The Evolution from Protein Chemistry to Proteomics

Basic Science to Clinical Application

The Evolution from Protein Chemistry to Proteomics

Basic Science to Clinical Application

Roger L. Lundblad

CRC Press
Taylor & Francis Group
Boca Raton London New York

CRC Press is an imprint of the
Taylor & Francis Group, an **informa** business
A TAYLOR & FRANCIS BOOK

First published 2006 by Taylor & Francis

Published in 2019 by
CRC Press
Taylor & Francis Group
6000 Broken Sound Parkway NW, Suite 300
Boca Raton, FL 33487-2742

© 2006 by Taylor & Francis Group, LLC
CRC Press is an imprint of Taylor & Francis Group, an Informa business

First issued in paperback 2019

No claim to original U.S. Government works

ISBN-13: 978-0-367-45399-2 (pbk)
ISBN-13: 978-0-8493-9678-6 (hbk)

Library of Congress Card Number 2005049929

Library of Congress Cataloging-in-Publication Data

Lundblad, Roger L.
　The evolution from protein chemistry to proteomics : basic science to clinical application / Roger L. Lundblad.
　　　p. cm.
　Includes bibliographical references and index.
　ISBN 0-8493-9678-6
　1. Proteomics. 2. Proteins--Chemical modification. I. Title.

QP551.L882 2005
572'.6--dc22   2005049929

Visit the Taylor & Francis Web site at
http://www.taylorandfrancis.com

and the CRC Press Web site at
http://www.crcpress.com

Preface

He thought he saw an Argument
that proved he was the Pope:
He looked again, and found it was
A Bar of Mottled Soap.
'A fact so dread,' he faintly said,
'Extinguishes all hope!'

Lewis Carroll

This book is intended to bring solution protein chemistry to proteomic technology and to provide a different perspective from other review works. Proteomics, while a diffuse area, is an extension of classical protein chemistry largely driven by major advances in the analytical capability of mass spectrometry. The content assembles together some disparate areas into a single volume. The author has also tried to use as much classical protein chemistry literature as possible as this serves as the intellectual basis for proteomics. The author also notes that recent literature fails to recognize prior contributions and is also limited with respect to technical detail. Effective commercialization of proteomics will require more diligence in these last two areas.

Roger L. Lundblad
Chapel Hill, North Carolina

Acknowledgments

I want to first express my thanks for the great assistance provided by the various libraries of the University of North Carolina at Chapel Hill. The author again acknowledges the support of Professor Bryce Plapp of the University of Iowa. Others who provided insight into some of the more complex issues include Professor Ralph Bradshaw of the University of California at Irvine and Professor Charles Craik of the University of California at San Francisco. The author also acknowledges the support of Dr. Judith Spiegel and her colleagues at Taylor and Francis.

Roger L. Lundblad
Chapel Hill, North Carolina

Contents

1 A Brief Discussion of Proteomics — Definition – Concepts – Illusions

CONTENTS

INTRODUCTION AND DEFINITION OF THE PROTEOME AND PROTEOMICS

Proteomics is an increasingly complex area of study[1,2] that is expected to yield results important for the development of therapeutics, diagnostics and for the emerging discipline of theranostics,[3,4] which emphasizes patient-specific therapeutics. What, however, exactly is proteomics? The term proteome dates back to 1995[5] when Humphrey-Smith and colleagues defined the proteome as "the total protein content of a genome." Genome is defined as "a complete single set of the genetic material of a cell or of an organism; the complete set of genes in a gamete."[6] It would follow that proteomics is the study of the proteome. A variety of other definitions have been proposed for proteomics. Morrison and coworkers[7] define the proteome as "the entire complement of proteins expressed by a cell at a point in time." In such cases, proteomics would be the study of the proteome; however, this definition would exclude extracellular collections of proteins such as those found in blood plasma,[8,9] urine[10,11] and lymphatic fluid.[12] These latter studies use some of the tools of proteomics such as two-dimensional electrophoresis and mass spectrometry but are clearly different from studies where isotope-coded affinity tag (ICAT) technology is used to study differential protein expression[13] and are used to identify biomarkers for diagnostics and therapeutics.

Whatever the precise definition, proteomics involves the study of complex mixtures of proteins and their interactions. This somewhat broader definition might be useful in that it extends the application of proteomics to diagnostics.[14,15] The technologies that underlie proteomics quite likely will improve sufficiently in analytical capability to be valuable in personalized medicine.[16,17] Ley and coworkers[18] have organized a

1

useful review of the applications of genomics and proteomics to the study of pathology and drug discovery. We will not be spending much time on the use of protein micro-arrays as a proteomic technology[19] as current technology presents more challenges than opportunities at this time. Challenges include the limited number of truly specific monoclonal antibodies or fragments, the limited stability of these targets and the high variance of on/off rates. Proteomics describes an experimental approach to the study of complex protein mixtures such as described by Dutt and Lee[20] and Dreger[21] and we will focus on the application of solution protein chemistry to proteomics and the limited success in some areas together with the challenges of reducing the research to practice.

Genealogy studies (Figure 1.1) would suggest that genomics[22] begat transcrip-tomics,[23] which begat proteomics, which begat... [24] Genomics is the study of the total genome of an organism (eukaryote, prokaryote or virus) and is generally depicted as the DNA sequence.[25,26] Where epigenetic information fits into this definition is not clear; the term "epigenome" has been proposed to include genomic aspects of methylation, for example.[27–31] Transcriptomics is the study of DNA expres-sion as measured by messenger RNA.[33–37] Protein expression, which should correlate with transcription, sometimes does and sometimes does not.[32,37] Functional genomics has been equated with proteomics.[38] Regardless of definition, the research must be of sufficient rigor to be useful.[39]

Proteomics appears to have replaced protein chemistry as a subdiscipline within biochemistry[40–45] but, as noted in another recent review, "Unfortunately, the word proteomics has come to mean virtually everything."[46] One of the goals of the current discussion is to dissect the various activities that appear to be contained with chem-ical proteomics and to firmly link some of the new nomenclature to more established disciplines. The development of new terms to describe old activities has resulted in a major nomenclature issue for anyone interested in performing a serious literature search in the area of proteomics; it also appears to make the intellectual property issue interesting and challenging.

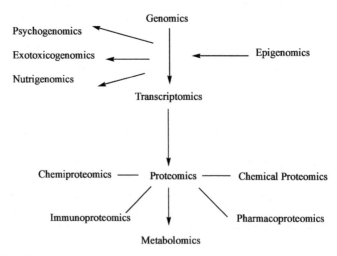

FIGURE 1.1 A genealogy for proteomics.

This chapter places proteomics in perspective with respect to its development, utility, and relationship to other more established disciplines. The derivation of proteomics is described above. Unfortunately, proteomics as a discipline has become incredibly diffuse and complicated with respect to nomenclature and applications. The nomenclature issue has been addressed in an elegant manner by Righetti and colleagues[24] and the diffuse nature of applications by several investigators who also note the potential useful interplay between the somewhat separate areas.[47-52] The term "omics" is a search term for PUBMED; a total of 185 citations were obtained using "omics" as the search term (April 2005) dating to 2002; and there is a journal entitled OMICS (Mary Ann Liebert) with most of the search items being citations to this journal. "Omics" is a suffix derived from the Greek *omes* which means "all" or "every" (as in genome, proteome). The use of "omics" as a suffix has enabled an explosion of terms (see Table 1). The use of genomics, transcriptomics and proteomics is useful but will require some discipline.[53-58] The various terms evolving for the use of "omics" as a suffix can be described as neologisms, where neologism is defined as a new word or term not infrequently greeted with derision; a secondary definition for neologism is a meaningless term coined by a psychotic.[59]

The overall intent of the current book is to address issues that are not discussed in detail by others and to avoid, where possible, redundancy in the coverage of information discussed in considerable detail in other sources.[60-66] The use of chemical modification in proteomics will be covered in great detail as will sample preparation and sample prefractionation. There is limited discussion of the specific separation technologies (two-dimensional gel electrophoresis, capillary electrophoresis and liquid chromatography) that result in the actual samples for mass spectrometry. There is little discussion of microarray technology other than chemistry associated with covalent linkage to a matrix. As noted above, microarray technology will only be of value when there is a better understanding of important analytes (biomarkers) and their importance to diagnosis and prognosis. Also, new technologies will be tied firmly to the concepts used in their development both to present the unique qualities of proteomics and to indicate that proteomics is not "magic" and that other, perhaps older technologies can be equally useful.[15,67] A danger exists both in failing to recognize the value of prior observations and technologies as well as lost opportunities. Success in the identification of tissue-based biomarkers will depend on the interplay of pathology and analytical biochemistry,[68] while the use of samples derived from serum or plasma will require the use of more traditional separation technologies prior to the analytical process. The emergence and success of proteomics depends more on the remarkable advances in mass spectrometry and rather less on advances in separation science.

While protein chemistry is a discipline, proteomics appears to be a somewhat undisciplined approach to the study of proteins and their function. In the least complex mode, proteomics can be nicely segmented into structural proteomics and functional proteomics.[48] A list of definitions useful in the study of proteomics is presented in Table 1.1. These definitions may be considered arbitrary but are useful in organizing our thoughts. These new terms should present major problems for

TABLE 1.1
Useful Definitions

Term	Definition
Activity-based proteomics	Identification of proteins in the proteome by the use of reagents that measure biological activity; most often used for enzymes where functional families of proteins can be identified.
Accurate mass tag (AMT)	A peptide of sufficiently distinctive mass and elution time from liquid chromatography that can be used as a single identifier of a protein.
Ampholyte	An amphoteric electrolyte. In proteomics, this term is used to describe a small multicharged organic buffer used to establish pH gradients in isoelectric focusing.
Amphoteric	Referring to a molecule such as a protein, peptide or amino acid capable of having a positive charge, negative charge or zero net charge. When at a zero net charge, it is also referred to as a zwitterion.
Affinity proteomics	The use of affinity reagents for the study of the proteome.
Algorithm	The underlying iterative method or mathematic theory for any particular computer programming technique; a precisely described routine process that can be applied and systematically followed through to a conclusion; a step-by-step procedure for solving a problem or accomplishing some end.
Arabidopsis thaliana	A small plant in the mustard family that is the model for studies of the plant genome (Meinke, D.W., Cheng, D.M., Dean, C., Rounsley, S.D. and Koorneeft, M., *Arabidopis thaliana*: A model plant for genome analysis, *Science*, 282, 662–682, 1998).
Bathochromic shift	A shift in the absorption/emission of light to a longer wavelength ($\lambda > \lambda_0$); a "red" shift.
Bioinformatics	The use of information technology to analyze data obtained from proteomic analysis. An example is the use of databases such as SWISSPROT to identify proteins from sequence information determined by the mass spectrometric analysis of peptides.
Biomarker	A change in response to an underlying pathology; current examples include C-reactive protein, fibrin D-dimer and troponin.
Bottom-up proteomics	Identification of unknown proteins by analysis of peptides obtained from unknown proteins by enzymatic (usually trypsin) hydrolysis.
Cytomics	The molecular analysis of heterogeneous cellular systems.
Chemical biology	The application of chemical techniques to problems in biology — the author has no idea how this differs from biochemistry or, perhaps more relevant, biological chemistry.
Chemical proteomics	Use of chemical modification to identify enzymes in the proteome.
Chemiproteomics	The use of small molecules as affinity materials for the discovery of specific binding proteins in the proteome.

Term	Definition
CHOmics	Study of the total carbohydrates in an organism/cell/tissue. See glycome and glycomics.
Classical proteomics	Proteomic analysis based on the direct analysis of the expressed proteome, such as an extract obtained from lysis of a cell; also referred to as forward proteomics as compared to reverse proteomics.
Clinomics	Application of oncogenomics to cancer care.
Cryosection	A tissue section cut from a frozen specimen; in this situation, ice is the supporting matrix.
Deconvolution	An algorithm used in electrospray mass spectrometry to translate the spectra of multiply charged ions into a spectrum of molecular species.
Desorption	Process by which molecules in solid or liquid form are transformed into a gas phase.
Electrophoresis/MS	Proteins are separated by one-dimension or more often two-dimensional gel electrophoresis. The separated proteins are subjected to *in situ* tryptic digestion and the peptides separated by liquid chromatography and identified by mass spectrometry (Nishihara, J.C. and Champion, K.M., Quantitative evaluation of proteins in one- and two-dimensional polyacrylamide gels using a fluorescent stain, *Electrophoresis*, 23, 2203–2215, 2002).
Embedding	Infiltration of a specimen with a liquid medium (paraffin) that can be solidified/polymerized to form a matrix to support the tissue for subsequent manipulation.
Epistasis	Masking of a phenotype caused by mutation of one gene by a mutation in another gene; epistasis analysis can be used to define order of gene expression in a genetic pathway.
Epitope	All epitopes present in the antigenic universe; also defined as an example, paradigm; a brief presentation or statement in most dictionaries.
Exotoxicogenomics	Study of the expression of genes important in adaptive responses to toxic exposures.
Expression profiling	The measurement or determination of DNA expression by the measurement RNA (transcriptomics); also used to refer to protein expression as determined by proteomic technology.
Functional genomics	Refers to establishing a verifiable link between gene expression and cell/organ/tissue function/dysfunction.
Functional proteomics	Study of changes in protein expression within the proteome; use of reactive chemical probes to identify enzymes in the proteome.
Genome	The complete gene complement of any organism, contained in a set of chromosomes in eukaryotes, a single chromosome in bacteria or a DNA or RNA molecule in viruses; the complete set of genes inside the cell or virus.
Genome-based proteomics	Gene-based analysis of the proteome.
Genomics	The study of the structure and function of the genome, including information about the sequence, mapping and expression and how genes and their products work in the organism; the study of the genetic composition of organisms.
Genotype	The internally coded, inheritable information carried by all living organisms; the genetic constitution of an organism.

(Continued)

TABLE 1.1
Useful Definitions (*Continued*)

Term	Definition
Global proteomics	Analysis of all proteins in a cell, tissue or organism.
Glycome	The total carbohydrates within an organism.
Glycomics	The study of the structure, function and interactions of carbohydrates within the glycome.
Hypsochromic	A shift of light absorption or emission to a shorter wavelength ($\lambda < \lambda_o$); a "blue" shift.
Immunomics	Study of the molecular functions associated with all immune-related coding and noncoding mRNA transcripts.
Immunoproteomics	Definition is a work in progress, varying from the screening of two-dimensional gels for reactive antibodies to the use of mass spectrometry to study targets of the immune system; use of proteomics to study the cellular and humoral immune systems.
Interactome	The protein–protein interactions within a proteome.
Isoelectric focusing (IEF)	An electrophoretic method for separating amphoteric molecules in pH gradients.
Kinomics	Analysis of all kinases in the proteome of a given organism.
Mass spectrometer	A device that assigns mass-to-charge ratios to ions based on their momentum, cyclotron frequency, time-of-flight or other parameters.
Metabolome	The total metabolites produced by the products of the genome.
Metabolomics	The study of the metabolome.
Modification-specific proteomics	The sum of all post-translational modifications of the proteome.
Nutrigenomics	Genomics of nutrition. The science of nutrigenomics seeks to provide a molecular understanding for how common dietary chemicals (i.e., nutrition) affect health by altering the expression and structure of an individual's genetic makeup (http://nutrigenomics.ucdavis.edu).
Oncogenomics	The use of molecular medicine tools such as DNA microarray and proteomics to study the oncology process; cancer genomics; study of oncogenes.
Organelle proteomics	Analysis of subcellular organelles such as mitochrondria, nucleus and the endocytotic apparatus by proteomic techniques.
Orthologues	Genes in different organisms that have similar functions.
Paralogues	Genes within the same genome that have evolved by gene duplication.
Peptidome	The peptide complement of a genome.[a]

Term	Definition
Pharmacoproteomics	The use of proteomics to predict individual reaction to a drug or drugs; related to personalized medicine, theranostics, pharmacogenomics.
Phenotype	The physical manifestation of the genes of an organism; the collection of structure and function expressed by the genotype of an organism; the visible properties of an organism that are produced by the interaction of a genotype and the environment.
Plasma/serum proteome	The identification and characterization of the proteins in the blood plasma/serum.
Poisson distribution	A probability density function that is an approximation to the biomodal distribution and is characterized by its mean being equal to its variance.
Post-translational modification	A covalent modification of a protein following translation of the RNA to form the polypeptide chain. Such modification may or may not be enzyme catalyzed (γ-carboxylation vs. nitration) and may or may not be reversible (phosphorylation vs. γ-carboxylation).
Protease proteomics	The proteases and protease substrates of the proteome.
Protein profiling	The use of algorithms to determine the relationship of multiple proteins as determined by mass spectrometric or liquid chromatographic analysis.
Proteometabolics	Pertaining to proteometabolism.
Proteome	The total expressed protein content of a genome.
Proteometabolism	Metabolism of the proteome.
Proteomics	The study of the proteome, not technologically limited; the qualitative and quantitative study of the proteome under various conditions including protein expression, modification, localization and function, and protein–protein interactions as a means of understanding biological processes.
Psychogenomics	The process of applying the tools of genomics, transcriptomics and proteomics to understand the molecular basis of behavioral abnormalities.
Quantitative proteomics	Determination of changes in the proteome following perturbation; quantitative measurement of proteins in the proteome following perturbation. Isotope-coded affinity tags (ICAT) have been useful in quantitative proteomics.
Reverse proteomics	Proteomic analysis where genomic sequence information is used to predict the resulting proteome, providing the basis for experiment design.
SELDI	Surface-enhanced laser/desorption ionization mass spectrometry; ProteinChip® (Tang, N., Tornatore, P. and Weinberger, S.R., Current developments in SELDI affinity technology, *Mass Spectrom. Rev*, 23, 34–44, 2004).
Shotgun proteomics	Identification of peptides (usually by mass spectrometry) obtained by the enzymatic or chemical digestion of the entire proteome. A naturally occurring protein mixture such as cell extract, blood plasma or other biological fluid is reduced, alkylated and subjected to tryptic hydrolysis. The tryptic hydrolysis is fractionated by liquid chromatography and analyzed by mass spectrometry (Wolters, D.A., Washburn, M.P. and Yates, J.R., III, An automated multidimensional protein identification technology for shotgun proteomics, *Anal. Chem.*, 73, 5683–5690, 2001; Liu, H., Sadygov, R.G. and Yates, J.R., III, A model for random sampling and estimation of relative protein abundance in shotgun proteomics, *Anal. Chem.*, 76, 4193–4201, 2004).

(Continued)

TABLE 1.1
Useful Definitions (*Continued*)

Term	Definition
Soft-ionization	Ionization techniques such as fast atom bombardment (FAB), electrospray ionization (ESI) or matrix-assisted laser desorption/ionization (MALDI) that initiate the desorption and ionization of nonvolative, thermally labile compounds such as proteins or peptides.
Structural proteomics	Study of the primary, secondary, and tertiary structure of the proteins in a proteome; functional predictions from primary structure.
Structural biology	Study of the secondary, tertiary and higher structures of proteins in the proteome including but limited to the use of crystallography, nuclear magnetic resonance and electron microscopy.
Structural genomics	Focuses on the physical aspects of the genome through the construction and comparison of gene maps and sequences as well as gene discovery, localization and characterization; determination of the three-dimensional structures of gene products; known previously as crystallography.
Surrogate marker	A biomarker which can be used in place of a clinical endpoint.
Synovial proteome	The total protein content of synovial fluid.
Systems biology	The integration of data at the genomic, transcriptomic, proteomic and metabolomic levels, including functional and structural data, to understand biological functions that can be described by a mathematical function.
Targeted proteomics	Analysis of a defined portion of a proteome such as a glycoproteome, phosphoproteome or ribosomal proteins.
Theranostics	The use of diagnostic laboratory tests to guide therapeutic outcomes. Current use has emphasized "real-time" PCR assays for the identification of pathogens.
Time-of-flight	Measured flight time of ions; lighter ions travel a greater distance that heavier ions with the mass proportional to the square of the time; converted to m/z by calibration with standards.
Top-down proteomics	Analysis of the intact unknown proteins by mass spectrometry; occasionally used to describe the analysis of cyanogen bromide peptides.
Topological proteomics	A technology which analyzes proteins on a single cell level; study of the toponome.
Transcriptomics	The total RNA transcripts produced by a genome; the complete set of RNA messages coded from the DNA within a cell.

a The distinction between peptide and protein can be arbitrary.

the literature searches that are necessary both for scholarly accuracy as well as intellectual property protection.

Proteomics, or more accurately proteomic techniques, have been used to study a broad variety of organs, tissues and cells, as illustrated by some examples in Table 1.2. A selected number of these studies that have focused on the identification of biomarkers for the development of diagnostics are discussed in the chapter on clinical proteomics.

DIVISION OF ACTIVITIES IN PROTEOMICS

ANALYTICAL PROTEOMICS

In considering the vast literature on proteomics, three general types of activities appear that are closely related to each other. The first is the elucidation of the proteome by analytical biochemistry, including various separation technologies, mass spectrometry

TABLE 1.2
Application of Proteomics to Organs, Tissues and Cells

Sample	Reference
Leukocytes	1
Whole human saliva	2, 5
Pancreatic juice	3
Membranes	4, 11
Human colon crypt	6
Human tears	7
Human mammary epithelial cells	8
Human cerebrospinal fluid	9, 18, 19
Biofilm	10
Seafood	12
Urine	13, 14
Monocyte/macrophages	15, 16, 21
Pulmonary edema fluid	17
CD4(+)T cells	20
Cell surface	22
Platelets	23–28
Podocyte (modified epithelial cells; visceral epithelial cells)	29
Human cilia	30
Bronchoalveolar lavage	31
Human seminal fluid	32
Proteosomes	33
Lymph	34
Pituitary	35–38
Human body fluids	39

REFERENCES FOR TABLE 1.2

1. Wang, X., Zhao, H. and Andersson, R., Proteomics and leukocytes: An approach to understanding potential molecular mechanisms of inflammatory responses, *J. Proteome Res.*, 3, 921–929, 2004.
2. Wilmorth, P.A., Riviere, M.A, Rustvold, D.L., Lauten, J.D., Madden, T.E. and David, L.L., Two-dimensional liquid chromatography study of the human whole saliva proteome, *J. Proteome Res.*, 3, 1017–1023, 2004.
3. Grønberg, M., Bunkenborg, J., Kristiansen, T.Z. et al., Comprehensive proteome analysis of human pancreatic juice, *J. Proteome Res.*, 3, 1042–1055, 2004.
4. Zhang, N.L., Li, N. and Li, L., Liquid chromatography MALDI MS/MS for membrane proteome analysis, *J. Proteome Res.*, 3, 719–727, 2004.
5. Messana, T., Cabras, T., Inzitari, D. et al., Characterization of the human salivary basic proline-rich protein complex by a proteomic approach, *J. Proteome Res.*, 3, 792–800, 2004.
6. Li, X.-M., Patel, B.B. and Blogoi, E.L., Analyzing alkaline proteins in human colon crypt proteome, *J. Proteome Res.*, 3, 821–833, 2004.
7. Zhou, L., Huang, L.Q., Beuerman, R.W. et al., Proteomic analysis of human tears: Defensin expression after ocular surface surgery, *J. Proteomic Res.*, 3, 410–416, 2004.
8. Jacobs, J.M., Hettaz, H.M. and Yu, L.-R., Multidimensional proteomic analysis of human mammary epithelial cells, *J. Proteome Res.*, 3, 68–75, 2004.
9. Wenner, B.R., Lowell, M.A. and Lynn, B.C., Proteomic analysis of human ventricular cerebrospinal fluid form neurologically normal, elderly subjects using two-dimensional LC-MS/MS, *J. Proteomic Res.*, 3, 97–103, 2004.
10. Vilain, S., Costette, P., Zimmerlin, I. et al., Biofilm proteins: Homogeneity or versatility? *J. Proteome Res.*, 3, 132–136, 2004.
11. Pedersen, S.K., Henry, J.L. and Sebastian, L., Unseen proteome: Mining below the tip fo the iceberg to find low-abundance and membrane proteins, *J. Proteome Res.*, 2, 303–311, 2003.
12. Piñero, C., Bãrros-Velásquez, J., Vãsquez, J., Figueras, A. and Gallards, J.M., Proteomics as a tool for the investigation of seafood and other marine products, *J. Proteome Res.*, 2, 127–135, 2003.
13. Kiernan, U.A., Tubbs, K.A., Nelelkov, D. et al., Comparative urine protein phenotyping using mass spectrometric immunoassay, *J. Proteome Res.*, 2, 191–197, 2003.
14. Pang, J.X., Ginanni, N., Dangree, A.R., Hefta, S.A. and Opitek, G.J., Biomarker discovery in urine by proteomics, *J. Proteome Res.*, 1, 161–169, 2002.
15. Gonzalez-Borderas, M., Gallego-Delgado, J., Mas, S. et al., Isolation of circulating monocytes with high purity for proteomic analysis, *Proteomics*, 4, 432–437, 2004.
16. Jin, M., Opalek, J.M., Marsh, C.B. and Wu, H.M., Proteome comparison of alveolar macrophages with monocytes reveals distinct protein characteristics, *Am. J. Respir. Cell. Mol. Biol.*, 31, 322–329, 2004.
17. Bowler, R.P., Duda, B., Chan, E.D. et al., Proteomic analysis of pulmonary edema fluid and plasma in patients with acute lung injury, *Am. J. Phys. Cell. Mol. Biol.*, 286, L1095–L1104, 2004.
18. Yuan, X., Russell, T., Wood, G. and Desiderio, D.M., Analysis of the human lumbar cerebrospinal fluid proteome, *Electrophoresis*, 23, 1185–1196, 2002.
19. Rohlff, C., Proteomics in molecular medicine: Applications in central nervous system disorders, *Electrophoresis*, 21, 1227–1234, 2000.
20. Ravtajok, K., Nyum, T.A. and Labesmaa, R., Proteome characterization of human T helper 1 and 2 cells, *Proteomics*, 4, 84–92, 2004.

21. Gadgil, H.S., Pabst, K.M., Giorgiumi, F. et al., Proteome of monocytes primed with lipopolysaccharide: Analysis of the most abundant proteins, *Proteomics*, 3, 1767–1780, 2003.
22. Jang, J.H. and Hanash, J., Profiling of the cell surface proteome, *Proteomics*, 3, 1947–1954, 2003.
23. Perrotta, P.L. and Bahou, W.F., Proteomics in platelet science, *Curr. Hemat. Rev.*, 3, 462–469, 2004.
24. Garcia, A., Zitzmann, N. and Watson, S.P., Analyzing the platelet proteome, *Sem. Thromb. Hemostas.*, 30, 485–489, 2004.
25. Marcus, K. and Meyer, H.E., Two-dimensional polyacrylamide gel electrophoresis for platelet proteomics, *Meth. Mol. Biol.*, 273, 421–434, 2004.
26. Garcia, A., Prabhaker, S., Brock, C.J. et al., Extensive analysis of the human platelet proteome by two-dimensional gel electrophoresis and mass spectrometry, *Proteomics*, 4, 656–658, 2004.
27. McRedmund, J.P., Park, S.D., Reilly, D.F. et al., Integration of proteomics and genomics in platelets: A profile of platelet proteins and platelet-specific genes, *Mol. Cell. Proteomics*, 3, 133–144, 2004.
28. Maguire, P.B. and Filtzgerald, D.J., Identification of the phosphotyrosine proteome from thrombin activated platelets, *Proteomics*, 2, 642–648, 2002.
29. Ramsom, R.F., Podocyte proteomics, *Contrib. Neph.*, 141, 189–211, 2004.
30. Ostrowski, L.E., Blackburn, K. and Radde, K.M., A proteomic analysis of human celia: Identification of novel components, *Mol. Cell. Proteomics*, 1, 451–465, 2002.
31. Noel-Georis, I., Bernard, A., Falmonge, P. and Wattiez, R., Database of bronchoalveolar lavage, *J. Chrom. A*, 771, 221–236, 2002.
32. Fung, K.Y., Glode, L.M., Green, S. and Duncan, M.W., A comprehensive characterization of the peptide and protein constituents of human seminar fluid, *Prostate*, 61, 171–181, 2004.
33. Utleg, A.G., Yi, E.C., Zie, T. et al., Proteomic analysis of human prostasomes, *Prostate*, 56, 150–161, 2003.
34. Leak, L.W., Liotta, L.A., Krutzsch, H. et al., Proteomic analysis of lymph, *Proteomics*, 4, 752–765, 2004.
35. Lloyd, R.V., Advances in pituitary pathology: Use of novel techniques, *Front Horm. Res.*, 32, 146–174, 2004.
36. Desiderio, D.M. and Zhan, X., The human pituitary proteome. The characterization of differentially expressed protein in an adenoma compared to a control, *Cell. Mol. Biol. (Noisy-le-grand)*, 49, 689–712, 2003.
37. Zhan, X. and Desiderio, D.M., Heterogeneity analysis of the human pituitary proteome, *Clin. Chem.*, 49, 1740–1751, 2003.
38. Beranova-Giorginanni, S., Giorgianni, F. and Desiderio, D.M., Analysis of the proteome in the human pituitary, *Proteomics*, 2, 534–542, 2002.
39. Kennedy, S., Proteomic profiling from human samples: The body fluid alternative, *Tox. Letters*, 120, 379–384, 2001.

and microarray platforms. In principle, the following two activities should follow the basic description of the proteome in that it is always useful to define normal before attacking the abnormal. The definition of normal is by no means a trivial issue since one must be concerned with low-abundance and high-abundance analytes. This issue is discussed in detail in the chapter on prefractionation (Chapter 5). The chapter on sample preparation addresses the critical issue of reproducibility and accuracy in analysis. As will be noted, differences in techniques between laboratories make it difficult to compare various sets of data. Developments in mass spectrometry with the concomitant developments in bioinformatics have provided the driving force for proteomics and experimental design and ideation have struggled to keep up.

EXPRESSION PROTEOMICS

The second general activity is expression proteomics. This term will be used to describe studies where a system, either *in vitro* or *in vivo*, is perturbed by a particular stimulus.[69–81] This activity represents the application of proteomic technology to study the physiology of a system; metabolomics[82–91] is a related area. In a literature search on metabolomics, surprisingly a large number of citations for plant biology appear as well as a large number of studies concerning issues in sample preparation for plant tissue. This topic is discussed in detail in the chapter on sample preparation. Expression profiling has used technologies such as isotope-coded affinity tags (ICAT)[12,73,92] and stable isotope labeling with amino acids in cell culture (SILAC).[68,71,93] While a few whole animal studies with expression profiling exist, most investigations have studied systems either in cell culture or fermentation, providing the opportunity to look at pathways (pathway proteomics) rather than individual proteins.[94] The techniques required for the study of expression proteomics are discussed in the chapter on chemical modification and aspects of application in the chapter on chemical proteomics.

BIOMARKER IDENTIFICATION

The third general area is biomarker identification, where the protein composition in tissue samples from certain disease conditions is compared to the composition from normal samples and biomarkers are identified.

Table 1.3 lists the relative popularity of diseases on citation frequency, and the chapter on clinical proteomics is devoted to a discussion of this activity. At the time of this writing, there is reason to be optimistic about the potential of

TABLE 1.3
Number of Proteomic Citations by Disease Category

Pathology	Number of Citations[a]
Cancer and proteomics	639[b]
Neurological disease and proteomics	111
Aging and proteomics	54
Pulmonary disease and proteomics	53
Gastrointestinal disease and proteomics	46
Bone and proteomics[c]	19

[a] Search performed on November 14, 2004.
[b] 543 if blood excluded from search; 488 if profiling excluded from search.
[c] No citations from orthopedics and proteomics.

proteomic technology to contribute the development of new diagnostic tests of value; however, as noted earlier and in the chapter on clinical proteomics, the mere fact that a biomarker is developed using proteomic technology does not imply added value and existing technologies should not be ignored.[15,67] Also of use would be the various investigators pursuing the identification of biomarkers clearly seeing the assay developed in terms of integration into existing clinical laboratories.

REFERENCES

1. Bradshaw, R.A., Proteomics — Boom or bust? *Mol. Cell. Proteomics*, 1, 177, 2002.
2. *Medical and Healthcare Marketplace Guide*, 17th Edition, Research Reports, Dorland's Biomedical, Philadelphia, PA, Vol. 1, I-584, 2001–2002.
3. Margez, P. and Shapiro, F., Theranostics: Turning tomorrow's medicine into today's reality, *Curr. Drug Disc.*, 23, September 2002.
4. Picard, F.J. and Bergeron, M.G., Rapid molecular theranostics in infectious diseases, *Drug Disc. Today*, 7, 1092, 2002.
5. Wasinger, V., Cordwell, S., Cerpa-Poljak, A., Yan, J., Gooley, A., Wilkins, M., Duncan, M., Harris, R., Williams, K. and Humphrey-Smith, I., Progress with gene-product mapping of the Mollicules: *Mycoplasma genitalium*, *Electrophoresis*, 16, 1090, 1995.
6. Stenesh, J., *Dictionary of Biochemistry*, Wiley-Interscience, New York, NY, 126, 1983.
7. Morrison, R.S., Kinoshita, Y., Johnson, M.D. and Conrads, T.P., Proteomics in the post genomic age, *Adv. Prot. Chem.*, 65, 1, 2003.
8. Anderson, N.L. and Anderson, N.G., The human plasma proteome. History, character, and diagnostic prospects, *Mol. Cell. Proteomics*, 1, 845, 2002.
9. Adkins, J.N., Varnum, S.M., Auberry, K.J. et al., Toward a human blood serum proteome. Analysis by multidimensional separation coupled with mass spectrometry, *Mol. Cell. Proteomics*, 1, 947, 2002.
10. Pieper, R., Gatlin, C.L., McGrath, A.M. et al., Characterization of the human urinary proteome: A method for high-resolution display of urinary proteins on two-dimensional electrophoresis gels with a yield of nearly 1400 distinct protein spots, *Proteomics*, 4, 1159, 2004.
11. Schaub, S., Wilkins, J., Weiler, T. et al., Urine protein profiling with surface-enhanced laser-desorption/ionization time-of-flight mass spectrometry, *Kidney Int.*, 65, 323, 2004.
12. Leak, L.V., Liotta, L.A., Krutzsch, M. et al., Proteomic analysis of lymph, *Proteomics*, 4, 753, 2004.
13. Gygi, S.P., Rist, B., Gerber, S.A. et al., Quantitative analysis of complex protein mixtures using isotope-coded affinity tags, *Nat. Biotech.*, 17, 994, 1999.
14. Petricoin, E.F., Zoon, K.C., Kohn, E.C. et al., Clinical proteomics: Translating bench-side promise into bedside reality, *Nat. Rev. Drug Disc.*, 1, 683, 2002.
15. Lundblad, R.L and Wagner, P.M., The potential of proteomics in developing diagnostics, *IVD Technology*, 11(2), 20, March 2005.
16. Califf, R.M., Defining the balance of risk and benefit in the era of genomics and proteomics, *Heath Affairs (Millwood)*, 23, 77, 2004.
17. Jain, K.K., Role of pharmacoproteomics in the development of personalized medicine, *Pharmacogenomics*, 5, 331, 2004.

18. Kley, N., Schmidt, S., Berlin, V. et al., Functional genomics and proteomics: Basics, opportunities and challenges, in *Molecular Nuclear Medicine: The Challenge of Genomics and Proteomics to Clinical Practice*, Springer-Verlag, Berlin, Ch. 3, 39, 2003.

19. Haab, B.B., Advances in protein microarray technology for protein expression and interaction profiling, *Curr. Opin. Drug Disc. Dev.*, 4, 116, 2001.

20. Dutt, M.J. and Lee, K.H., Proteomic analysis, *Curr. Opin. Biotech.*, 11, 176, 2000.

21. Dreger, M., Emerging strategies in mass-spectrometry based proteomics. *Eur. J. Biochem.*, 270, 569, 2003.

22. Honoré, B., Østergaard, M. and Vorum, H., Functional genomics studied by proteomics, *BioEssays*, 26, 901, 2004.

23. Hegde, P.S., White, I.R. and Debouck, C., Interplay of transcriptomics and proteomics, *Curr. Opin. Biotech.*, 14, 647, 2003.

24. Righetti, P.G., Castagna, A., Antonucci, F. et al., The proteome: *Anno Domini* 2002, *Clin. Chem. Lab. Med.*, 41, 425, 2003.

25. Umar, A., Applications of bioinformatics in cancer detection: A lexicon of bioinformatics terms, *Ann. N.Y. Acad. Sci.*, 1020, 263, 2004.

26. Matros, E., Wang, Z.C., Richardson, A.L. and Iglehart, J.D., Genomic approaches in cancer biology, *Surgery*, 136, 511, 2004.

27. Kohn, E.C., Lu, Y., Wang, H. et al., Molecular therapeutics: Promises and challenges, *Seminars in Oncology*, 31(1), Suppl. 3, 39, 2004.

28. Burchiel, S.W., Knoll, C.M., Davis, J.W., II, Paules, R.S., Boggs, S.E. and Afshan, C.A., Analysis of genetic and epigenetic mechanisms of toxicity: Potential roles of toxicogenomics and proteomics in toxicology, *Tox. Sci.*, 59, 193, 2001.

29. Miklos, C.L. and Moleszko, R., Integrating molecular medicine with functional proteomics, *Proteomics*, 1, 30, 2001.

30. Sakatoni, T. and Onyango, P., Oncogenomics: Prospects for the future, *Expert Rev. Anticancer Ther.*, 3, 891, 2003.

31. Yang, H.H. and Lee, M.P., Application of bioinformatics in cancer epigenetics, *Ann. N.Y. Acad. Sci.*, 1020, 67, 2004.

32. Hegde, P.S., White, I.R. and Debouck, C., Interplay of transcriptomics and proteomics, *Curr. Opin. Biotech.*, 14, 667, 2003.

33. Chan, G., Gharib, T.G., Huang, C.C. et al., Discordant protein and MRNA expression in lung adenocarcinomas, *Mol. Cell. Proteomics*, 1, 304, 2002.

34. Urbanczyk-Wachniak, E., Luedemann, A., Kopka, J. et al., Parallel analysis of transcript and metabolic profiles: A new approach in systems biology, *EMBO Reports*, 4, 989, 2003.

35. Kellner, U., Steinhert, R., Seibert, V. et al., Epithelial cell preparations for prosteomics and transcriptomic analysis in human pancreatic tissue, *Path. Res. Prac.*, 200, 155, 2004.

36. Kim, C.H., Kim, do K, Cho, S.J. et al., Proteomics and transcriptomic analysis of interleukin-1 beta treated lung carcinoma cell lines, *Proteomics*, 3, 2454, 2003.

37. Ideker, T., Tharsson, V., Ranish, J.A. et al., Integrated genomic and proteomic analysis of a systematically perturbed metabolic network, *Science*, 292, 929, 2001.

38. Hieter, P. and Boguski, M., Functional genomics: It's all how you read it, *Science*, 278, 601, 1997.

39. Weinstein, J.N., Fishing expeditions, *Science*, 282, 628, 1998.

40. Yates, J.R., III, Mass spectrometry. From genomics to proteomics, *Trends Gen.*, 16, 5, 2000.

41. Mann, M., Hendrickson, R.C. and Pandey, A., Analysis of proteins and proteomics by mass spectrometry, *Ann. Rev. Biochem.*, 70, 437, 2001.

42. Liebler, D.C., Proteomic approaches to characterize protein modifications: New tools to study the effects of environmental exposures, *Env. Health Pers.*, 110, Suppl. 1, 3, 2002.

43. Norin, M. and Sundstrom, M., Structural proteomics: Developments in structure-to-function predictions. *Trends Biotech.*, 20, 79, 2002.

44. Chakravarti, D.N., From the decline and fall of protein chemistry to proteomics, *BioTechniques*, 2, Suppl. March 2002.

45. Marshall, T. and Williams, K.M., Proteomics and its impact upon biomedical science, *Brit. J. Biomed. Sci.*, 59, 47, 2002.

46. Petach, H. and Gold, L., Dimensionality is the issue: Use of photoaptamers in protein microarrays, *Curr. Opin. Biotech.*, 13, 309, 2002.

47. Aardema, M.J. and MacGregor, J.T., Toxicology and genetic toxicology in the new era of "toxicogenomics": Impact of "-omics" technologies, *Mut. Res.*, 29, 13, 2002.

48. Lefkovits, I., Functional and structural proteomics: A critical appraisal, *J. Chrom. B*, 787, 1, 2003.

49. Ge, H., Walhout, A.J. and Vidal, M., Integrating 'omic' information: A bridge between genomic and systems biology, *Trends Gen.*, 19, 551, 2003.

50. Berman, H.M. and Westbrook, J., The need for dictionaries, ontologies, and controlled vocabularies, *OMICS*, 7, 9, 2003.

51. Werner, T., Proteomics and regulomics: The yin and yang of functional genomics, *Mass Spec. Rev.*, 23, 25, 2004.

52. Kleno, T.G., Kiehr, B., Baunsgaard, D. and Sidelmann, U.G., Combination of 'omics' data to investigate the mechanism(s) of hydrazine-induced hepatoxicity in rats and to identify potential biomarkers, *Biomarkers*, 9, 116, 2004.

53. Genomic Glossary, http://www.genomicsglossaries.com/conent/proteomic_categories.asp.

54. Free Dictionary, http://www.thefreedictionary.com/proteomics.

55. Mednet Dictionary, http://www.medterms.com.

56. Redei, C.P., *Encyclopedic Dictionary of Genetics, Genomics, and Proteomics*, Wiley-VCH, Hoboken, NJ, 2003.

57. Kahl, G., *The Dictionary of Gene Technology, Genomics, Transcriptomics, Proteomics*, 3rd ed., Wiley, Hoboken, NJ, 2004.

58. Dictionary of Linguistics, http://www.explore-dictionary.com/liguistics/0/-omics.html.

59. *Webster's Ninth New Collegiate Dictionary*, Merriam-Webster, Springfield, MA, 1987.

60. Simpson, R.J., Ed., *Proteins and Proteomics: A Laboratory Manual*, Cold Spring Harbor Laboratory Press, Cold Spring Harbor, NY, 2003.

61. Simpson, R.J., Ed., *Purifying Proteins for Proteomics: A Laboratory Manual*, Cold Spring Harbor Laboratory Press, Cold Spring Harbor, NY, 2004.

62. Arrigan, G.G. and Goodacre, R., Eds., *Metabolic Profiling: Its Role in Biomarker Discovery and Gene Function Analysis*, Kluwer, Boston, MA, 2003.

63. Cullis, C.A., *Plant Genomics and Proteomics*, John Wiley & Sons, Hoboken, NJ, 2004.

64. Hamadeh, H.K. and Asfari, C.A., Eds., *Toxicogenomics: Principles and Applications*, Wiley-Liss, Hoboken, NJ, 2004.

65. Strege, M.A. and Lagu, A.L., Eds., *Capillary Electrophoresis of Proteins and Peptides*, Humana Press, Towata, NJ, 2004.

66. Pechkova, E., *Proteomics and Nanocrystallography*, Kluwer Academic/Plenum Publishers, New York, NY, 2003.

67. Longtin, R., Out of fashion: Non-proteomic research vying for attention, *J. Natl. Cancer Inst.*, 95, 1034, 2003.

68. Melle, C., Ernst, G., Schimmel, B. et al., Biomarker discovery and identification in laser microdissected head and neck squamous cell carcinoma with ProteinChip® technology, two-dimensional electrophoresis, tandem mass spectrometry and immunohistocytochemistry, *Mol. Cell. Proteomics*, 2, 443, 2003.

69. Vogt, J.A., Schroer, K., Hölzer, K. et al., Protein abundance quantification in embryonic stem cells using incomplete metabolic labelling with ^{15}N amino acids, matrix-assisted laser desorption/ionization time-of-flight mass spectrometry, and analysis of relative isotopologue abundance of peptides, *Rapid Comm. Mass Spec.*, 17, 1273, 2003.

70. Hecker, M. and Volker, U., Towards a comprehensive understanding of *Bacillus subtilis* cell physiology by physiological proteomics, *Proteomics*, 4, 3727, 2004.

71. Wang, Y.K., Ma, Z., Quinn, D.F. and Fu, E.W., Inverse ^{15}N-metabolic labeling/mass spectrometry for comparative proteomics and rapid identification of protein markers/targets, *Rapid Comm. Mass Spec.*, 16, 1389, 2002.

72. Zhu, H., Pan, S., Gu, S., Bradbury, E.M. and Chen, X., Amino acid residue specific stable isotope labeling for quantitative proteomics, *Rapid Comm. Mass Spec.*, 16, 2115, 2002.

73. Tao, W.A. and Aebersold, R., Advances in quantitative proteomics via stable isotope tagging and mass spectrometry, *Curr. Opin. Biotech.*, 14, 110, 2003.

74. Harris, R.D., Nindl, G., Balcavage, W.X., Weiner, W. and Johnson, M.T., Use of proteomics methodology to evaluate inflammatory protein expression in tendonitis, *Biomed. Sci. Instr.*, 39, 493, 2003.

75. Krijgsveld, J., Ketting, R.F., Mahmoudi, T. et al., Metabolic labeling of *C. elegans* and *D. melanogaster* for quantitative proteomics, *Nat. Biotech.*, 21, 927, 2003.

76. Chei, F.Y, Eipper, B.A., Mains, R.E. and Fricker, L.D., Quantitative peptidomics of pituitary gland from mice deficient in copper transport, *Cell. Mol. Biol. (Noisy-legrand)*, 49, 713, 2003.

77. Ibarrola, N., Kalume, D.E., Gronborg, M., Iwahori, A. and Pandey, A., A proteomic approach for quantitation of phosphorylation using stable isotope labeling in cell culture, *Anal. Chem.*, 75, 6043, 2003.

78. Everly, P.A., Krijgsveld, J., Zetter, B.R. and Gygi, S.P., Quantitative cancer proteomics: Stable isotope labeling with amino acids in cell culture (SILAC) as a tool for prostate cancer research, *Mol. Cell. Proteomics*, 3, 729, 2004.

79. Hardwidge, P.R., Rodriguez-Escudero, I., Goode, D. et al., Proteomic analysis of the intestinal epithelial cell response to enteropathogenic *Escherichia coli*, *J. Biol. Chem.*, 279, 20127, 2004.

80. Shiiio, Y., Donohoe, S., Yi, E.C., Gooodlett, D.R., Aebersold, R. and Eisnemann, R.N., Quantitative proteomic analysis of Nyc oncoprotein function, *EMBO J.*, 21, 5088, 2002.

81. Wright, M.E., Eng, J., Sherman, J. et al., Identification of androgen-coregulated protein networks from the microsomes of human prostate cancer cells, *Genome Biol.*, 5, R4, 2003.

82. Bino, R.J., Hall, R.D., Fiehn, O. et al., Potential of metabolomics as a functional genomics tool, *Trends Plant Sci.*, 9, 418, 2004.

83. Whitfield, P.D., German, A.J. and Noble, P.J., Metabolomics: An emerging postgenomic tool for nutrition, *Br. J. Nutrition*, 92, 549, 2004.

84. Schmidt, C., Metabolomics takes its place as latest up-and-coming "omic" science, *J. Natl. Cancer Inst.*, 96, 732, 2004.

85. Kell, D.B., Metabolomics and systems biology: Making sense of the soup, *Curr. Opin. Microbiol.*, 7, 296, 2004.
86. Jansen, J.J., Hoefsloot, H.C., Boelens, H.F., van der Greef, J. and Smilde, A.K., Analysis of longitudinal metabolomics data, *Bioinformatics*, 20, 2438, 2004.
87. Weckwerth, W., Metabolomics in systems biology, *Ann. Rev. Plant Biol.*, 54, 669, 2004.
88. Castrillo, J.I., Hayes, A., Mohammed, S., Gaskell, S.J. and Oliver, S.G., An optimized protocol for metabolome analysis in yeast using direct infusion electrospray mass spectrometry, *Phytochemistry*, 62, 929, 2003.
89. Phelps, T.J., Palumbo, A.V. and Beliaev, A.S., Metabolomics and microarrays for improved understanding of phenotypic characteristics controlled by both genomics and environmental constraints, *Curr. Opin. Biotechl.*, 13, 20, 2002.
90. Flehn, O., Metabolomics — The link between genotypes and phenotypes, *Plant Mol. Biol.*, 48, 155, 2002.
91. Reo, N.V., NMR-based metabolomics, *Drug Chem. Tox.*, 25, 375, 2002.
92. Gygi, S.P., Rist, B. et al., Quantitative analysis of complex protein mixtures using isotope-coded affinity tags, *Nat. Biotech.*, 17, 994, 1999.
93. Ong, S.E., Blagoev, B., Kratchmarova, I. et al., Stable isotope labeling by amino acids in cell culture, SILAC, as a simple and accurate approach to expression proteomics, *Mol. Cell. Proteomics*, 1, 376, 2002.
94. Graumann, J., Dunipace, L.A., Seol, J.H. et al., Applicability of tandem affinity purification MudPiT to pathway proteomics in yeast, *Mol. Cell. Proteomics*, 3, 226, 2004.

STRUCTURAL AND FUNCTIONAL GENOMICS

Bogyo, M. and Hurley, J.H., Proteomics and genomics, *Curr. Opin. Chem. Biol.*, 7, 2–4, 2003.
Förster, J., Gombert, A.K. and Nielsen, J., A functional genomics approach using metabolomics and *in silico* pathway analysis, *Biotech. Bioeng.*, 79, 703–712, 2002.
Machius, M., Structural biology: A high-tech tool for biomedical research, *Curr. Opin. Neph. Hyper.*, 12, 431–438, 2003.
Pédelacq, J.-D., Piltch, E., Liong, E.C. et al., Engineering soluble proteins for structural genomics, *Nat. Biotech.*, 20, 927–932, 2002.
Sørlie, T., Børresen-Dale, A., Lønning, P.E., Brown, P.O. and Botstein, D., Expression profiling of breast cancer: From molecular portraits to clinical utility, in *Oncogenomics: Molecular Approaches to Cancer*, C. Brenner and D. Duggan, Eds., John Wiley & Sons, Ch. 5, 77–100, 2004.

TRANSCRIPTOMICS

Kellner, U., Steinhert, S., Seibert, U. et al., Epithelial cell preparations for proteomic and transcriptomic analysis in human pancreatic tissue, *Path. Res. Prac.*, 200, 155–163, 2004.
Hegde, P.S., White, I.R. and Debouck, C., Interplay of transcriptomics and proteomics, *Curr. Opin. Biotech.*, 14, 647–651, 2003.
Kiechle, F.L. and Holland-Staley, C.A., Genomics, transcriptomics, proteomics, and numbers, *Arch. Path. Lab. Med.*, 127, 1089–1097, 2003.

Ragno, S., Romano, M., Howell, S. et al., Changes in gene expression in macrophages infected with *Mycobacterium tuberculosis*: A combined transcriptomic and proteomic approach, *Immunology*, 104, 99–108, 2001.

Shi, Y. and Shi, Y., Metabolic enzymes and coenzymes in transcription — A direct link between metabolism and transcription, *Trends Gen.*, 20, 445–452, 2004.

Urbanczyk-Wochniak, E., Luedemann, A., Kopka, J., Selbig, J., Roessner-Tunali, U., Willmitzer, L. and Fernie, A.R., Parallel analysis of transcript and metabolic profiles: A new approach in systems biology, *EMBO Reports*, 4, 989–993, 2003.

GENERAL

Agataon, C., Uhlén, M. and Hober, S., Genome-based proteomics, *Electrophoresis*, 25, 1280–1289, 2004.

Arnesano, F., Banci, L., Bertini, I. and Martinelli, M., Ortholog search of proteins in copper delivery to cytochrome c oxidase and functional analysis of paralogs and gene neighbors by genomic context, *J. Proteome Res.*, 4, 63–70, 2005.

Baggerly, K.A., Morris, J.S. and Coombes, K.R., Reproducibility of SELDI-TOF protein patterns in serum: Comparing datasets from different experiments, *Bioinformatics*, 20, 777–785, 2004.

Blander, G. and Guarente, L., The Sir2 family of protein deacetylases, *Ann. Rev. Biochem.*, 73, 417–435, 2004.

Brunet, S., Thibault, P., Gagnon, E. et al., Organelle proteomics: Looking at less to see more, *Trends Cell Biol.*, 13, 629–638, 2003.

Domon, B. and Broder, S., Implications of new proteomics strategies for biology and medicine, *J. Proteome Res.*, 3, 253–260, 2004.

Dorland's Illustrated Medical Dictionary, 30th Edition, Saunders (Elsevier), Philadelphia, PA, 2003.

Ge, H., Walhout, A.J.M. and Vidal, M., Integrating 'omic' information. A bridge between genomics and systems biology, *Trends Gen.*, 19, 551–560, 2003.

Glassel, J.A. and Deutscher, M.P., Eds., *Introduction to Biophysical Methods for Protein and Nucleic Acid Research*, Academic Press, San Diego, CA, 1995.

Hicks, J., Genome, proteome, and metabolome: Where are we going? *Ultrastructural Path.*, 27, 289–294, 2003.

Johnson, E.S., Protein modification by SUMO, *Ann. Rev. Biochem.*, 73, 355–382, 2004.

Koonin, E.V., An apology for orthologs — or brave new names, *Genome Biol.*, 2(4), Comment 1005.1–1005.2, 2001.

Righetti, P.G., Bioanalysis: Its past, present, and some future, *Electrophoresis*, 25, 2111–2127, 2004.

Righetti, P.G., Castagna, A., Antonucci, F. et al., The proteome: *Anno Domini* 2002, *Clin. Chem. Lab. Med.*, 41, 425–538, 2003.

Schubert, W., Topological proteomics, toponomics, MELK-technology, *Adv. Biochem. Eng. Biotech.*, 83, 189–209, 2003.

Speicher, D.W., Overview of proteome analysis, in *Proteome Analysis, Interpreting the Genome*, D.W. Speicher, Ed., Elsevier BV, Amsterdam, Ch. 1, 1–18, 2004.

Wadsworth, J.F., Somers, K.D., Cazares, L.H. et al., Serum protein profiles to identify head and neck cancer, *Clin. Cancer Res.*, 10, 1625–1632, 2004.

FUNCTIONAL PROTEOMICS

Greenbaum, D., Baruch, A., Hayrapetian, L. et al., Chemical approaches for functionally probing the proteome, *Mol. Cell. Proteomics*, 1, 60–68, 2002.

Lefkovits, I., Functional and structural proteomics: A critical appraisal, *J. Chrom. B.*, 787, 1–10, 2003.

Resing, K.A., Analysis of signaling pathways using functional proteomics, *Ann. N.Y. Acad. Sci.*, 971, 608–614, 2002.

Ruoppolo, M., Orru, S., D'Amato, A. et al., Analysis of transglutaminase protein substrates by functional proteomics, *Protein Sci.*, 12, 1290–1297, 2003.

CHEMICAL PROTEOMICS

Adam, G.C., Sorensen, E.J. and Cravatt, B.F., Trifunctional chemical probes for the consolidated detection and identification of enzyme activities from complex proteomes, *Mol. Cell. Proteomics*, 1, 828–835, 2002.

Kozarich, J.W., Activity-based proteomics: Enzyme chemistry redux, *Curr. Opin. Chem. Biol.*, 7, 78–83, 2003.

INDIVIDUAL "OMICS"

Bredemeyer, A.J., Lewis, R.M., Malone, J.P., Davis, A.E., Gross, J., Townsend, R.R. and Ley, T.J., A proteomic approach for the discovery of protease substrates, *Proc. Natl. Acad. Sci. USA*, 101, 11785–11790, 2004.

Brenner, C. and Duggen, D., Eds., *Oncogenomics: Molecular Approaches to Cancer*, Wiley-Liss, Hoboken, NJ, 2004.

Cambridge Health Insitute, http://www.genomicsglossary.com.

Dasuri, K., Antonovici, M., Chen, K. et al., The synovial proteome: Analysis of fibroblast-like synoviocytes, *Arthritis Res. Ther.*, 6, R161–R168, 2004.

Figeys, D., Novel approaches to map protein interactions, *Curr. Opin. Biotech.*, 14, 119–125, 2003.

Haas, G., Karaali, G., Ebermayer, K. et al., Immunoproteomics of *Helicobacter pylori* infection and relation to gastric disease, *Proteomics*, 2, 313–324, 2002.

Jain, K.K., Role of pharmacoproteomics in the development of personalized medicine, *Pharmacogenomics*, 5, 331–336, 2004.

Janossy, G., Clinical flow cytometry, a hypothesis-driven discipline of modern cytomics, *Cytometry (Part A)*, 58A, 87–97, 2004.

Krah, A. and Jungblut, P.R., Immunoproteomics, *Meth. Mol. Med.*, 94, 19–32, 2004.

Mann, M. and Jensen, O.N., Proteomic analysis of post-translational modifications, *Nat. Biotech.*, 21, 255–261, 2003.

Moore, J.M.R., Galicia, S.J., McReynolds, A.C., Nguyen, N.-H., Scanlan, T.S. and Guy, R.K., Quantitative proteomics of the thyroid hormone receptor-coregulator interactions, *J. Biol. Chem.*, 279, 27584–27590, 2004.

Nestler, E.J., Psychogenomics: Opportunities for understanding addiction, *J. Neuro.*, 21, 8324–8327, 2001.

Norin, M. and Sundstrom, M., Structural proteomics: Developments in the structure-to-function predictions, *Trends Biotech.*, 20, 79–84, 2002.

Purcell, A.W. and Gorman, J.J., Immunoproteomics. Mass spectrometry-based methods to study the targets of the immune response, *Mol. Cell. Proteomics*, 3, 193–208, 2004.

Saito, R., Suzuki, H. and Hayashizaki, Y., Global insights into protein complexes through integrated analysis of the reliable interactome and knockout lethality, *Biochem. Biophys. Res. Comm.*, 301, 633–640, 2003.

Valet, G.K. and Tárnok, A., Cytomics in predictive medicine, *Cytometry (Part B)*, 53B, 1–3, 2003.

Von Hoof, D.D., Han, H. and Beass, D., Clinomics: Postgenomic cancer care, in *Oncogenomics: Molecular Approaches to Cancer*, C. Brenner, and D. Duggen, Eds., John Wiley & Sons, Hoboken, NJ, Ch. 15, 341–355, 2004.

Weinstein, J.N., Pharmacogenomics — Teaching old drugs new ideas, *New Eng. J. Med.*, 343, 1408–1409, 2000.

Wolters, D.A., Washburn, M.P. and Yates, J.R., III, An automated multidimensional protein identification technology for shotgun proteomics, *Anal. Chem.*, 73, 5683–5690, 2001.

Zijlstra, A., Testa, J.E. and Quigley, J.P., Targeting the proteome/epitome, implementation of subtractive immunization, *Biochem. Biophys. Res. Comm.*, 303, 733–744, 2003.

PLASMA/SERUM PROTEOME

Adkins, J.N., Varnum, S.M., Auberry, K.J. et al., Toward a human blood serum proteome. Analysis by multidimensional separation coupled with mass spectrometry, *Mol. Cell. Proteomics*, 1, 947–955, 2002.

Anderson, N.L. and Anderson, N.G., The human plasma proteome. History, character, and diagnostic prospects, *Mol. Cell. Proteomics*, 1, 845–867, 2002.

Cristea, I.M., Gaskell, S.J. and Whetton, A.D., Proteomics techniques and their application to hematology, *Blood*, 103, 3624–3634, 2004.

Fugii, K., Nakano, T., Kawamura, T. et al., Multidimensional protein profiling technology and its application to human plasma proteome, *J. Proteome Res.*, 3, 712–718, 2004.

Marshall, J., Jankowski, A., Furesz, S. et al., Human serum proteins preseparated by electrophoresis or chromatography followed by tandem mass spectrometry, *J. Proteome Res.*, 3, 364–382, 2004.

Marshall, J., Kupchak, P., Zhu, W. et al., Processing of serum proteins underlies the mass spectral fingerprinting of myocardial infarction, *J. Proteome Res.*, 2, 361–372, 2004.

Omenn, G.S., The human proteome organization plasma proteome project pilot phase: Reference specimens, technology platform comparisons, and standardized data submissions and analyses, *Proteomics*, 3, 1235–1240, 2004.

Reddy, K.S. and Perrotta, P.L., Proteomics and transfusion medicine, *Transfusion*, 44, 601–604, 2004.

Villanueva, J., Philip, J., Entenberg, D. et al., Serum peptide profiling by magnetic particle-assisted, automated sample processing and MALDI-TOF mass spectrometry, *Anal. Chem.*, 76, 1560–1570, 2004.

Wu, S.-L., Choudhary, G., Ramström, M., Bergquist, J. and Hancock, W.S., Evaluation of shotgun sequencing for proteomic analysis of human plasma using HPLC coupled with either ion trap or Fourier transform mass spectrometry, *J. Proteome Res.*, 2, 383–393, 2003.

SYSTEMS BIOLOGY

Kitono, H., Systems biology: A brief overview, *Science*, 295, 1662–1664, 2002.

Oltvai, Z.N. and Barbási, A.-L., Life's complexity pyramid, *Science*, 298, 763–764, 2002.

van der Greef, J., Stroobant, P. and van der Heijden, R., The role of analytical sciences in medical systems biology, *Curr. Opin. Chem. Biol.*, 8, 559–565, 2004.

Weston, A.D. and Hood, L., Systems biology, proteomics, and the future of health care: Toward predictive, preventive, and personalized medicine, *J. Proteome Res.*, 3, 179–196, 2004.

2 An Overview of the Chemical Modification of Proteins

CONTENTS

INTRODUCTION

The primary purpose of this chapter is to provide an overview of the current status of the specific chemical modification of proteins. The approach is intended to be general but particular importance is given to certain reagents and reactions that provide the basis for chemical modifications useful for proteomics. Specific applications to proteomics such as chemical proteomics, two-dimensional fluorescence difference gel electrophoresis (DIGE) and isotope-coded affinity tags (ICAT) are discussed in Chapter 3. Also included is a limited discussion of *in vivo* chemical modifications of proteins separate from translational processing as such modifications are of major interest to investigators using proteomics to study toxicology and environmental health. The reader is directed to several reviews for experimental detail[1-3] on the specific chemical modification of proteins.

SPECIFIC CHEMICAL MODIFICATION OF PROTEINS

This section will focus on selected recent studies on the specific chemical modification of proteins. The specific chemical modification of proteins is defined as the modification of one or more specific amino acid residues in a protein, for example, the modification of cysteine residues or the modification of lysine residues. Site-specific modification infers that the modification occurs at a single amino acid residue in a protein. Non-specific modification infers that the modification occurs at several different amino acid residues such as modification of lysine and cysteine by a single reagent. Chemical modification of proteins is accomplished with exogenous reagents such as acetic anhydride, iodoacetamide, tetranitromethane and phenylglyoxal, all of which are discussed below. Most of the modifications utilize common chemical reactions in organic chemistry such as those described in Figure 2.1. Chemical modification of proteins also occurs with zendogenous reagents such as peroxynitrite derived from nitric oxide (Figure 2.2) or biological aldehydes such as methyl glyoxal and 4-hydroxy-2-nonenal (Figure 2.3).[4] Figure 2.3 shows the action of peroxynitrite as a reagent for the nitration of tyrosine or the oxidation of lysine. The reaction of aldehydes with lysine can be complex, as demonstrated by the complex reactions involved in the cross-linking of collagen.[5]

The reactivity of a functional group in a protein is largely dependent upon two factors. The first is the nucleophilicity of the amino acid residue and the second is the accessibility of the functional group to the chemical reagent. The nucleophilicity of a functional group is in turn a function of the inherent chemistry of the functional group and the environment surrounding the residue. The environments of the various amino acid residues in a protein are not identical. Due to this lack of homogeneity, a variety of surface polarities surrounds the various functional groups. The physical and chemical properties of any given functional group is strongly influenced by the nature (e.g., polarity) of the local microenvironment. For example, consider the effect of the addition of an organic solvent, ethyl alcohol, on the pKa of acetic acid. In 100% H_2O, acetic acid has a pKa of 4.70. The addition of 80% ethyl alcohol results in an increase of the pKa to 6.9. In 100% ethyl alcohol, the pKa of acetic acid is 10.3. This result is particularly important when considering the reactivity of nucleophilic groups such as amino groups, cysteine, carboxyl groups and the phenolic hydroxyl group. In the case of the primary amines present in protein, these functional groups are not reactive except in the freebase form. In other words, the proton present at neutral pH must be removed from the ε-amino group of lysine before this functional group can function as an effective nucleophile. A listing of the "average" pKa values for the various functional groups present in proteins is given in Table 1.

Some modification reactions take advantage of differences in pKa values in similar chemical groups. For example, the difference in pKa values between an α-amino group and an ε-amino group makes it possible to selectively modify the α-amino group in a protein. Another example is the selective modification of the γ-carboxyl groups on a protein without modification of the γ-carboxyl groups since the protonated form of the carboxylic acid is required for successful reaction. Other factors that

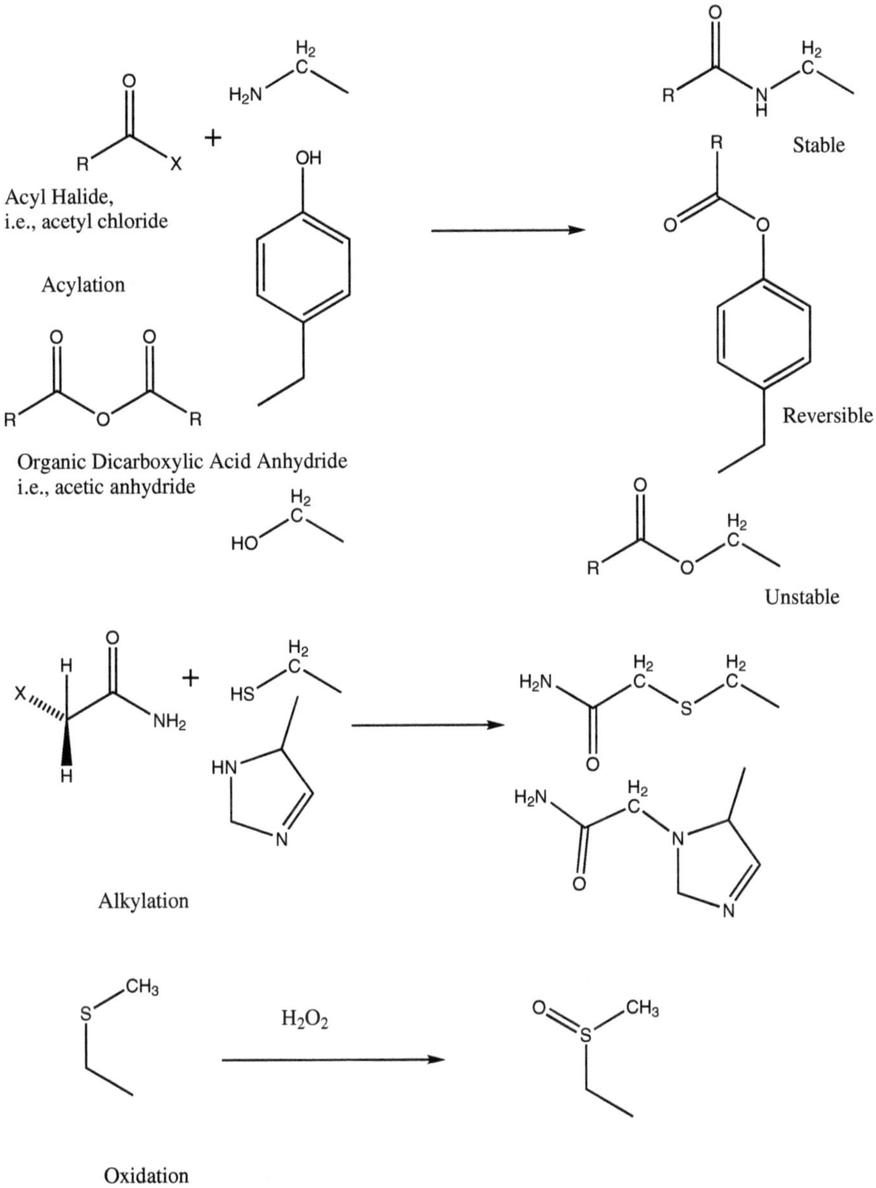

FIGURE 2.1 Some common chemical reactions used to modify functional groups in proteins. The great majority of reactions are either acylation reactions or alkylation reactions. Alkylation reactions can either be the substitution reactions (S_N2) as shown or insertion reactions such as those seen with Michael addition reactions as shown in Figure 10.

can influence the pKa of a functional group in a protein include hydrogen binding with an adjacent functional group, the direct electrostatic effect of a charged group's presence in the immediate vicinity of a potential nucleophile and direct steric effects on the availability of a given functional group. An excellent example of the effect

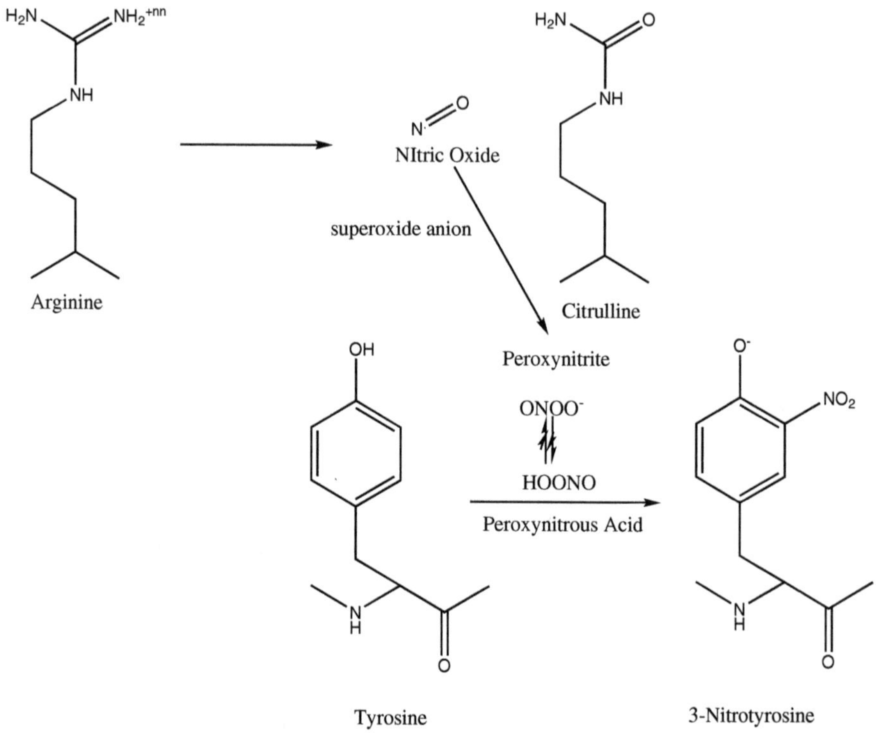

FIGURE 2.2 Some reaction of peroxynitrite with functional groups on proteins. Shown is the pathway responsible for the formation of peroxynitrite from nitric oxide and superoxide and the subsequent modification of tyrosine by nitration. Not shown are the oxidation reactions seen with peroxynitrite and cysteine, methionine or lysine.

of a neighboring group on the reaction of a specific amino acid residue is provided by the comparison of the rates of modification of the active-site cysteinyl residue by chloroacetic acid and chloroacetamide in papain.[6,7] A rigorous evaluation of the effect of pH and ionic strength on the reaction of papain with chloroacetic acid and chloroacetamide demonstrated the importance of a neighboring imidazolium group in enhancing the rate of reaction at low pH. The essence of the experimental observations is that the plot of the pH dependence of the second-order rate constant for the reaction with chloroacetic acid is bell-shaped with an optimum at about a pH of 6.0 while that of chloroacetamide is S-shaped approaching maximal rate of reaction at a pH of 10.0. Other excellent examples of the effect of neighboring functional groups is provided by the reaction of 2,4-dinitrophenyl acetate with a lysine residue at the active site of phosphonoacetaldehyde hydrolase where the pKa of the lysine residue is decreased to 9.3 as a result of a positively charged environment[8] and the effect of remote sites on the reactivity of histidine residues in ribonuclease A.[9] We would be remiss to not mention the seminal observations of Schmidt and Westheimer[10] on the reaction of 2,4-dinitrophenyl propionate with the active site lysine of acetoacetate decarboxylase demonstrating a pKa of 5.9 for this

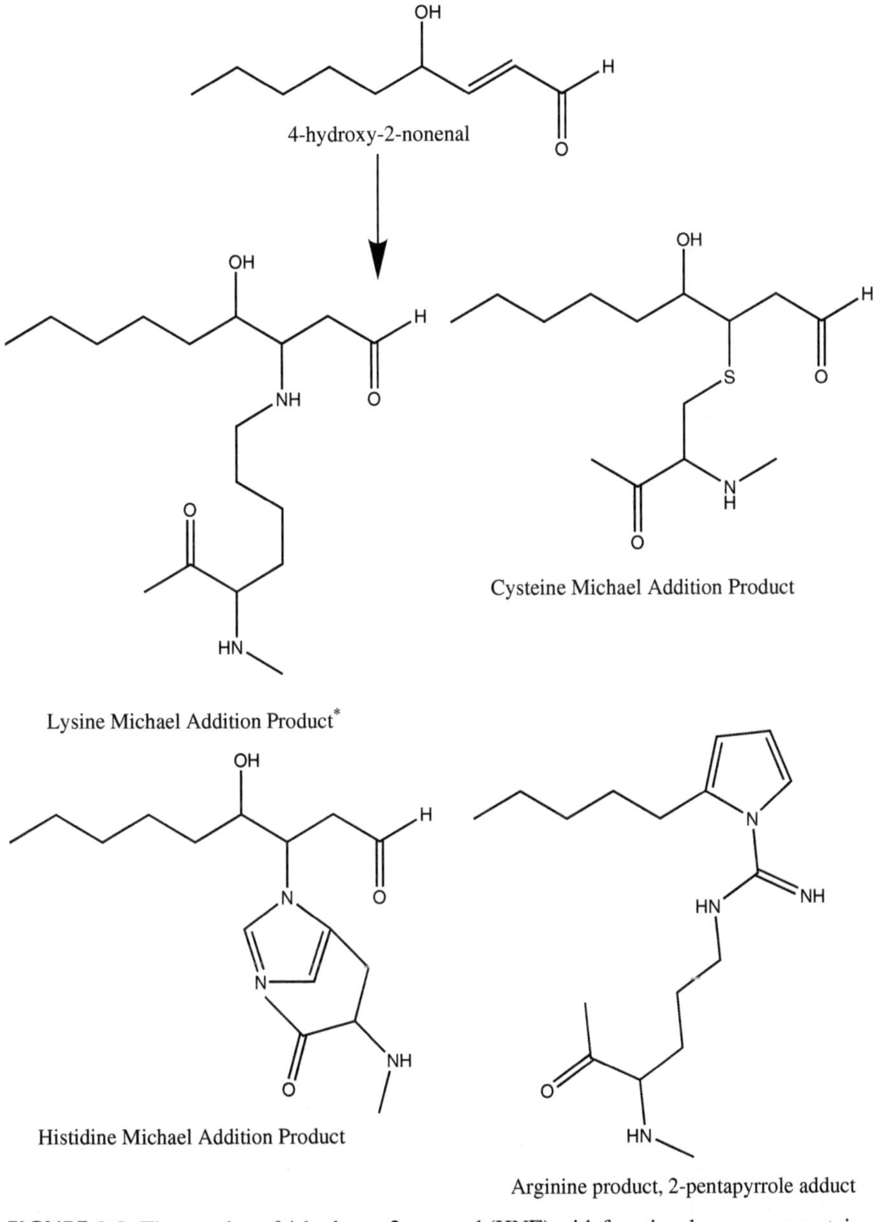

FIGURE 2.3 The reaction of 4-hydroxy-2-nonenal (HNE) with functional groups on proteins. HNE is derived from the peroxidation of lipid fatty acids such as linoleic acid (Uchida, K. and Stadtman, E.R., Covalent attachment of 4-hydroxynonenal to glyceraldehydes-3-phosphate dehydrogenase, *J. Biol. Chem.*, 266, 6388–6393, 1993).

TABLE 2.1
Dissociation Constants for Nucleophiles in Proteins

Potential nucleophile	pKa
γ-Carboxyl (Glutamic acid)	4.25
β-Carboxyl (Aspartic acid)	3.65
A-Carboxyl (Isoleucine)	2.36
Sulfhydryl (Cysteine)	10.46
α-Amino (Isoleucine)	9.68
Phenolic hydroxyl (Tyrosine)	10.13
ε-Amino (Lysine)	10.79
Imidazole (Histidine)	6.00
Guanidino (Arginine)	12.48

Data for Table 1 taken from Mooz, E.D., Data on the naturally occurring amino acids, in *Practical Handbook of Biochemistry and Molecular Biology*, G.D. Fasman, Ed., CRC Press, Boca Raton, Florida, 1989.

residue. These examples clearly demonstrate the effect of electrostatic effects in the reactivity of amino acid residues in proteins.

Another consideration exists that can, in a sense, be considered either a cause or consequence of microenvironmental polarity and involves the environment immediately around the residue modified. These are the "factors" that can cause a "selective" increase (or decrease) in reagent concentration in the vicinity of a potentially reactive species. The most clearly understood example of this action is the process of affinity labeling.[11] Yet another consideration is the partitioning of a reagent such as tetranitromethane between the polar aqueous environment and the nonpolar interior of the protein. Tetranitromethane is an organic compound and, in principle, can react equally well with exposed and buried tyrosyl residues.[12,13] Thus, caution must be taken when interpreting results in terms of residue availability to solvent.

Since most of the data obtained for the reactivity of amino acid residues in proteins has been derived from structure-function studies with native proteins, little data exists on the reactivity of functional groups in denatured proteins but the prolonged time course required for complete modification of sulfydryl groups in a protein[14] and the observation that substances such as detergents influence reactivity[15] suggest that there are microenvironmental effects in denatured proteins as well. A considerable amount of information exists on the "exposure" of functional groups such as the sulfydryl group of cysteine as judged by chemical reactivity.[16,17]

A limited listing of reactions for the specific chemical modification of proteins is presented in Table 2.2. This list is not meant to be exclusive. As cited above, most chemical modifications can be assessed by mass spectrometry. The following section describes the chemical modification of individual amino acids in some detail. Chemical mutagenesis has proved useful in the study of protein structure. In the current context, chemical mutagenesis is defined as a reaction or reactions

TABLE 2.2
Reagents for Chemical Modification in Proteomics

Residue	Reagent	Specificity[a]	Ease[b]	MS[c]	Ref.
Arginine	phenylglyoxal	high	yes	yes	1
Lysine	acetic anhydride	medium	yes	yes	2
Histidine	diethylpyrocarbonate	medium	yes	yes	3
Cysteine	alkyl methanethiosulfonates	high	yes	yes	4
Cysteine	α-haloalkyl derivatives	low	yes	yes	5
Cysteine	N-alkylmaleimides	high	yes	yes	6
Cystine	phosphine[d]	high	yes	yes	7
Carboxyl	carbodiimide[e]	medium	yes	yes	8
Tryptophan	N-bromosuccinimide	medium	yes	yes	9
Tyrosine	tetranitromethane	medium	yes	yes	10

[a] High: little, if any reaction, at other amino acid residues. Medium: reaction does occur at other amino acid residues but is either infrequent or controlled by appropriate reaction conditions. Low: considerable reaction described at other amino acid residues.

[b] The chemical modification is performed without the need for special conditions or precautions.

[c] Analysis of modification by mass spectrometry has been described.

[d] Includes tris(2-carboxyethyl)phosphine and tributylphosphine.

[e] Includes N,N′-dicyclohexylcarboiimide and 1-ethyl-[3-(dimethylamino)propyl]-carbodiimide.

REFERENCES FOR TABLE 2.2

1. Sheng, J. and Preiss, J., Arginine[294] is essential for the inhibition of *Anabaena* PCC 7120 ADP-glucose pyrophosphorylase by phosphate, *Biochemistry*, 36, 13077–13084, 1997.
2. Smith, C.M., Gafken, P.R., Zhang, Z., Gottschling, D.E., Smith, J.B. and Smith, D.L., Mass spectrometric quantification of acetylation at specific lysine within the amino-terminal tail of histone H4, *Anal. Biochem.*, 316, 23–33, 2003.
3. Jin, X.R., Abe, Y., Li, C.Y. and Hamasaki, N., Histidine-834 of human erythrocyte band 3 has an essential role in the conformational changes that occur during the band 3-mediated anion exchange, *Biochemistry*, 42, 12927–12932, 2003.
4. Britto, P.J., Knipling, L., McPhie, P. and Wolff, J., Thiol/disulfide interchange in tubulin: Kinetics and the effect on polymerization, *Biochem. J.*, 389, 549–588, 2005.
5. Pasquarello, C., Sanchez, J.-C., Hochstrasser, D.F. and Corthals, G.L., *N*-t-Butyl-iodoacetamide and iodoacetanilide: Two new cysteine alkylating reagents for relative quantitation of proteins, *Rapid Comm. Mass Spec.*, 18, 117–127, 2004.
6. Niwayama, S., Kurono, S. and Matsumoto, H., Synthesis of d-labeled *N*-alkylmaleimides and application to quantitative peptide analysis by isotope differential mass spectrometry, *Bioorg. Med. Chem. Letters*, 11, 2257–2261, 2001.
7. Thevis, M., Loo, R.R.O. and Loo, J.A., Mass spectrometric characterization of transferrins and their fragments derived by reduction of disulfide bonds, *J. Am. Soc. Mass Spec.*, 14, 635–647, 2003.
8. Izumi, S., Kaneko, H., Yamazaki, T., Hirata, T. and Kominami, S., Membrane topology of guinea pig cytochrome P450 17 alpha revealed by a combination of chemical modifications and mass spectrometry, *Biochemistry*, 42, 14663–14669, 2003.

9. McDermott, L., Moore, J., Brass, A. et al., Mutagenic and chemical modification of the ABA-1 allergen of the nematode *Ascaris*: Consequences for structure and lipid binding properties, *Biochemistry*, 40, 9918–9926, 2001.

10. Willard, B.B., Ruse, C.I., Keightley, J.A., Bond, M. and Kinter, M., Site-specific quantitation of protein nitration using liquid chromatography/tandem mass spectrometry, *Anal. Chem.*, 75, 2370–2376, 2003.

that change the quality of an amino acid functional group, such as the conversion of cysteine to a lysine analogue by reaction with aziridine or bromoethyl amine (Figure 2.4) or the insertion of sulfydryl function by the modification of lysine with 2-iminothiolane or *N*-succinimidyl *S*-acetylthioacetate (Figure 2.5). A partial listing of these reactions is presented in Table 2.3.

ARGININE

The modification of arginyl residues in proteins continues to be based on the reaction of vicinal dicarbonyl compounds such as phenylglyoxal, 2,3-butanedione and 1,2-cyclohexanedione to form an adduct (Figure 2.6). Interest also exists in the reaction of methylglyoxal with arginine residues in proteins[18–20] as well as 2-hydroxy-4-nonenol (Figure 2.3).[4] A review of the literature from 1997 to 2003 shows that phenylglyoxal was the most extensively used reagent for the specific modification of arginine residues in proteins, with somewhat less use of 2,3-butanedione and little use of 1,2-cyclohexanedione. Modification of proteins with phenylglyoxal and *p*-hydroxy-phenylglyoxal (Figure 2.7) is used to identify arginyl residues involved in binding anionic allosteric modifiers[21] and substrates.[22] More extensive use of phenylglyoxal is found in the modification of membrane transport.[23–26] Since the adducts formed between the vicinal dicarbonyl compounds are not covalent, stability can be an issue. Notwithstanding this potential issue, identification of the chemical

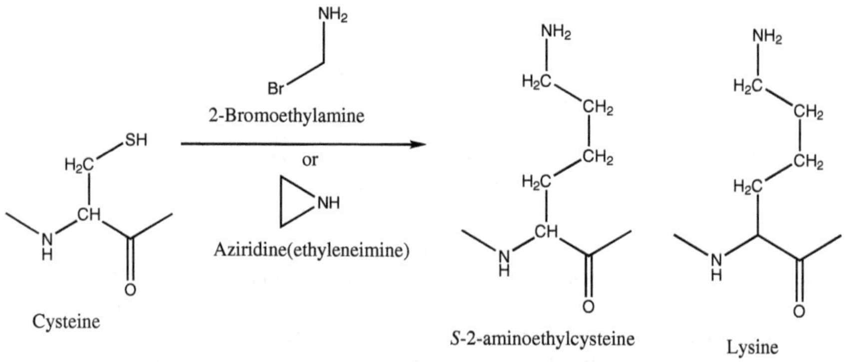

FIGURE 2.4 The chemical mutation of cysteine to lysine. The reaction of cysteine with 2-bromoethylamine or aziridine (ethyleneimine) results in the formation of *S*-2-aminoethyl-cysteine, an analog of lysine. This allows cleavage by trypsin and forms the basis for chemical rescue (Hopkins, C.E., O'Connor, P.B., Allen, K.N. et al., Chemical modification rescue assessed by mass spectrometry demonstrates that gamma-thia-lysine yields the same activity as lysine in aldolase, *Protein Sci.*, 11, 1591–1599, 2002).

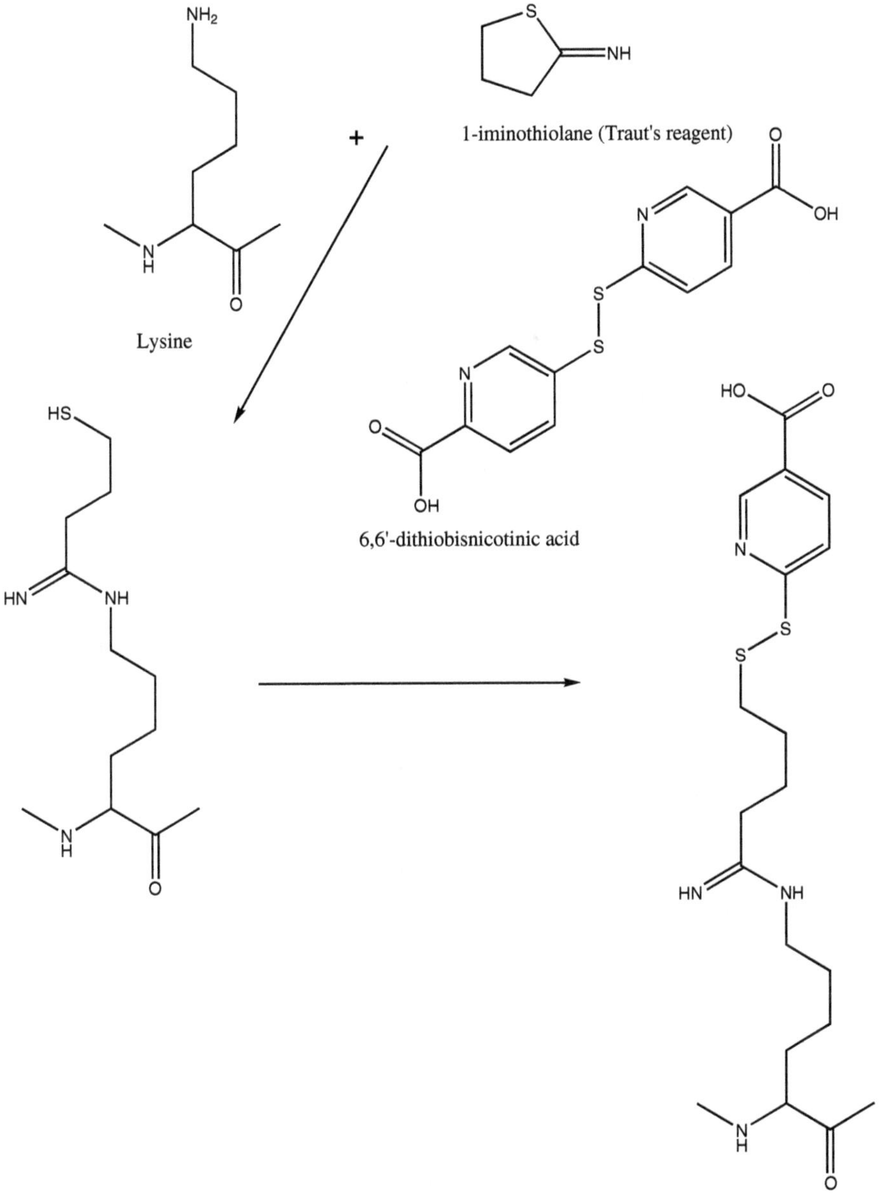

FIGURE 2.5 Method for the insertion of a reactive sulfydryl group by modification of lysine. Shown is the reaction of lysine with 2-iminothiolane (methyl-4-mercaptobutyrimidate, Traut's reagent) to yield a reactive sulfydryl group which is modified *in situ* with 6,6′ -dithiobis-nicotinic acid (Meunier, L., Bourgerie, S. and Mayer, R., Optimized conditions to couple two water-soluble biomolecules through alklyamine thiolation and thioetherification, *Bioconjugate Chem.*, 10, 206–212, 1999).

TABLE 2.3
The Chemical Mutagenesis of Proteins

Amino Acid/Reagent	Product/Analogue	Ref.
Cysteine/N-succinimidyl-S-acetylthio-acetate/NH_2OH	S-thioacetylcysteine	1
	lysine	
Cysteine/ethyleneimine (aziridine) or	S-aminoethylcysteine	2
bromoethylamine	lysine	
Cysteine/iodomethane	S-methylcysteine	4[a]
	methionine	
Cysteine/2-bromoethanol	S-hydroxyethylcysteine	4[a]
	serine	
Cysteine/(2-sulfonatoethyl)methanethiosulfonate	S-sulfonatoethyl-cysteine	5
	glutamic acid	
Lysine/citraconic anhydride	Citraconylated lysine[b,c]	6
	charge change/glutamic, asparic	
Lysine/succinic anhydride	succinylated lysine[c]	7
	change change/glutamic, aspartic	
Lysine/N-succinimidyl S-acetylthioacetate	Mercaptoacetyl lysine[c]	8
	cysteine	
Lysine/2-iminothiolane (Traut's reagent)	1-Mercapto-3-iminobutyl-lysine	9
	cysteine	
Lysine/$HCHO$/$NaCNBH_3$ (reductive methylation)	Methyl-lysine	10
Lysine/O-methylisourea (guanidation)	Homoarginine	11
	arginine	
Tyrosine/$C(NO_2)_4$/$NaCNBH_3$	3-Aminotyrosine	12
	introduction of amino group	

[a] This approach uses the alkylation of cysteine with substituted haloalkane to produce analogues of a wide variety of amino acids.
[b] Citraconylation can be reversed by mild acid treatment.
[c] Reaction also occurs at α-amino groups.

REFERENCES FOR TABLE 2.3

1. Konigsberg, R.J., Godtel, R., Kissel, T. et al., The development of IL-2 conjugated liposomes for therapeutic purposes, *Biochim. Biophys. Acta*, 1370, 243–251, 1998.
1a. Duncan, R.J., Weston, P.D. and Wrigglesworth, R., A new reagent which may be used to introduce sulfydryl groups into proteins, and its use in the preparation of conjugates for immunoassay, *Anal. Biochem.*, 132, 68–73, 1983.
1b. Visser, C.C., Voowinden, L.H., Harders, L.R. et al., Coupling of metal containing homing devices to liposomes via a maleimide linker: Use of TCEP to stabilize thiol-groups without scavenging metals, *J. Drug Target.*, 12, 569–573, 2004.
2. Hollenbaugh, D., Aruffo, A. and Senter, P.D., Effects of chemical modification on the binding activities of P-selectin mutants, *Biochemistry*, 34, 5678–5684, 1995.
3. Cook, L.L. and Gafni, A., Protection of phosphoglycerate kinase against *in vitro* aging by selective methylation, *J. Biol. Chem.*, 263, 13991–13993, 1988.

4. Schindler, J.F. and Viola, R.E., Conversion of cysteinyl residues to unnatural amino acid analogues. Examination in a model system, *J. Protein Chem.*, 15, 737–742, 1996.
5. Chen, S., Hartmann, H.A. and Kirsch, G.E., Cysteine mapping in the ion selectivity and toxin binding region of the cardiac NaA$^+$ channel pore, *J. Membrane Biol.*, 155, 11–25, 1997.
6. Kadlík, V., Strohalm, M. and Kodíek, M., Citraconylation — A simple method for high protein sequence coverage in MALDI-TOF mass spectrometry, *Biochem. Biophys. Res. Comm.*, 305, 1091–1093, 2003.
7. Kim, S.-H., Lee, J.-H., Yun, S.-Y. et al., Reaction monitoring of succinylation of collagen with matrix-assisted laser desorption/ionization mass spectrometry, *Rapid Comm. Mass Spec.*, 14, 2125–2128, 2003.
8. Duncan, R.J.S., Weston, P.D. and Wrigglesworth, R., A new reagent which may be used to incorporate sulfydryl groups into proteins, and its use in the preparation of conjugates for immunoassay, *Anal. Biochem.*, 132, 68–73, 1983.
9. Masuko, T., Minami, A., Iwasaki, N. et al., Thiolation of chitosan. Attachment of proteins via thioether formation, *Biomacromolecules*, 6, 880–884, 2005.
10. Hsu, J.L., Huang, S.Y., Shiea, J.T. et al., Beyond quantitative proteomics: Signal enhancement of the al ion as a mass tag for peptide sequencing with dimethyl labeling, *J. Proteome Res.*, 4, 101–108, 2005.
11. Brancia, F.L., Butt, A., Benyon, R.J. et al., A combination of chemical derivatisation and improved bioinformatic tools optimise protein identification for proteomics, *Electrophoresis* 22, 552–559, 2001.
12. Beckingham, J.A., Housden, N.G., Muir, N.M. et al., Studies on a single immunoglobulin-binding domain of protein L from *Peptostreptococcus magnus*: The role of tyrosine-53 in the reaction with IgG, *Biochem. J.*, 353, 395–401, 2001.

modification of arginine by mass spectrometry is possible[4,27] and the reactions are reasonably specific for arginine residues.

CYSTEINE

Cysteine is an attractive target for specific modification in proteins reflecting the relative ease of reaction (Figure 2.8) without concomitant modification of other nucleophiles such as the ε-amino group of lysine or the imidazole ring of histidine. As a result, a large number of reagents are available for the modification of cysteine and the modification of cysteine is used more in proteomics than modification of other amino acids. A consideration of the recent literature suggests that alkyl methanethiosulfonate derivatives (Figure 2.9)[28–30] are the most commonly used reagents for the study of cysteine in the structure-function in proteins. N-Alkylmaleimide derivatives (Figure 2.10)[31,32] are also extensively used as are α-haloalkyl compounds (Figure 2.8); iodoacetamide and iodoacetic acid are used for the modification of cysteine residues after the reduction of proteins.[33,34] The development of isotope-coded affinity tags (ICAT) as described in Chapter 3 is based on α-haloalkyl chemistry.

The alkyl methanethiosulfonates react with cysteine to form a mixed disulfide much in the same manner that sodium tetrathionate reacts with cysteine to form the S-sulfonate derivative (Figure 2.11),[35] while α-haloketo compounds such as

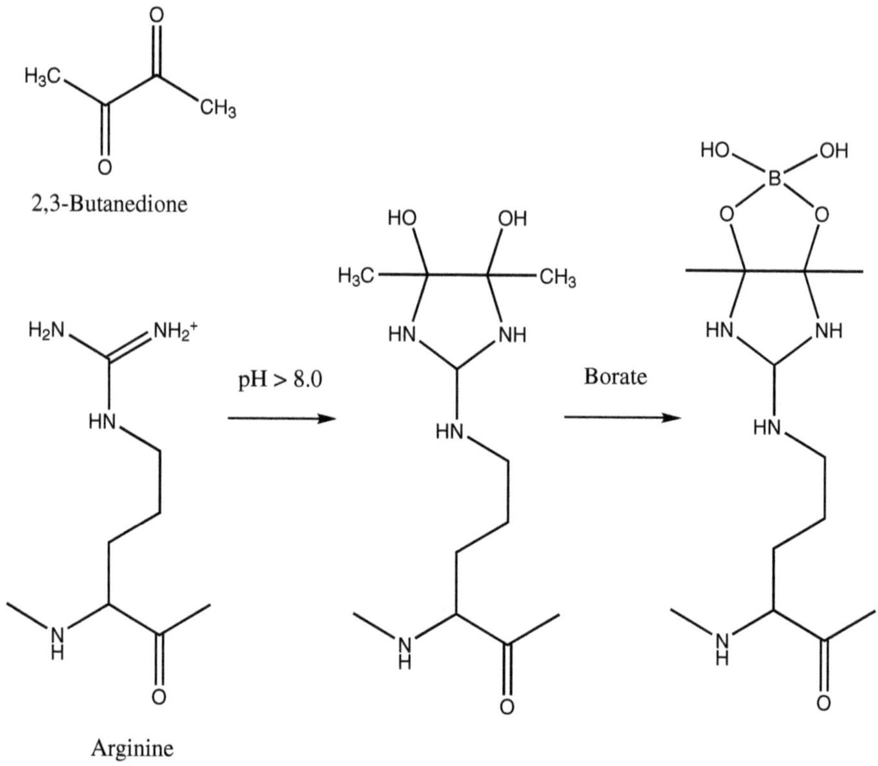

FIGURE 2.6 Some reactions of vicinyl dicarbonyl compounds with arginine. Shown as an example is the reaction of arginine with 2,3-butanedione. A similar reaction is observed with 1,2-cyclohexanedione.

iodoacetamide and N-alkylmaleimides react with cysteine via an S_N2 reaction to form a stable thioether linkage (see below). The mixed disulfides formed with reaction with the alkyl methanethiosulfonates are somewhat labile, although stability studies have not been formed on the various derivatives. Marked differences exist in the rate of reaction of disparate methanthiosulfonate derivatives with sulfydryl groups.[3] Wolff and coworkers[37] reported on the rates of modification of cysteine in tubulin with iodoacetamide, N-ethylmaleimide, and IAEDANS ([5-((((2-iodoacetyl)amino)ethyl) amino) naphthalene-1-sulfonic acid], Figure 2.12). This reference is an excellent paper and highly recommended for serious study. For the purpose of the current discussion, this study provides two important points critical for understanding cysteine reactivity in proteins. First, the local electrostatic environment is an important consideration in individual residues reactivity. Second, while both iodoacetate and N-ethylmaleimide react with cysteine via an S_N2 process (Figure 2.13), iodoacetate is a nucleophilic displacement reaction while N-ethylmaleimide is a nucleophilic addition reaction. In this study, the reaction of sulfhydryl groups with N-ethylmaleimide was faster and more extensive than that observed for iodoacetamide. This would suggest that the maleimide derivatives might be more suitable for modification of cysteine residues in proteins. Maleimide derivatives are likely more specific than the α-haloalkyl

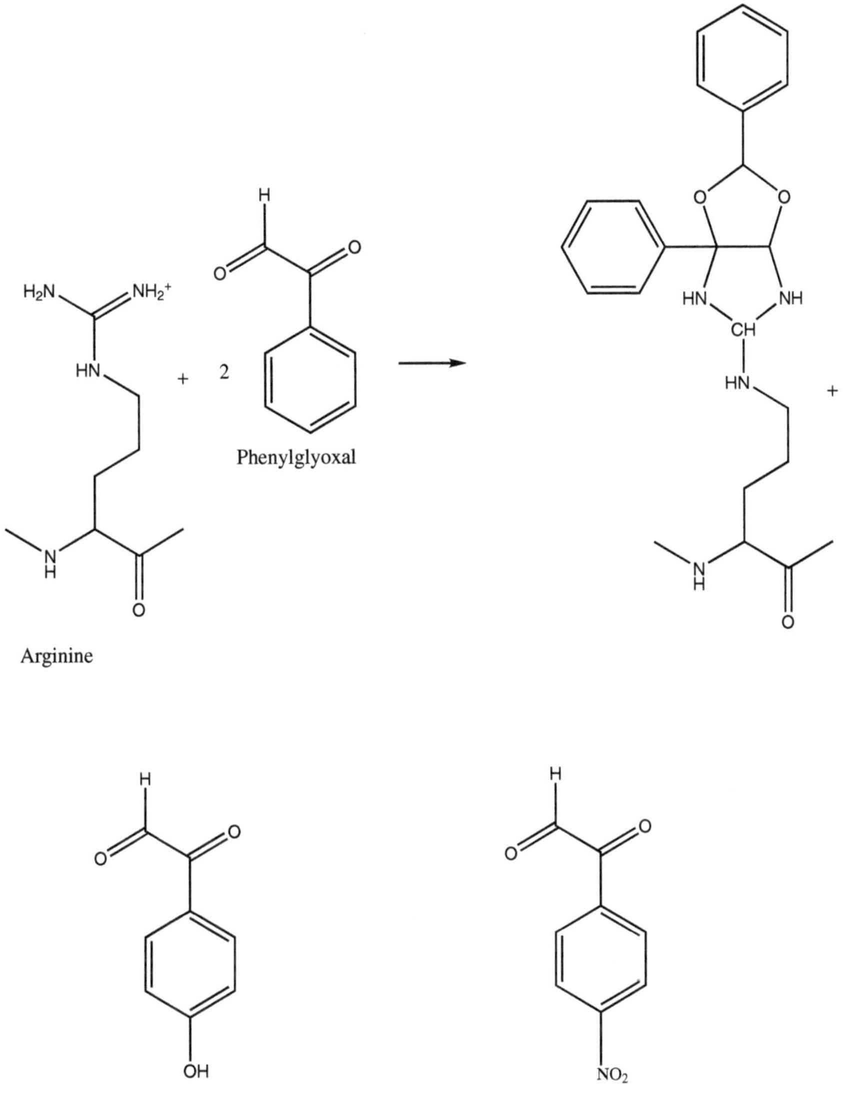

p-hydroxyphenylglyoxal *p*-nitrophenylglyoxal

FIGURE 2.7 The reaction of phenylglyoxal with arginine. Also shown are two useful derivatives of phenylglyoxal-*p*-hydroxyphenylglyoxal and *p*-nitrophenylglyoxal.

compounds since the authors have been unable to find any reference to reaction of maleimide compounds with functional groups other than thiol groups in proteins; α-haloalkyl compounds react with imidazole, amino, methionine and carboxyl groups in addition to cysteine. Maleimides do not react with thiourea included in solvents for the preparation of proteins for two-dimensional electrophoresis; other reagents such as acrylamide, which react via a Michael addition mechanism, do not react with thiourea. Vinyl pyridine is another example of a reagent reacting with sulfydryl groups

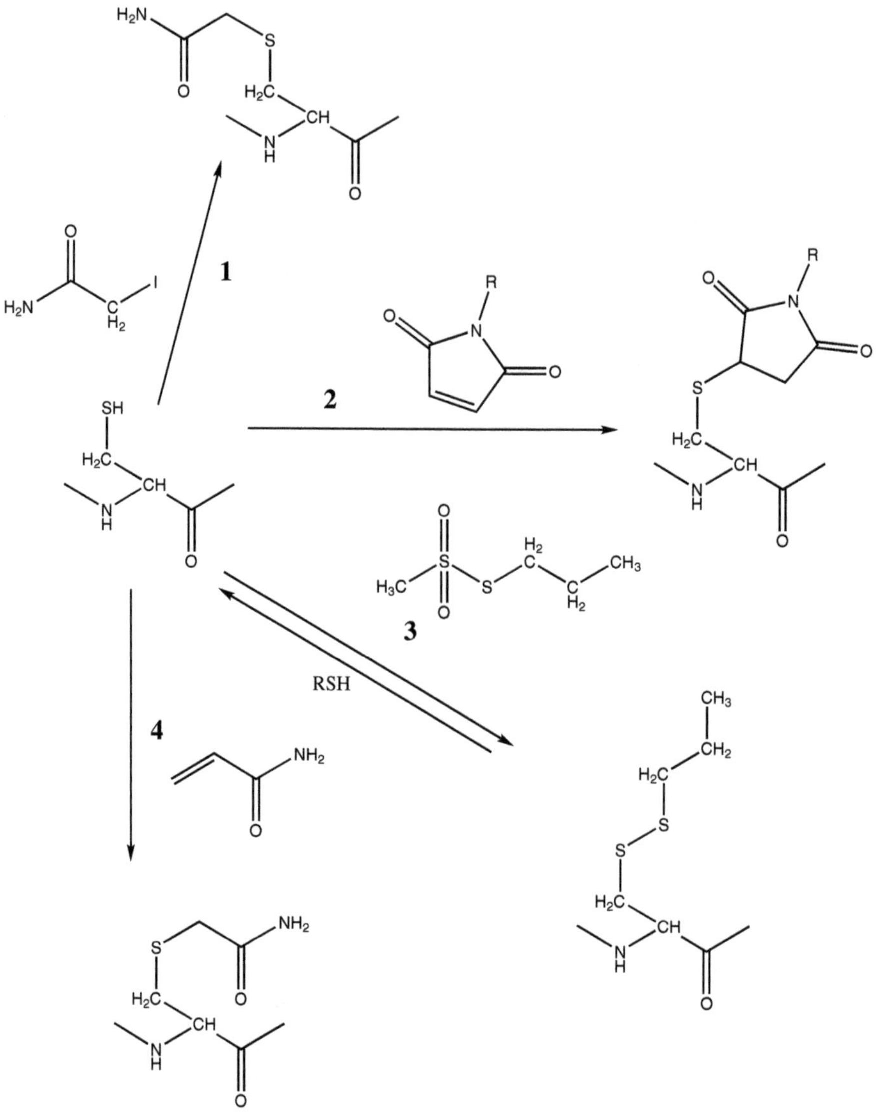

FIGURE 2.8 Multiple pathways for the chemical modification of cysteine residues in proteins. Shown is (1) carboxymethylation (alkylation) with iodoacetamide; (2) a Michael addition reaction with an N-alkylmaleimide; (3) the reversible reaction with an alkyl-methanethiosulfonate derivative forming a disulfide addition product; and (4) a Michael addition with acrylamide.

via a Michael addition (Figure 2.14);[38] studies suggest that vinyl pyridine can effectively alkylate cysteine residues at lower pH than iodoacetic acid.[39]

Cysteine residues can be modified with bromoethylamine or ethyleneimine (aziridine, Figure 2.4) to produce a lysine analogue that is susceptible for cleavage by trypsin. Regnier and colleagues have described the modification of cysteine with

Methyl Methanethiosulfonate
Neutral

2-Sulfoethyl Methanethiosulfonate
Negatively Charged

[2-(trimethylamonnium)ethyl] methanethiosulfonate
Positively Charged

Benzyl Methanethiosulfonate
Aromatic

FIGURE 2.9 Various alkyl and aryl methanethiosulfonate derivative which can be used to modify cysteine residues in proteins. Shown are neutral, charged and aromatic derivatives.

(3-acrylamidopropyl)trimethylammonium chloride (Figure 2.15).[40] Cysteine residues can be modified with reagents such as bromoethylamine to produce an analogue of lysine (Figure 2.5) that can be cleaved in proteins; this unique amino acid can also be modified with O-methylisourea to yield a homoarginine analogue (Figure 2.16).[41]

CYSTINE

The reduction and alkylation of proteins under denaturing conditions is a prerequisite for the determination of structure.[42,43] Also possible is to use the selective reduction of disulfide bonds in proteins to study structure-function relations in proteins for some time[44] where reduction is performed in the absence of chaotropic agents. Some

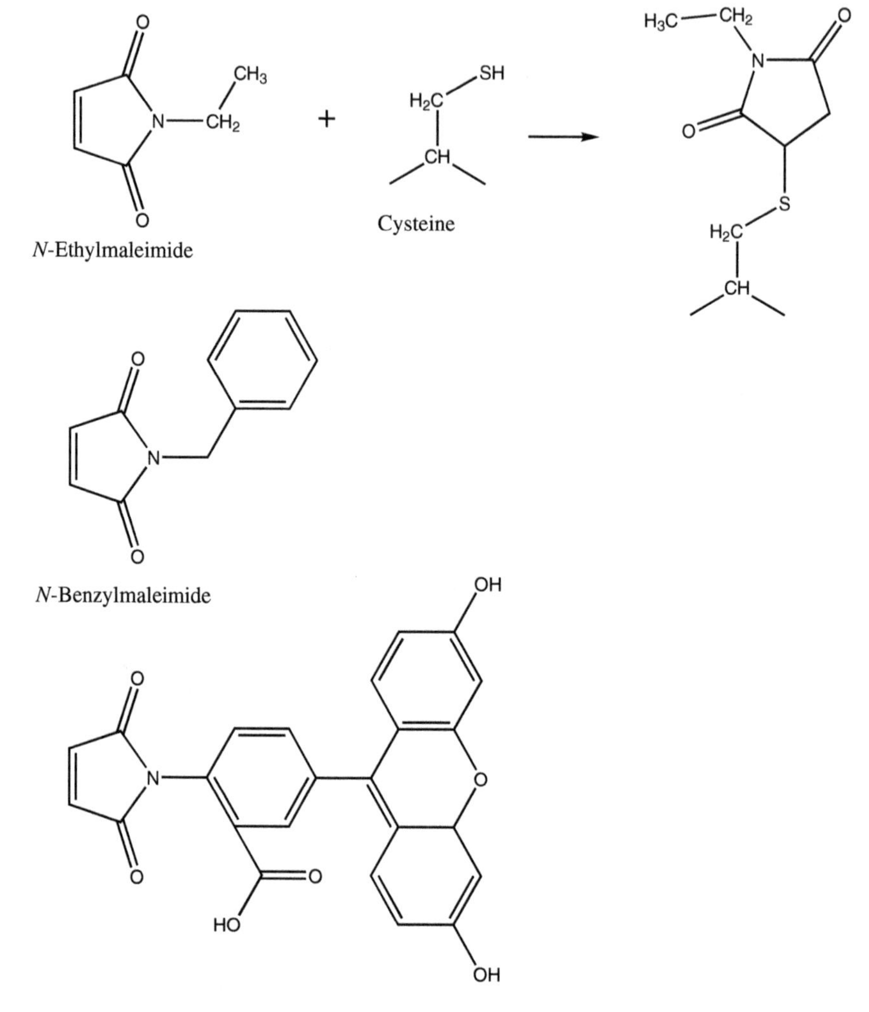

N-Ethylmaleimide

Cysteine

N-Benzylmaleimide

N-Fluorosceinmaleimide

FIGURE 2.10   Some maleimide derivatives used to modify cysteine residues in proteins.

recent studies also exist on the disulfide proteome.[45] The reduction of a disulfide bond in α-lactalbumin by ultraviolet light has been reported.[46] In work directly related to proteomics, phosphine derivatives appear to be more useful than thiol compounds in the reduction of proteins prior to two-dimensional electrophoresis (Figure 2.17).[47,48] Cleavage at S-cyanocysteine residues after disulfide bond reduction with tris(2-carboxyethyl) phosphine (Figure 2.18) and reaction with 1-cyano-4-dimethy-lamino pyridinium tetrafluoroborate is useful in structural analysis.[49] Cystine trisulfide (thiocystine, Figure 2.19)[50,51] is formatted as a result of protein fractionation in an multicompartment electrolyzer.[52]

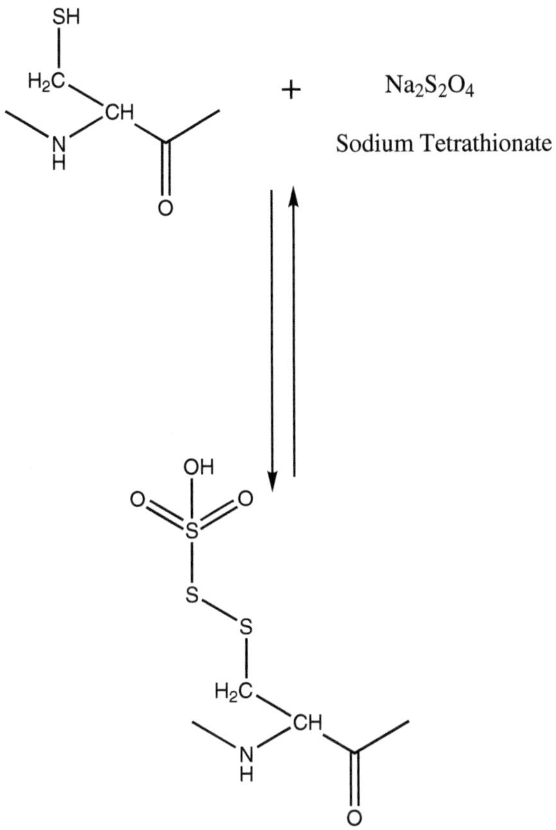

FIGURE 2.11 The reaction of sodium tetrathionate with cysteine. This is a reversible reaction and can be used to protect cysteine residues from modification.

CARBOXYL GROUPS

Both water soluble and water-insoluble carbodiimides have been used to modify carboxyl groups in proteins (Figure 2.20).[53,54] Woodward's reagent K (Figure 2.21) has been used to modify a specific glutamic acid residue in human erythrocyte band 3 protein.[55] Both Woodward's reagent K and carbodiimides modify carboxyl groups in the absence of an added nucleophile. In principle, proteins modified with Woodward's reagent K can undergo slow hydrolysis of the activated ester intermediate while carbodiimide-modified proteins undergo slow rearrangement of the O-acylisourea to a stable N-acylurea (Figure 2.22). The addition of a nucleophile such as glycine methyl ester is required to establish the extent of modification. N-(3-(Dimethylamino)propyl)-N′ -ethylcarbodiimide has been used to form an intramolecular crosslink protein between a lysine residue and a glutamic acid residue in a process referred to as zero-length cross-linking and is used frequently to determine areas of protein contact; the chemistry of protein cross-linking is

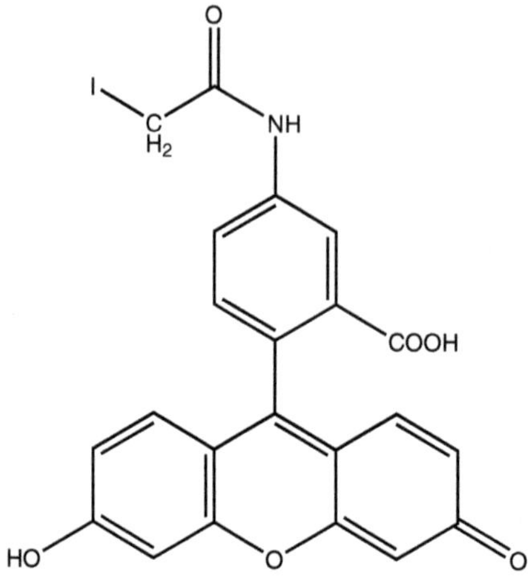

5-(iodoncetoamido)fluorescein

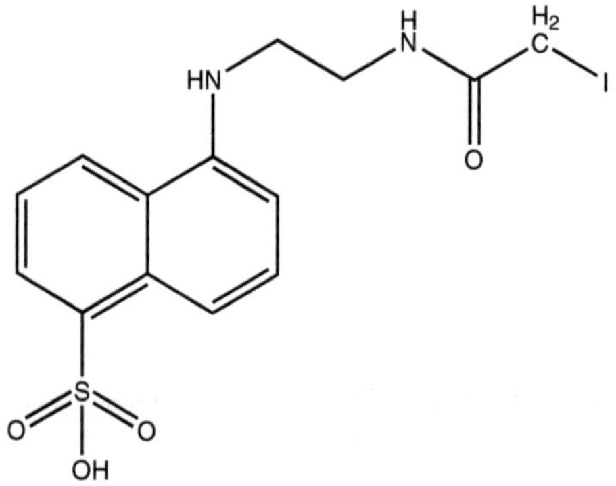

1,5 IAEDANS (*N*-iodoacetaminoethyl)-1-naphthylamine-5-sulfonic acid

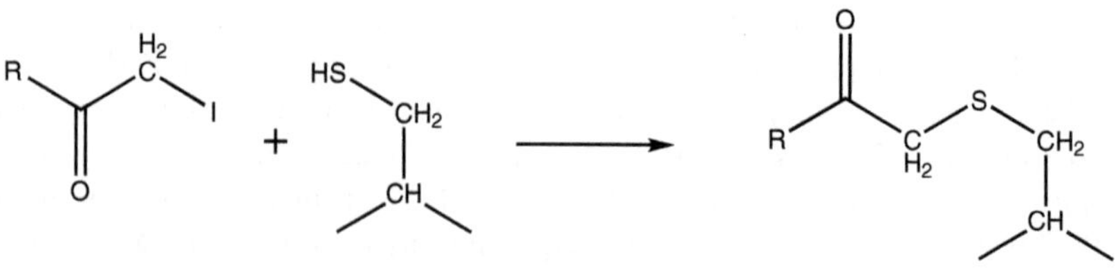

FIGURE 2.12 Fluorescent labels used to modify cysteine residues in proteins.

S_N2; Substitution, Nucleophile, Second Order

$R\text{-}Cl + H_2O \longrightarrow ROH + HCl$

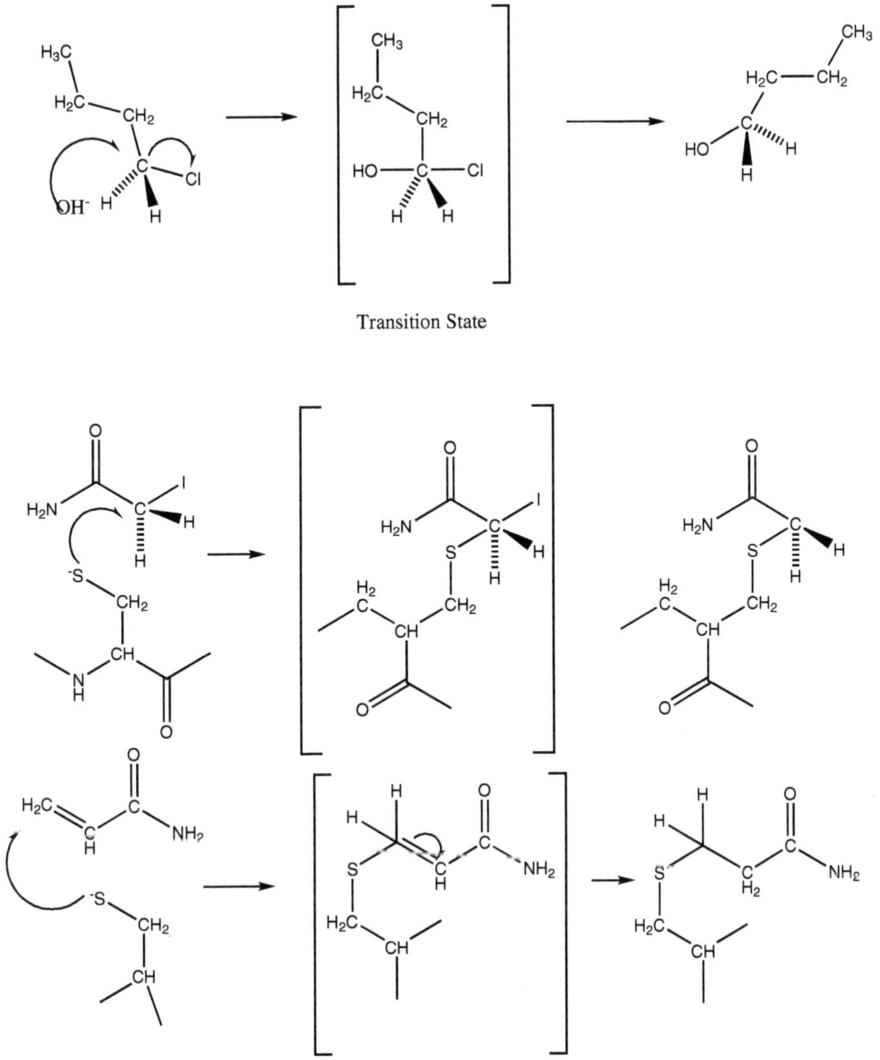

Transition State

FIGURE 2.13 Nucleophilic addition and substitution reactions, S_N2 (substitution, nucleophilic, biomolecular) reaction. This reaction is characterized by the formation of a transition state intermediate.

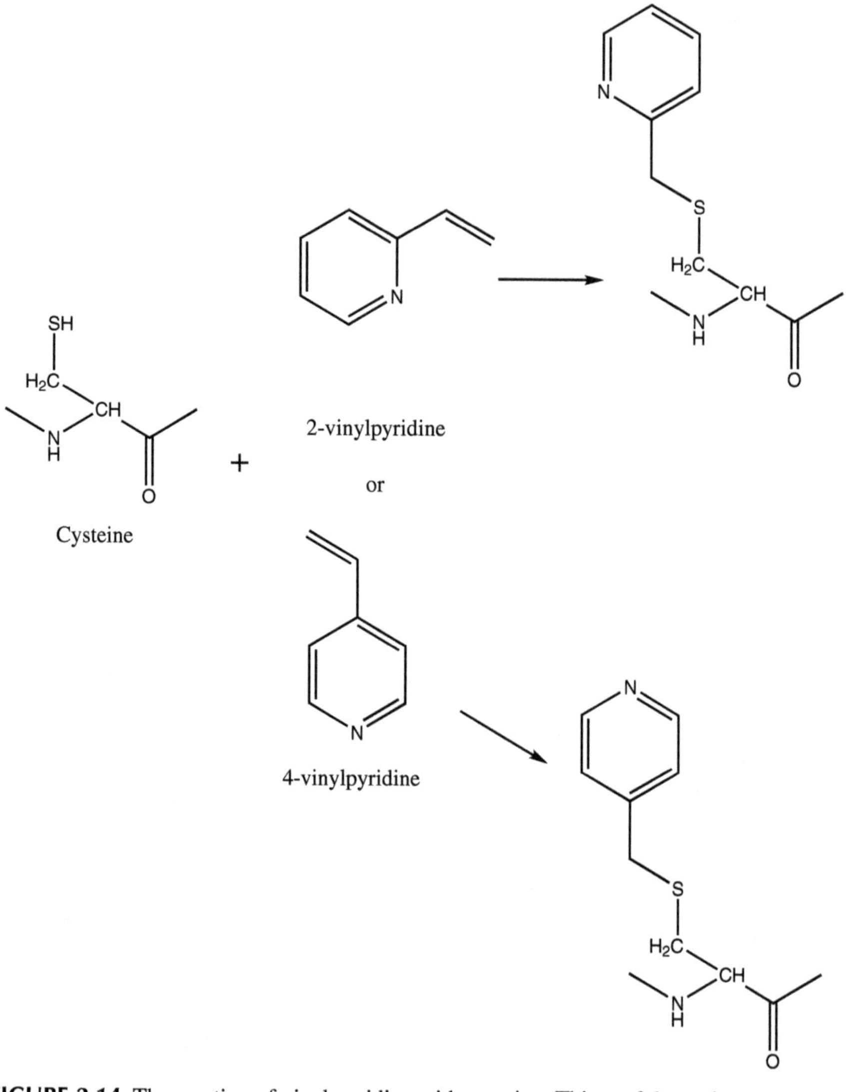

FIGURE 2.14 The reaction of vinyl pyridine with cysteine. This useful reaction can be used in place of carboxymethylation

discussed in greater detail below. Although not considered a chemical modification, it would be remiss to exclude the incorporation of ^{18}O into terminal peptide carboxyl groups during proteolytic digestion.[56]

HISTIDINE

Diethylpyrocarbonate continues to be the reagent of choice of the modification of histidyl residues in proteins (Figure 2.23).[1] Mono- and di-substitution reactions occur with di-substitution resulting in irreversible modification of the histidine

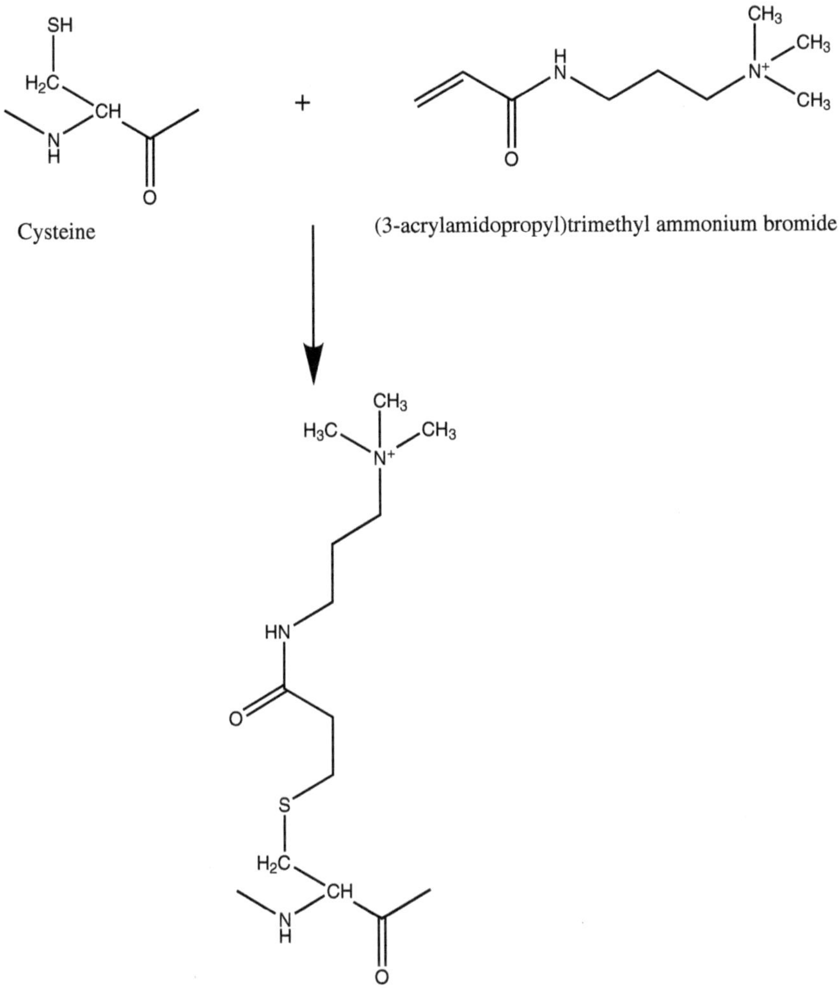

FIGURE 2.15 The modification of cysteine with (3-acrylamidopropyl)trimethyl ammonium bromide.

residues. Di-substitution reactions usually require a high molar excess of reagent. The mono-substitution reaction is reversible under alkaline conditions or in the presence of a suitable nucleophile such as hydroxylamine. Modification can occur at tyrosine residues; this reaction is associated with a change in the UV spectrum and is reversible under the above conditions. Reaction can also occur at lysine residues and that reaction is essentially irreversible. Mass spectrometry has identified the sites of reaction of diethylpyrocarbonate in proteins,[57–59] allowing the identification of both histidine and tyrosine residues modified by this reagent. The reaction of histidine with 4-hydroxy-2-nonenal[60] is of some interest (Figure 2.3), as is the oxidation of the imidazole ring of histidine (Figure 2.24).[61,62] Histidine residues also react with the α-haloalkyl compounds used for the modification of cysteine[63] but such residues are usually located at the enzyme active site.[64]

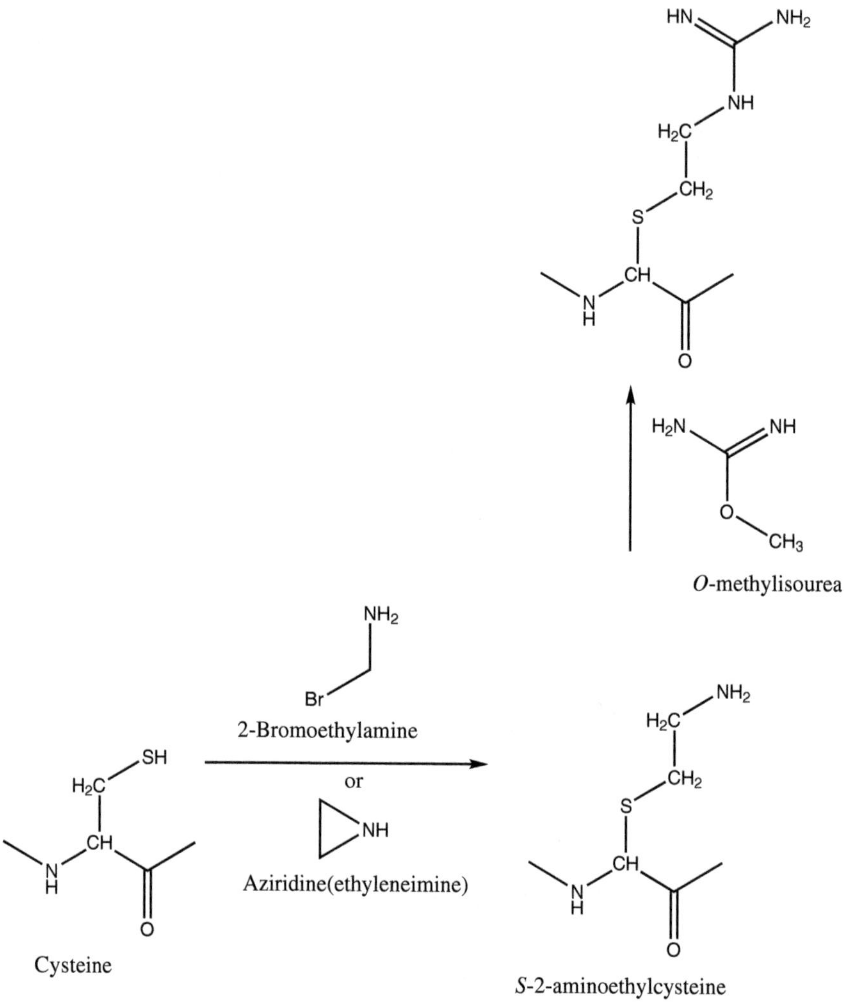

FIGURE 2.16 A reaction scheme for the modification of cysteine to yield a homoarginine analogue in proteins.

Carboxymethylation of a histidine residue in interferon was observed with (soccinimidyl casbonyl PEG) SC-PEG.[65] Histidine may also be methylated with methyl-p-nitrobenzenesulfonate (Figure 2.25).[66]

AMINO GROUPS

Proteins contain two types of amino groups, the ε-amino group of lysine and the N-terminal α-amino group. As shown in Table 1, these groups generally have different pKa with ε-amino groups having a lower value than ε-amino groups, which should make the α-amino groups more reactive; however, the α-amino groups are

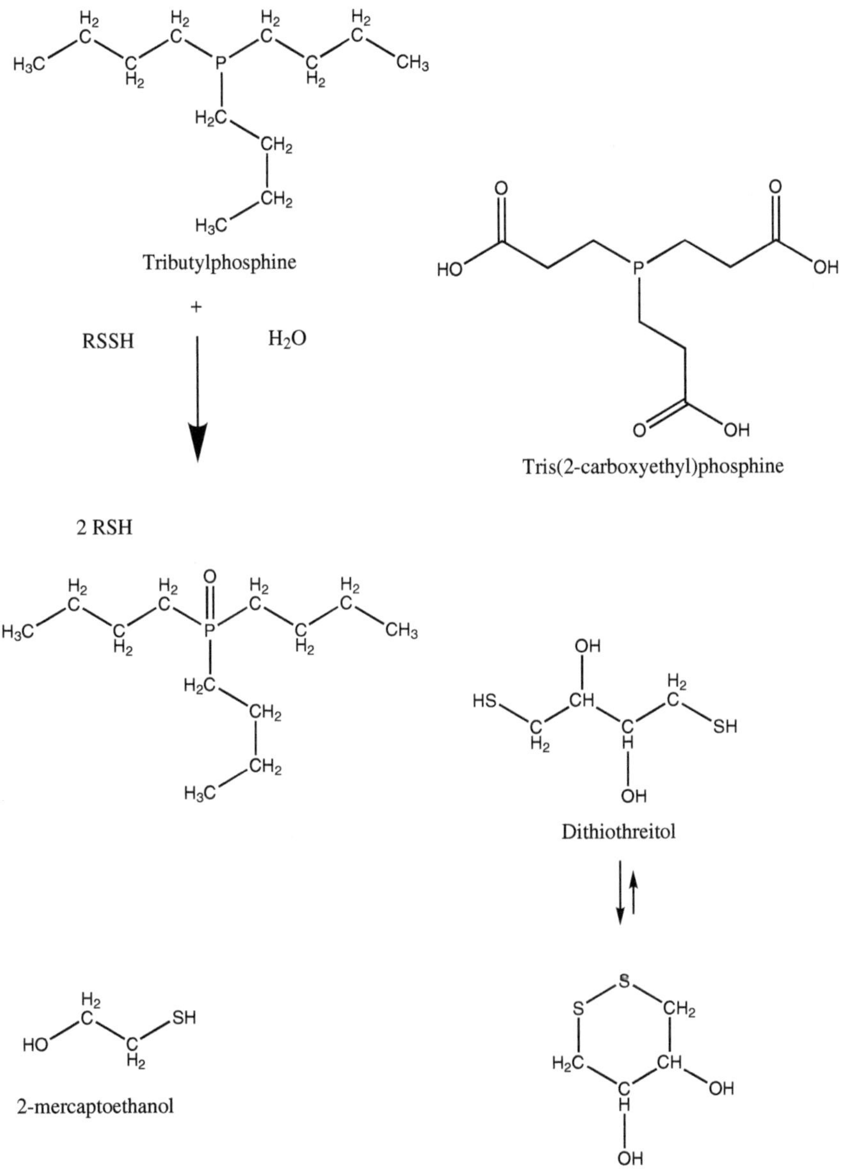

FIGURE 2.17 The reduction of cystine residues in protein. Shown are several reagents used for the reduction of disulfide bonds in proteins.

frequently not available. Some of the most frequently used modifications of amino groups in proteins are shown in Figure 2.26 with the ε-amino group as an example. The reaction of α-amino groups is important in the binding of proteins to surfaces in the manufacture of protein microarrays (see Chapter 3),[67] the attachment of spectral probes (Figure 2.27)[68] and the cross-linking of proteins.[69] α-Amino groups are the site of attachment of poly (ethylene) glycol (Figure 2.28) used for the

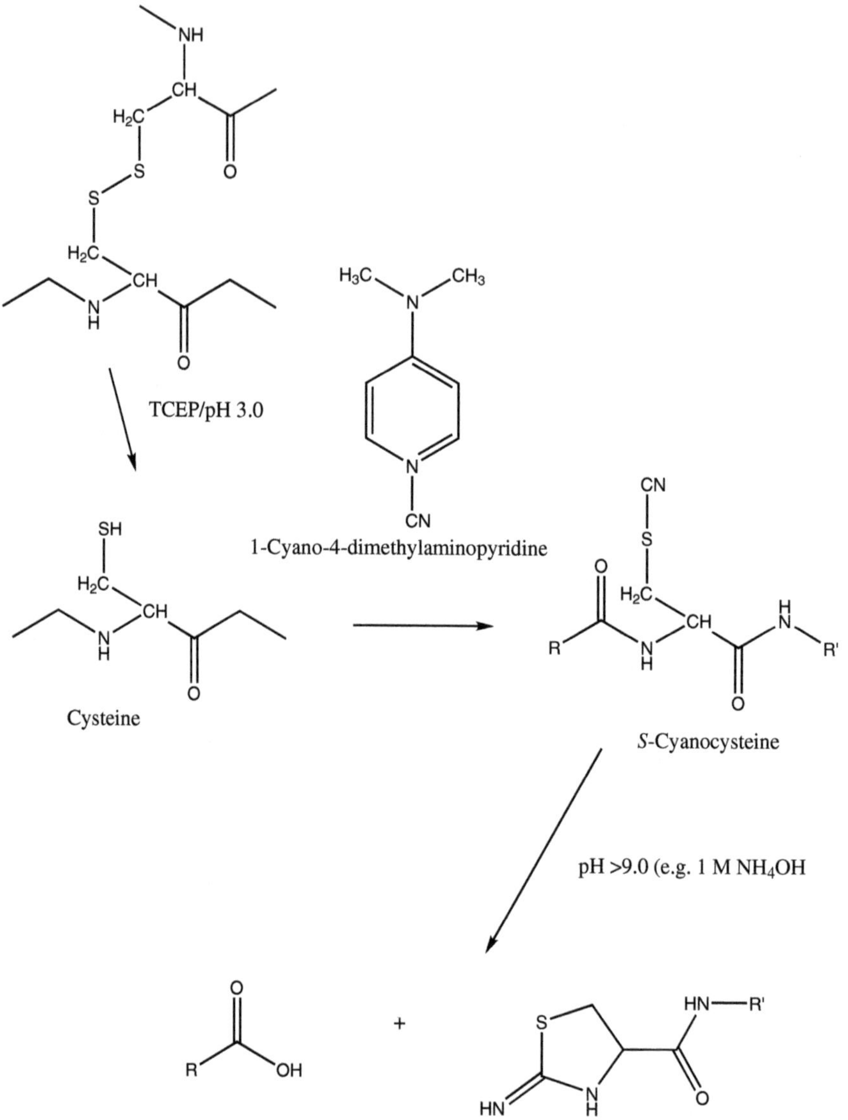

FIGURE 2.18 The reduction of cystine, cyanylation of the cysteine residues and subsequent peptide bond cleavage. In the example shown, reduction of the disulfide bond is accomplished by tris(2-carboxyethyl)phosphine (TCEP) at pH 3.0, which yields a partially reduced species useful for structural analysis (Gray, W.R., Echistatin disulfide bridges: Selective reduction and linkage assignment, *Protein Sci.*, 2, 1749–1755, 1993).

manufacture of biopharmaceuticals. α-Amino groups are also the major site of modification in the glycation of proteins and the oxidation of proteins to form complex cross-linked products. Chemical modification can be used to generate "lysine" residues in proteins from cysteine residues as described above; lysine residues can be converted into sulfhydryl groups by modification with reagents such

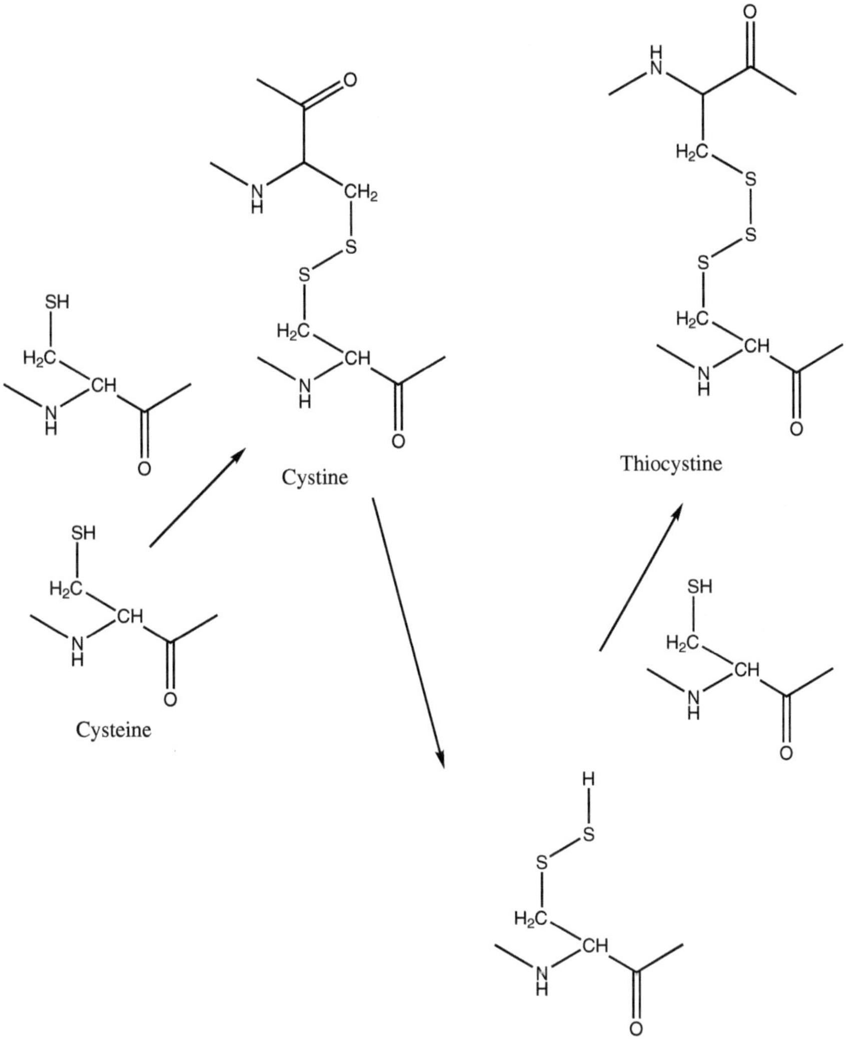

FIGURE 2.19 Trisulfide bonds in proteins. These unique derivatives may provide analytical issues in the mass spectrometric analysis of proteins.

as *N*-hydroxysuccinimide *S*-acetylthioacetate[70] or 2-iminothiolane (Traut's reagent).[71] Amino groups can be acylated with a variety of organic acid anhydride including acetic anhydride, succinic anhydride, maleic anhydride and citraconic anhydride as well as by derivatives of *N*-hydroxysuccinimide (Figure 2.29).[1] The modification with citraconic anhydride is important as it is a reversible reaction (Figure 2.30).[72] The reversible modification of proteins with citraconic anhydride has been used to reverse the effects of formalin fixation (reaction with formaldehyde) as a universal method to recover antigen activity in tissue sections.[73] Modification of lysine residues in proteins with *O*-methylisourea yields homoarginine

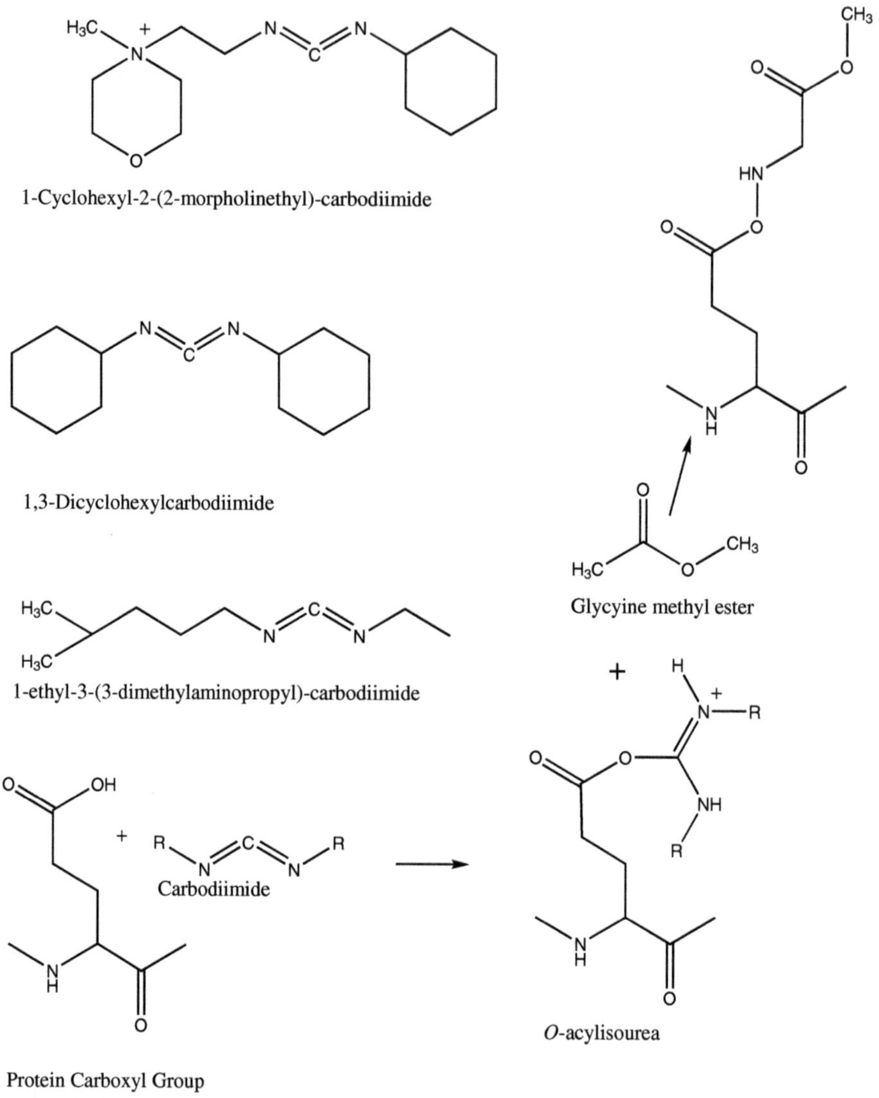

1-Cyclohexyl-2-(2-morpholinethyl)-carbodiimide

1,3-Dicyclohexylcarbodiimide

1-ethyl-3-(3-dimethylaminopropyl)-carbodiimide

Glycyine methyl ester

Carbodiimide

Protein Carboxyl Group

O-acylisourea

FIGURE 2.20 Carbodiimides and the modification of carboxyl groups in proteins. Shown are water-soluble derivatives (1-ethyl-3-(3-dimethylaminopropyl)-carbodiimide, EDC) and water-insoluble derivatives (1,3-dicyclohexylcarbodiimide, DCC) as well as 1-cyclohexyl-2-(morpholinethyl)-carbodiimide.

residues[74] which adds sensitivity to mass spectrometric analysis of tryptic peptides (see Chapter 6).[75] Modification of lysine with N-hydroxysuccinimide S-acetylthioacetate or N-hydroxysuccinimide-S-acetylthiopropionate yields thiol derivatives (Figure 2.31).[76] Lysine can also be modified with 2-S-thiuroniummethanesulfonate to yield homoarginine (Figure 2.32).[77]

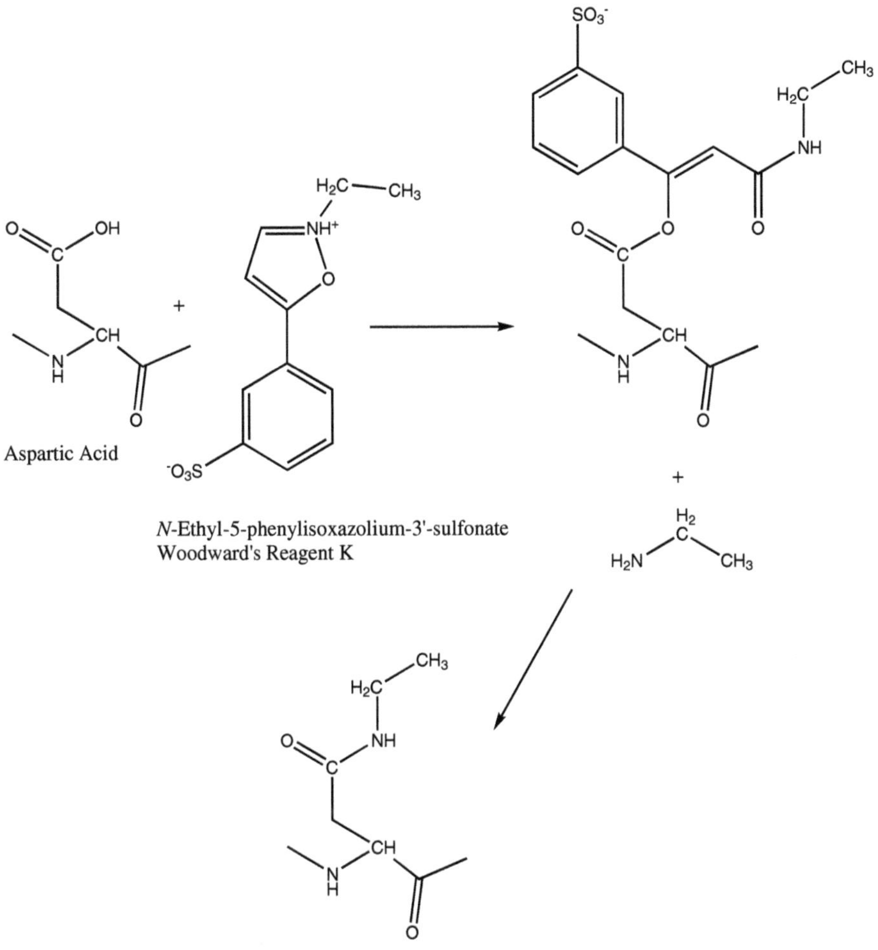

FIGURE 2.21 Woodward's reagent K and the modification of carboxyl groups in proteins.

METHIONINE

Methionine residues in proteins can be alkylated with iodoacetamide (Figure 2.33).[78-81] This reaction is not preferred under solvent conditions generally used for the alkylation of protein (pH $\geq$ 7) and usually is accomplished at low pH, which allows specificity of modification. Methionine can also be modified by bromoethylamine[82] providing a site for modification with reagents reacting with amino groups (Figure 2.34). Methionine can be oxidized initially to the sulfone and subsequently to the sulfoxide by reagents such as chloramine T (Figure 2.35).[83]

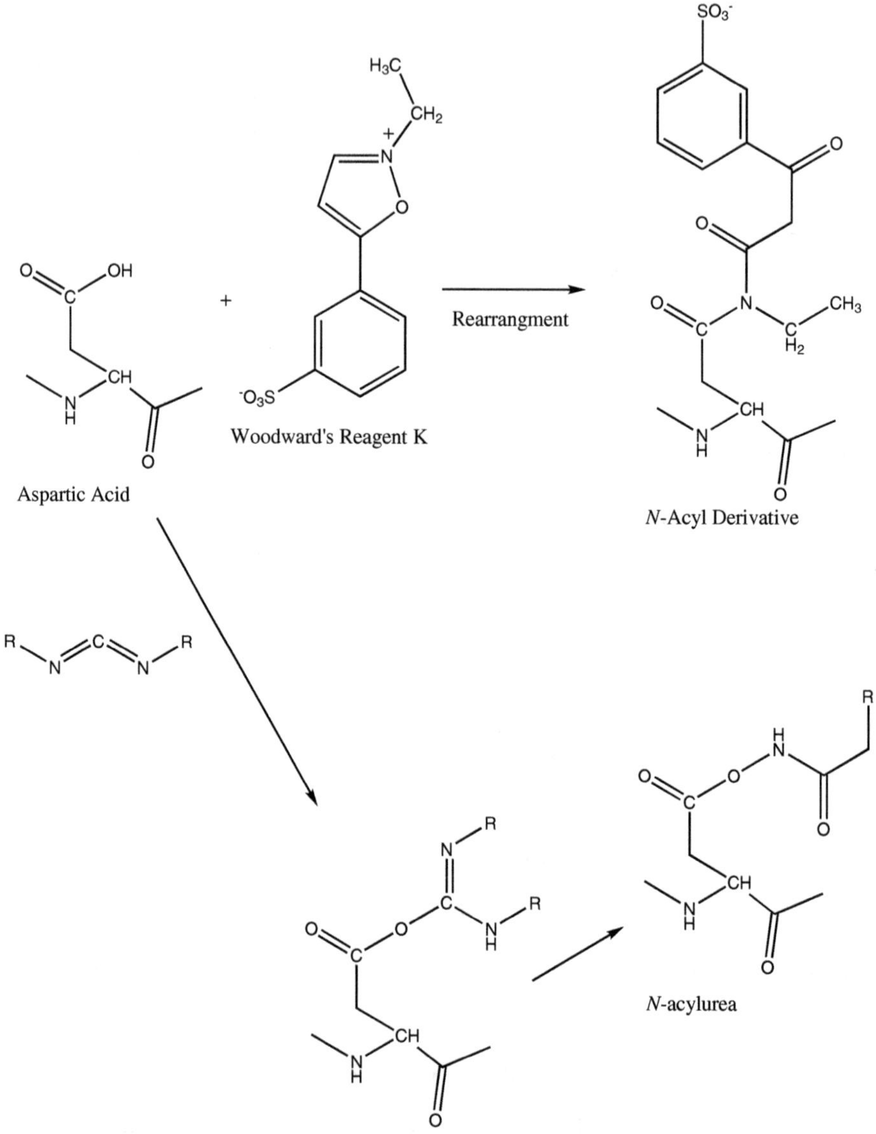

FIGURE 2.22 The modification of carboxyl groups in proteins without nucleophile.

TYROSINE

Tetranitromethane (Figure 2.36)[84,85] and N-acetylimidazole (Figure 2.36)[86] can be used for the modification of tyrosyl residues. 3-Nitrotyrosine, the product of the reaction of tyrosine with tetranitromethane, can be reduced with sodium dithionite to 3-aminotyrosine, which can be alkylated with a probe such as dansyl chloride (Figure 2.37).[87] Acetylation with N-acetylimidazole is a reversible reaction; acetylation can also occur at α-amino groups and on the imidazole group of histidine.

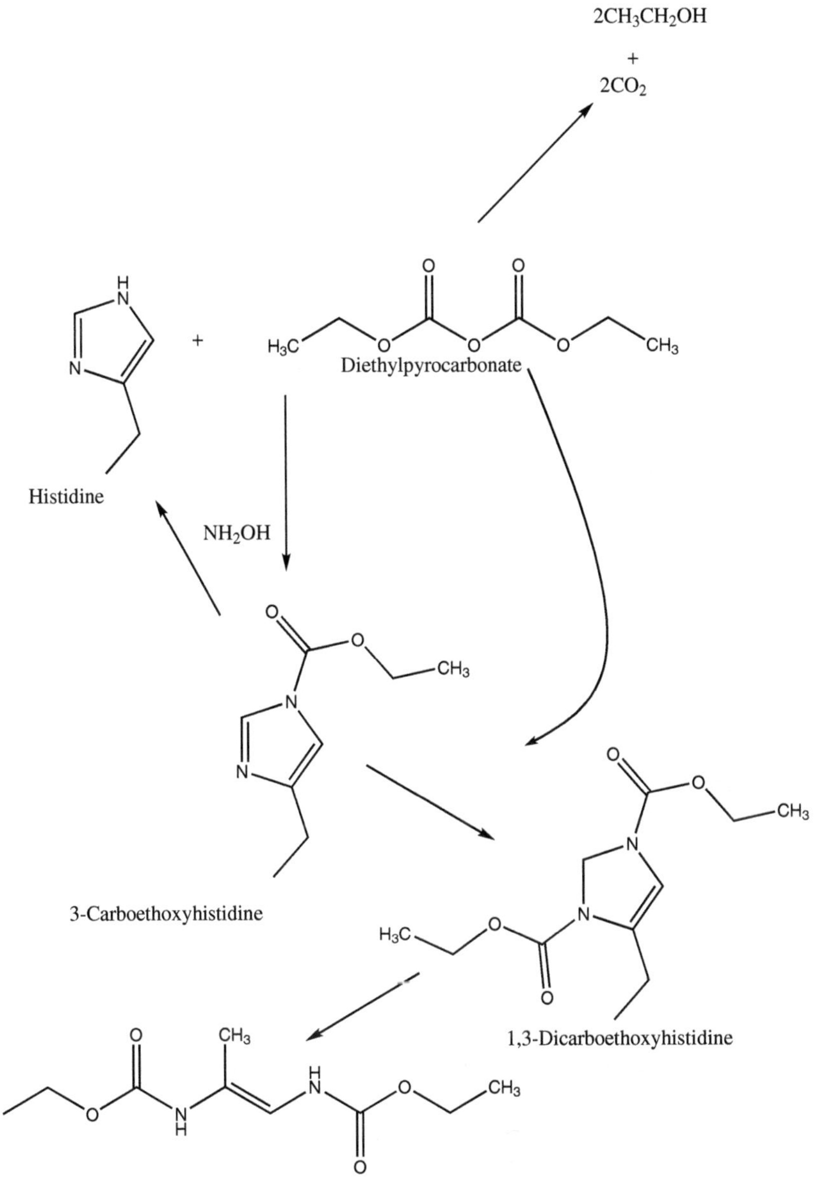

FIGURE 2.23 The modification of histidine with diethylpyrocarbonate (ethoxyformic anhydride).

Cross-linking via the formation of dityrosine can be accomplished with hydrogen peroxide/copper ions[88] or as a result of reaction with tetranitromethane or peroxynitrite (Figure 2.38).[89] Of significant interest are recent reports that phenylmethylsulfonyl fluoride, a compound developed for the modification of "active-site" serine residues, reacts with a tyrosyl residue in a superoxide dismutase resulting in inactivation of the enzyme (Figure 2.39).[90] Some earlier reports exist of the phosphorylation of

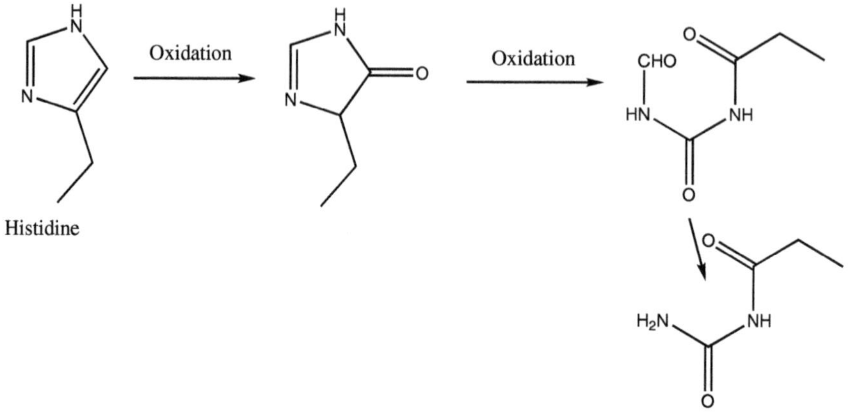

FIGURE 2.24 The oxidation of histidine in proteins. Oxidizing agents include hypochlorous acid and oxygen radicals.

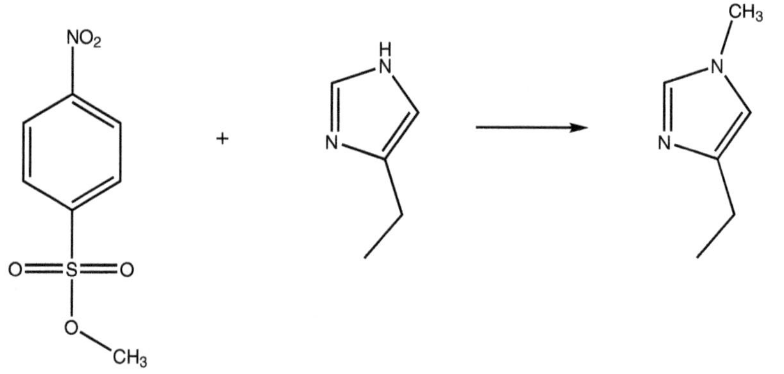

Methyl-p-nitrobenzenesulfonate N^3-methylhistidine

FIGURE 2.25 The methylation of histidine with methyl-p-nitrobenzenesulfonate.

tyrosine in human serum albumin with diisopropylphosphorofluoridate.[91] Peptide chloromethylketones, reagents developed for the modification of histidine residues at the active site of serine proteinases, do react with other amino acid residues in disparate proteins.[92] This unexpected broad spectrum of reactivity no doubt provides an explanation for the effect of peptide chloromethylketones in complex systems.[93,94]

TRYPTOPHAN

Most recent studies on the modification of tryptophan have used N-bromosuccinimide (Figure 2.40).[1,95] Fewer examples exist of the use of 2-hydroxy-5-nitrobenzyl bromide (Figure 2.41),[96] 3-bromo-3-methyl-2-(2-nitrophenylmercapto)-3H-indole (BPNS-skatole, Figure 2.42)[97] or 2-nitrobenzylsulfenyl chloride (Figure 2.43).[98–100] Under appropriate reaction conditions — usually mild acid — specific modification of tryptophanyl residues can be obtained. The product obtained on reaction of

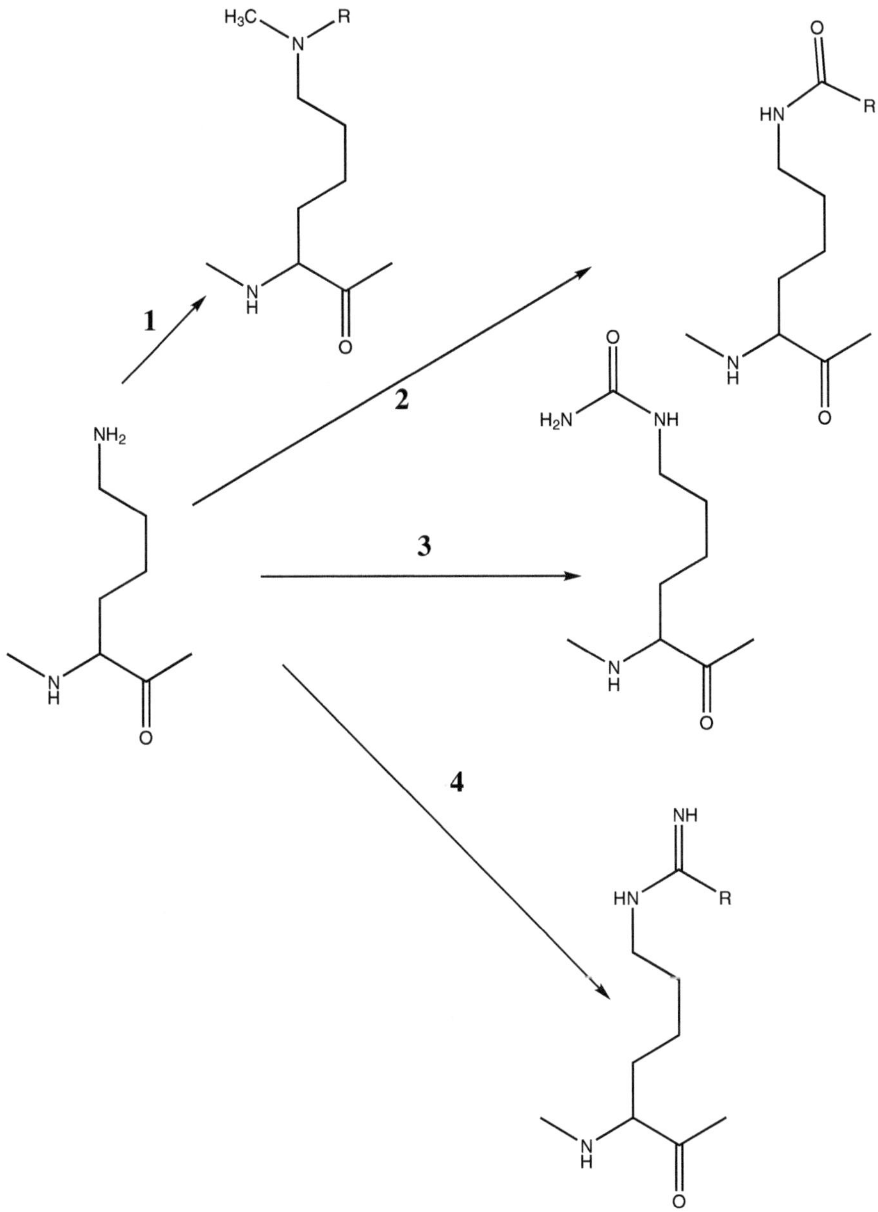

FIGURE 2.26 Example of mechanisms for the modification of amino groups in proteins. The ε-amino group of lysine is used as an example. Shown is (1) reductive methylation as with formaldehyde and sodium cyanoborohydride; (2) acylation as with acetyl chloride — the same reaction product would be obtained with acetic anhydride; (3) amidation as with methyl acetamidate; and (4) guanidation as with O-methylisourea.

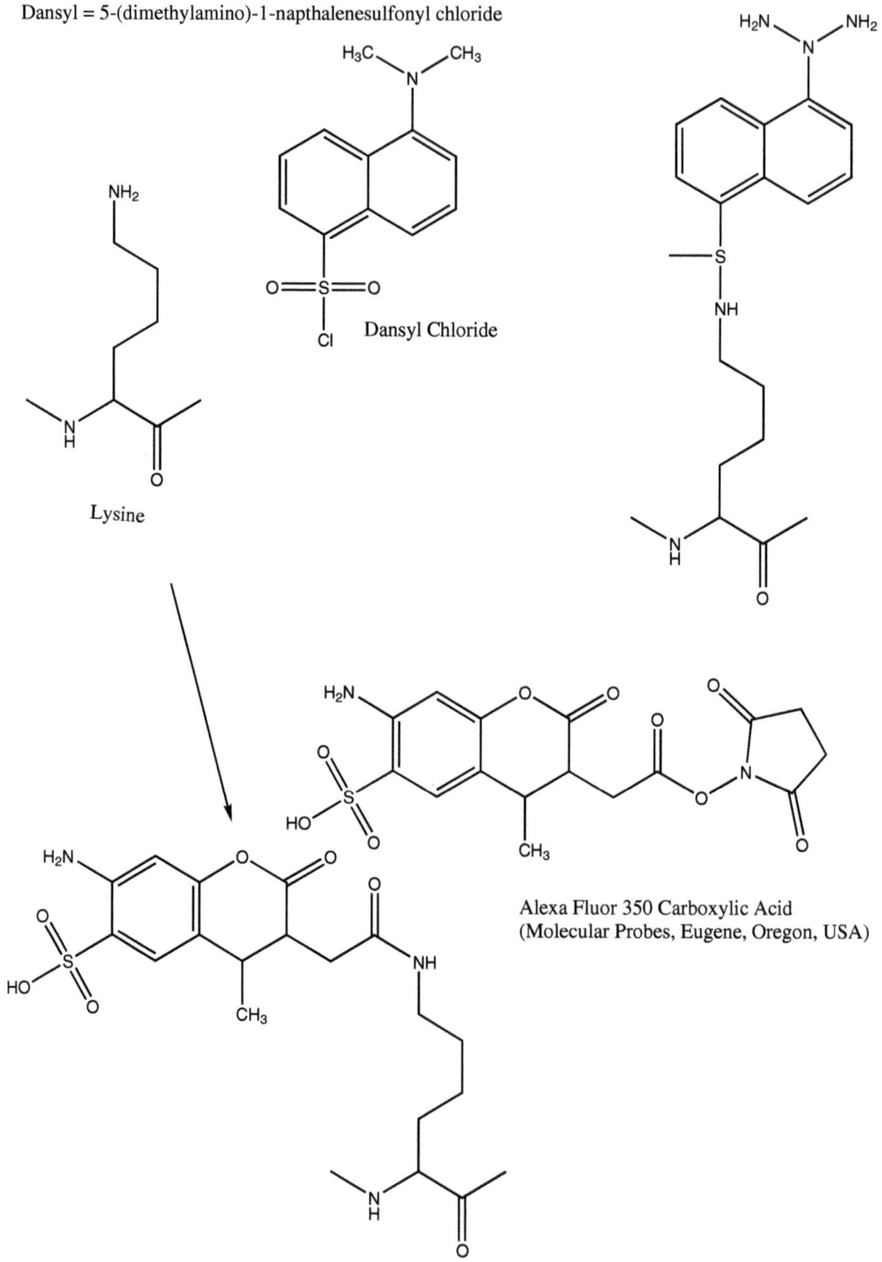
Dansyl = 5-(dimethylamino)-1-napthalenesulfonyl chloride
Dansyl Chloride
Lysine
Alexa Fluor 350 Carboxylic Acid
(Molecular Probes, Eugene, Oregon, USA)

FIGURE 2.27 The modification of lysine residues in proteins with fluorescent reagents.

H-(CH$_2$CH$_2$)$_n$-OH

poly(ethylene)glycol

CH$_3$-(CH$_2$CH$_2$)$_n$-OH

monomethoxypoly(ethylene)glycol

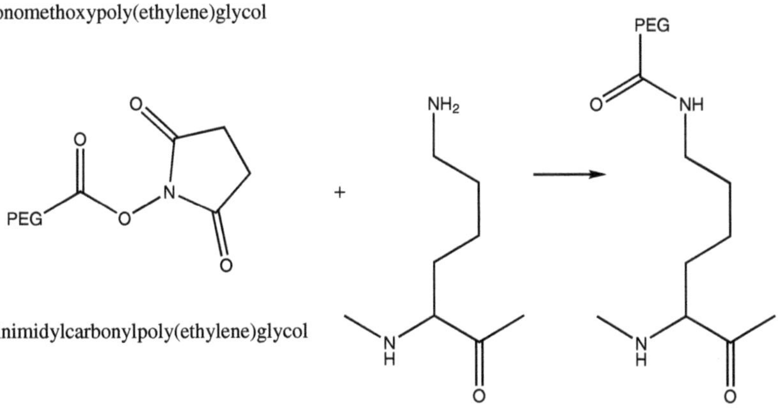

N-succinimidylcarbonylpoly(ethylene)glycol

Lysine PEG-Protein

FIGURE 2.28 The modification of lysine with poly(ethylene) glycol. Various chemistries can be used for the modification as shown in Figure 2.26 and Figure 2.27. The specific example shown here uses an N-hydroxysuccinimide mediated acylation process.

tryptophanyl residues with 2-nitrophenylsulfenyl chloride can be reduced with the production of thiol tryptophan, which can then be modified with sulfydryl reagents such as iodoacetamide.[101] The reaction of most of these reagents is not specific for tryptophan and reaction at cysteine residues, if present, should be expected; modification at other residues cannot be excluded and should be evaluated.

CHEMICAL CLEAVAGE OF PEPTIDE BONDS

Cleavage of methionine-containing peptide bonds with cyanogen bromide (CNBr)[102–105] is likely the most widely used method for specific chemical cleavage of peptide bonds. The reaction cleaves a peptide bond (Figure 2.44) in which methionine contributes the carboxyl moiety. Methionine is converted into homoserine lactone and homoserine during this process with the loss of methyl thiocyanate. The reaction is reasonably quantitative although, as indicated below, variable amounts of CNBr might be required. The methionine content of most proteins is low enough[106] that a reasonably small number of fragments are obtained, providing a distinct advantage in primary structure analysis. The major advances in mass spectrometry have resulted in the development of a "top-down" strategy for protein structure analysis which utilizes CNBr fragmentation.[107]

The chemistry of this reaction is straightforward, involving the nucleophilic attack of the thioether sulfur on the carbon in CNBr followed by cyclization to form the iminolactone, which is hydrolyzed by water and results in cleavage of the peptide bond (Figure 2.44). At acid pH, this reaction in and by itself does not

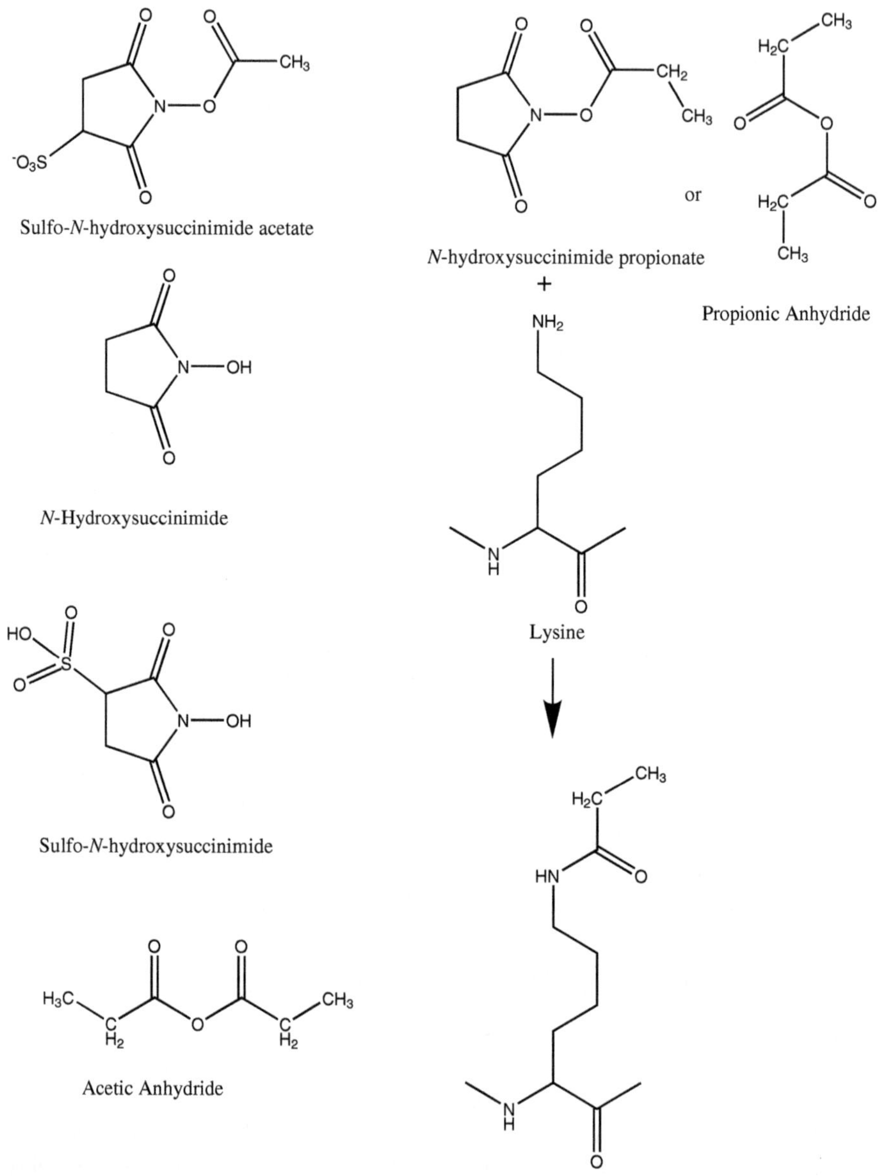

FIGURE 2.29 The acylation of lysine residues with organic acid anhydrides or activated acyl derivatives of N-hydroxysuccinimide.

generally affect any other amino acid with the exception of cysteine, which is converted to cysteic acid. Cleavage of peptide chains at methionine with CNBr proceeds best with a fully denatured protein in mild acid. Most recent studies have used 70% formic acid, trifluoroacetic acid[108] or an equal mixture of formic acid and trifluoroacetic acid. The reaction proceeds effectively with a 20- to 100-fold molar excess of CNBr (added either as a solid to the protein or peptide dissolved

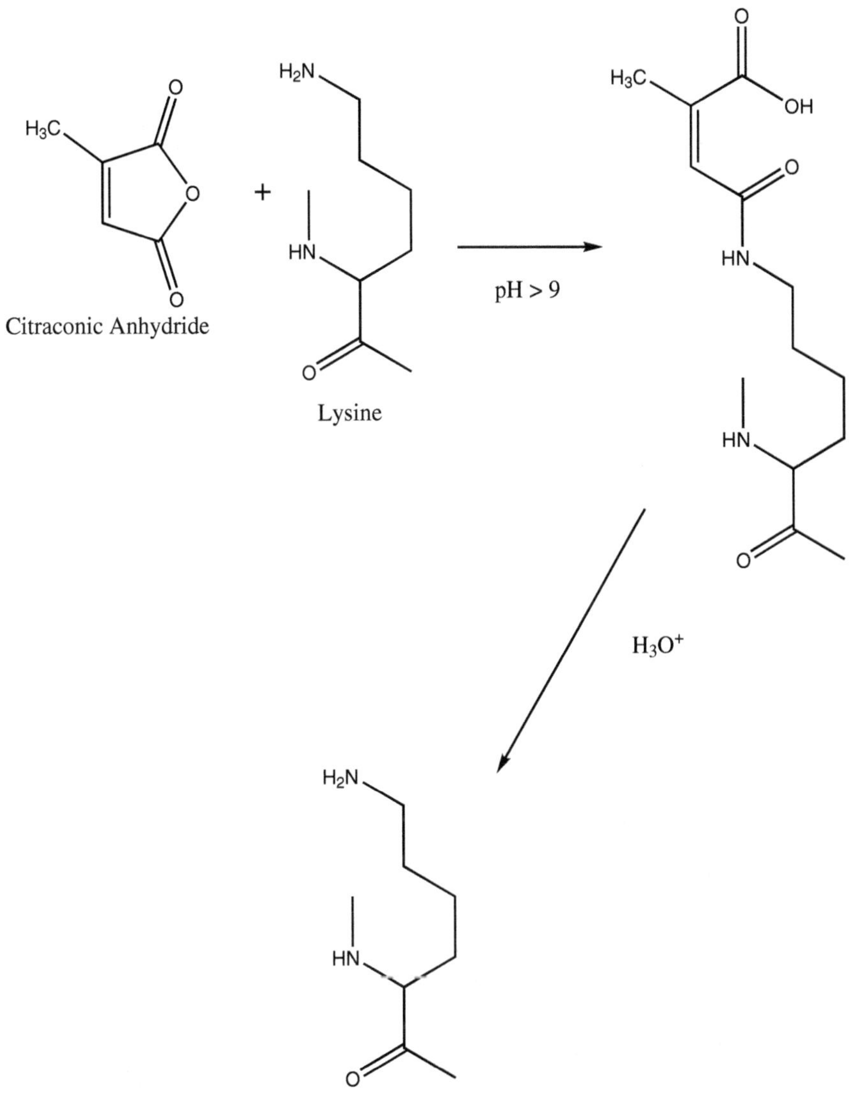

FIGURE 2.30 The reversible modification of protein amino groups with citraconic anhydride (methyl maleic anhydride).

in the solvent of choice). The molar ratio of CNBr to methionine residues should be established for each peptide and protein; it was necessary to use a 3000-fold molar excess to cleave a particular methionine-serine peptide bond in pancreatic deoxyribonuclease.[109]

Sokolov et al.[110] describe a modified method for direct CNBr cleavage directly within the polyacrylamide gel. After identification of the proteins by either staining or autoradiography, the gel is sliced and dried. The dried gel is taken into 30 μl CNBr (200 mg/ml in 70% HCOOH). The reaction is allowed to proceed for 16 hours

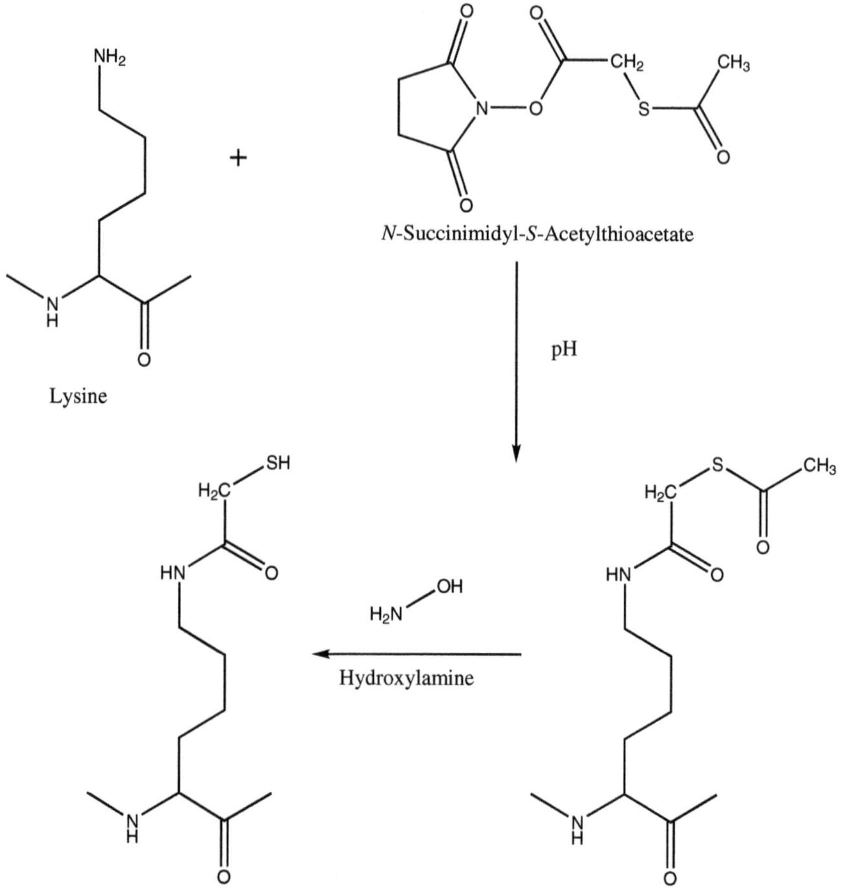

FIGURE 2.31 The reaction of lysine with *N*-succinimidyl-*S*-acetylthioacetate (SATA). The same scheme would applied to *N*-succinimidly-*S*-acetylthiopropionate (SATP). As shown, this allows the introduction of a protected sulfydryl group onto the protein.

at ambient temperature or 3 hours at 37°C. The gel is then dried and the reaction products are separated by a second electrophoretic step. The drying of the gel prior to the cleavage step is critical to avoid protein loss during this procedure. An alternative approach has been developed by Jahnen et al.[111] These investigators isolated the fragments from the CNBr cleavage of proteins on polyacrylamide gel (the gel slices were dried by lyophilization prior to the CNBr cleavage step) by either a second electrophoretic step or by HPLC after elution. The electrophoretic step is recommended over the HPLC step. More recent work by Robillard and coworkers[112] has provided methods for the in-gel cleavage of integral membrane proteins for subsequent analysis by MALDI-TOF mass spectrometry. The CNBr cleavage reaction was performed in 70% TFA (one small CNBr crystal was dissolved in 200 to 300 μl 70% trifluoroacetic acid added to the gel slice) for 14 hours in the dark at room temperature. The digested gel piece was sonicated for 5 minutes and then

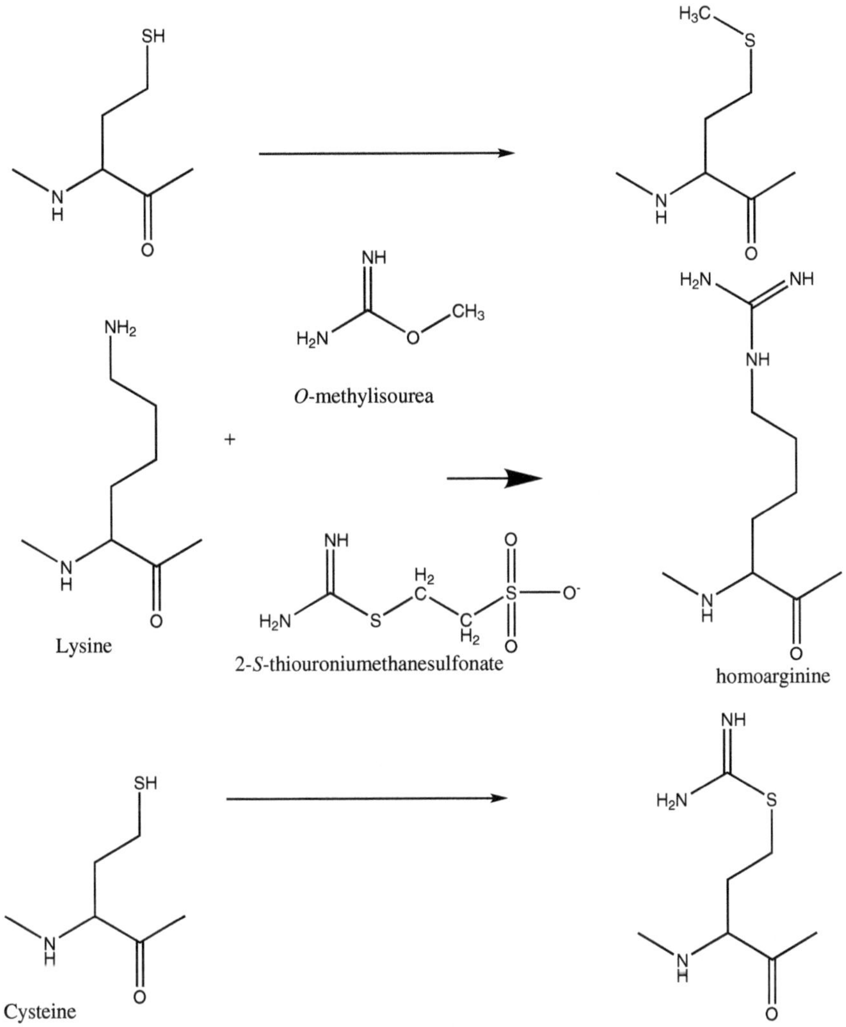

FIGURE 2.32 The reaction of lysine with 2-S-thiuroniummethanesulfonate to yield homoarginine. Also shown is the reaction of lysine with O-methylisourea to yield homoarginine. Homoarginine has proved to be an advantage in mass spectrometric analysis.

extracted twice with sonication with 30 μl 60% acetonitrile/1% TFA and concentrated *in vacuo* prior to analysis.

Partial acid hydrolysis is the oldest of the various chemical approaches to the cleavage of specific peptide bonds. The general principle of partial acid hydrolysis is based on the use of dilute acid at a pH just adequate to maintain the β-carboxyl group of aspartic acid in the protonated form. Under these conditions, peptide bonds in which the carboxyl moiety is contributed by aspartic acid are cleaved 100-fold more rapidly than other peptide bonds. The use of 0.03 N HCl *in vacuo* at 105°C for 20 hours has been found to be satisfactory for the partial acid hydrolysis of

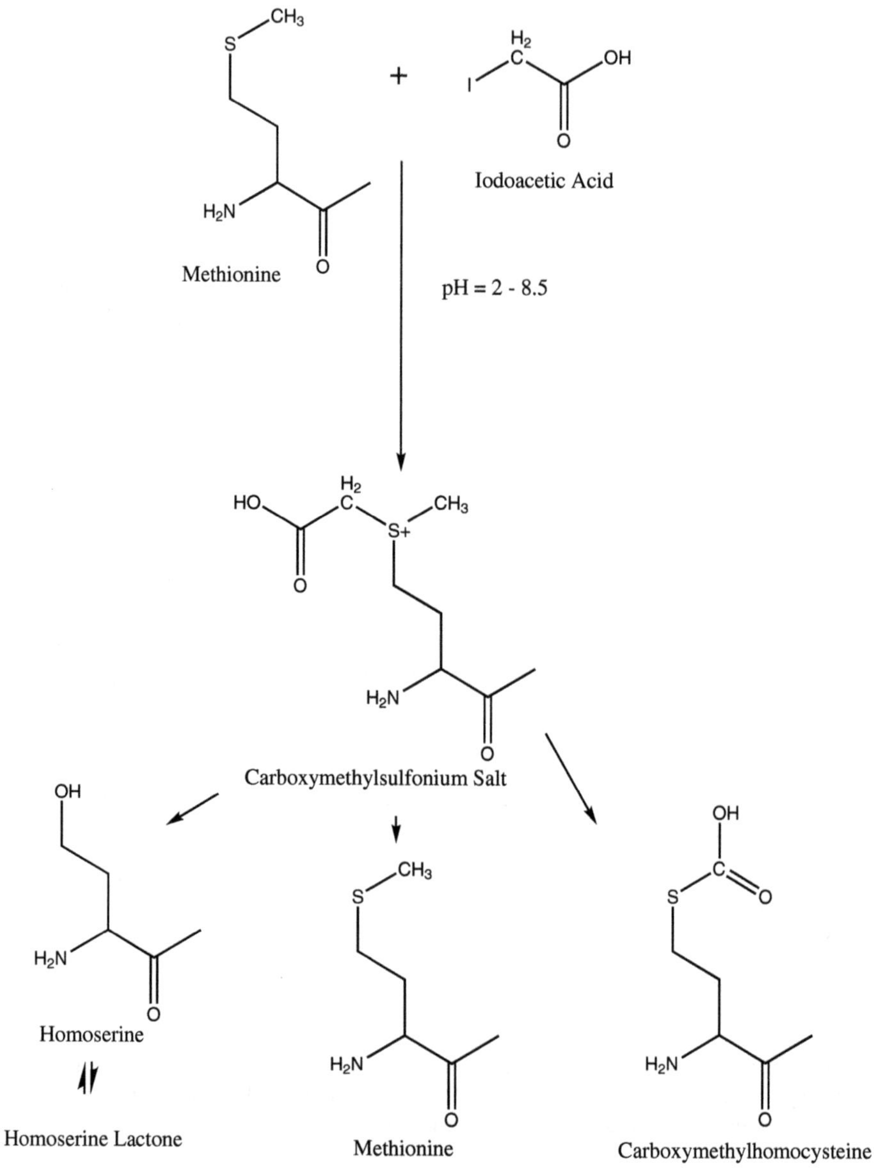

FIGURE 2.33 The carboxymethylation of methionine.

proteins. Hulmes and Pan[113] demonstrated that gas phase trifluoroacetic acid preferentially cleaves peptide chains at the amino-termini of threonine and serine (22°C, 1 to 15 days or 45°C for 2 to 3 days). The degradation of antiflammin 2 under acidic conditions has been studied.[114] The reaction was more rapid at a pH of 3.0 (37°C, sodium citrate) than at a pH of 2.37 (37°C, sodium phosphate) or a pH of 4.10 (37°C, sodium citrate).

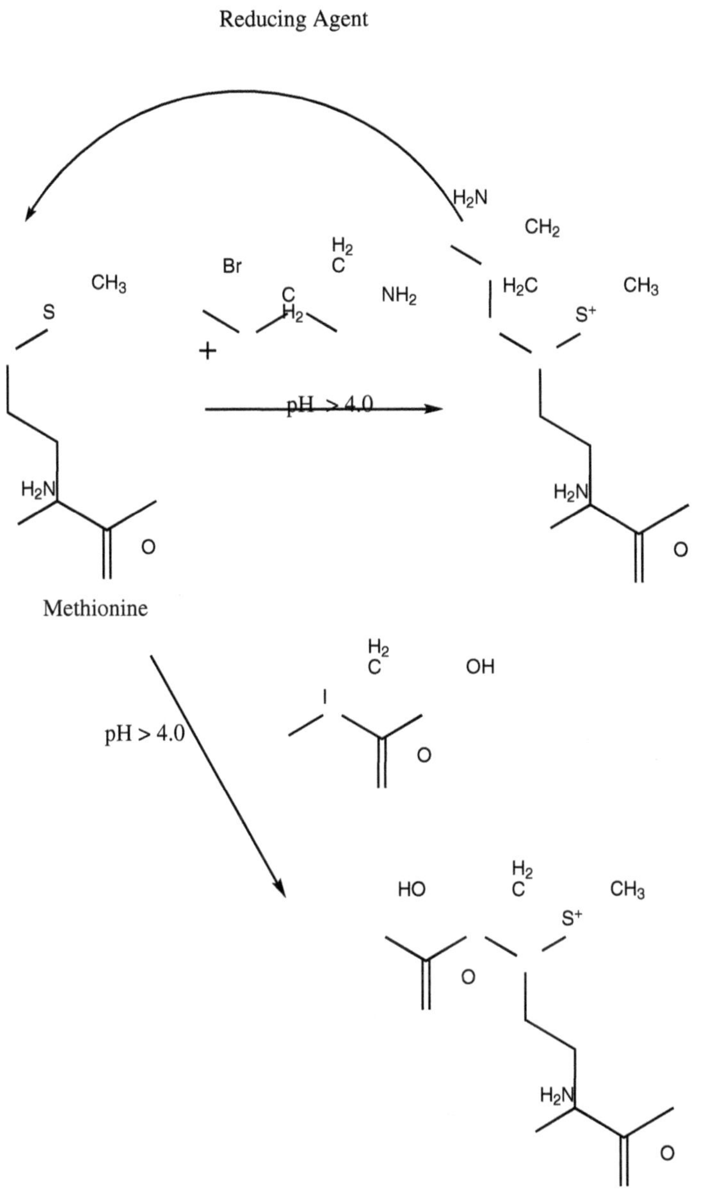

FIGURE 2.34 The alkylation of methionine with bromoethylamine. This reaction can be used for the selective modification of methionine at low pH and removal of methionine peptide. The reaction is reversed by reduction.

The cleavage of peptide bonds containing cystine (disulfide groups) has been examined in some detail. S-Cyanocysteine is obtained by reaction of cysteine or cystine with 2-nitro-5-thiocyano-benzoic acid (Figure 2.17). Cleavage of the S-cyano-cysteine is achieved by incubation in 0.1 M sodium borate, 6 M guanidine, pH 9.0 at 37° C with the formation of 2-iminothiazolidine-4-carboxyl peptides. Lu and Gracy[115]

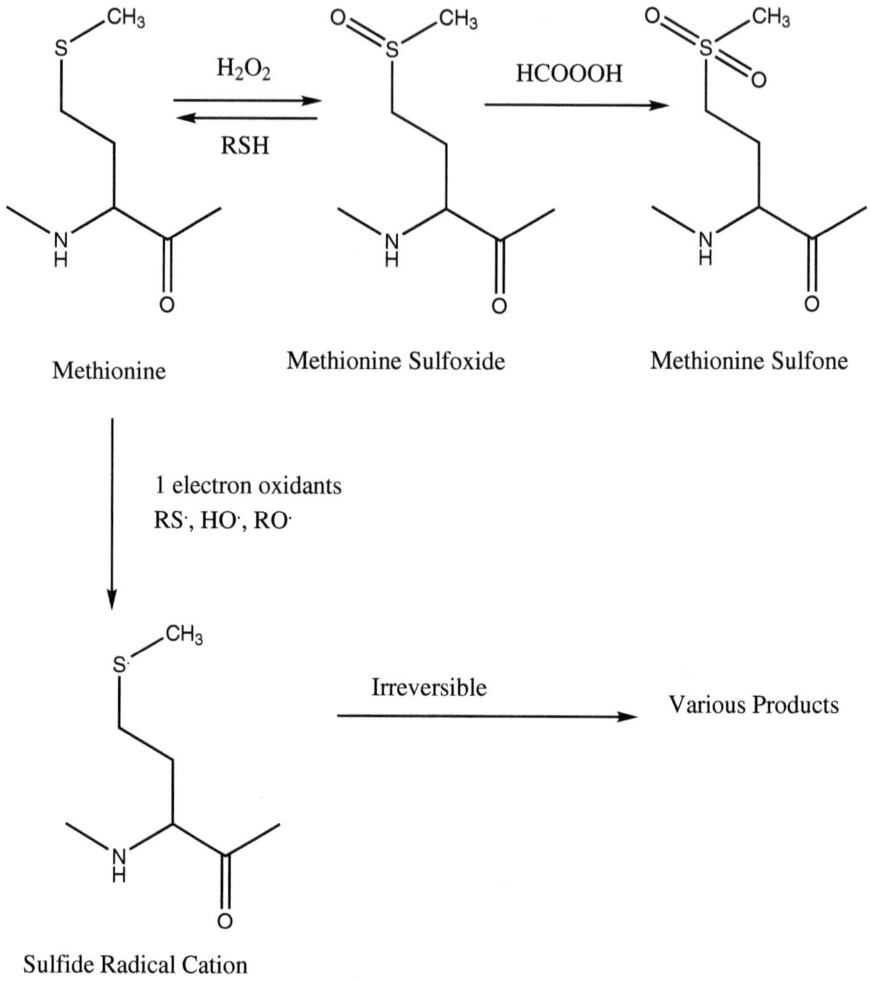

FIGURE 2.35 The oxidation of methionine and the formation of methionine sulfoxide and methionine sulfone.

have employed 2-nitro-5-thiocyanobenzoic acid to convert the cysteinyl residues in human placental glucosephosphate isomerase to *S*-cyanocysteine, followed by cleavage at the modified cysteine residues. Conversion to the *S*-cyanocysteinyl derivative was accomplished with a five- to tenfold molar excess of 2-nitro-5-thiocyanobenzoic acid in 0.2 M Tris-acetate, 6 M guanidinium chloride, pH 9.0 (protein previously incubated with adequate dithiothreitol — fourfold molar excess over the sulfhydryl groups in the protein) for 5 hours at 37°C. The modified protein was dialyzed extensively against 10% acetic acid and lyophilized. Cleavage of *S*-cyanylated protein was achieved by incubation in 0.2 M Tris-acetate, 6 M guanidium chloride, pH 9.0 at 37°C for 2 hours. Watson and colleagues[116] recently used cyanylation combined with mass spectrometric analysis to determine the disulfide structure of sillucin. These investigators used a combination of partial reduction and CN-induced

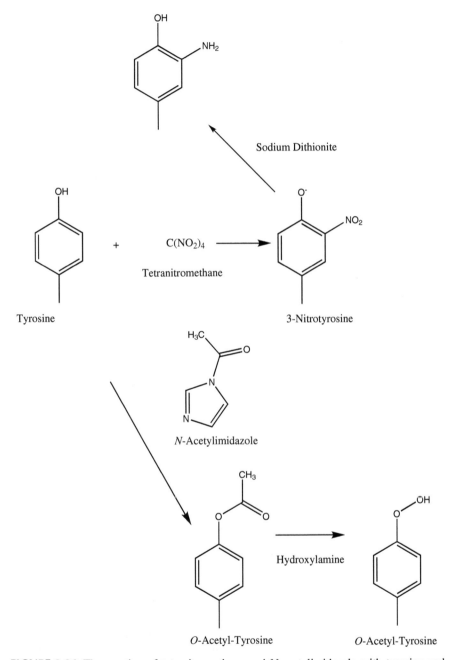

FIGURE 2.36 The reaction of tetranitromethane and N-acetylimidazole with tyrosine and the formation of 3-nitrotyrosine and O-acetyltyrosine.

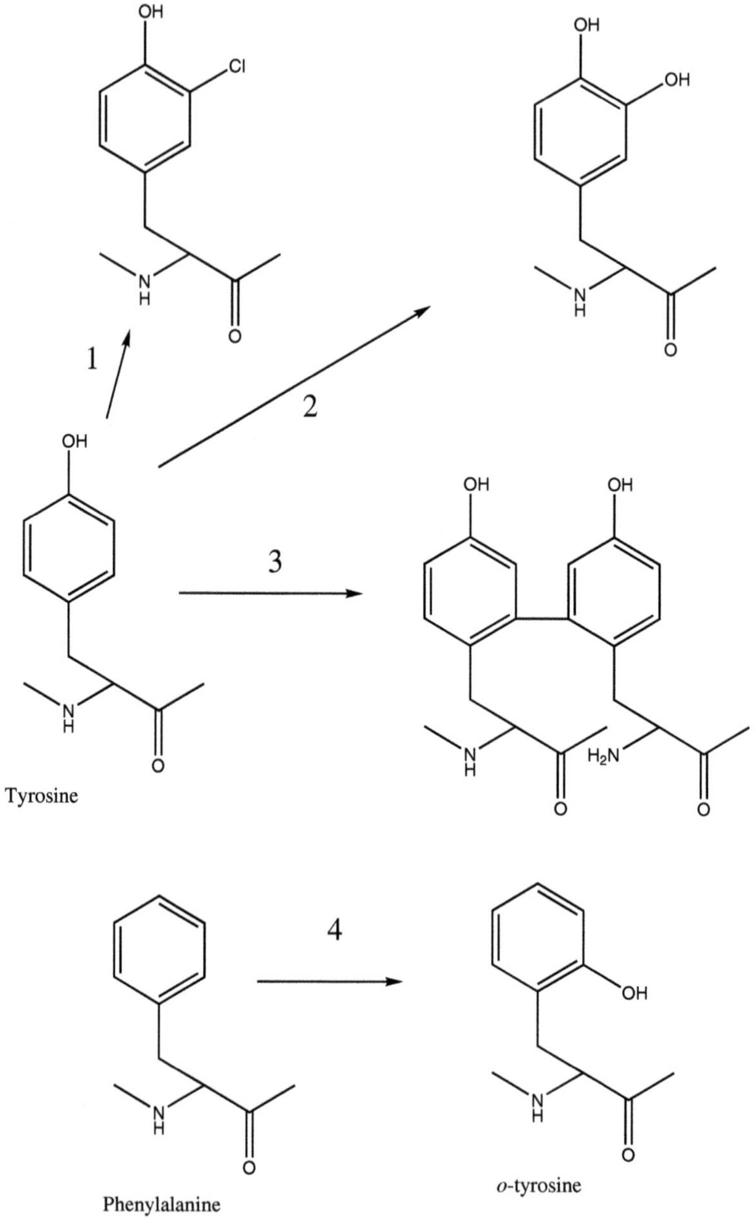

FIGURE 2.37 The reduction of 3-nitrotyrosine to form 3-aminotyrosine and the subsequent reaction with dansyl chloride. Also shown is the side reaction of tetranitromethane with tyrosine to form a dimer.

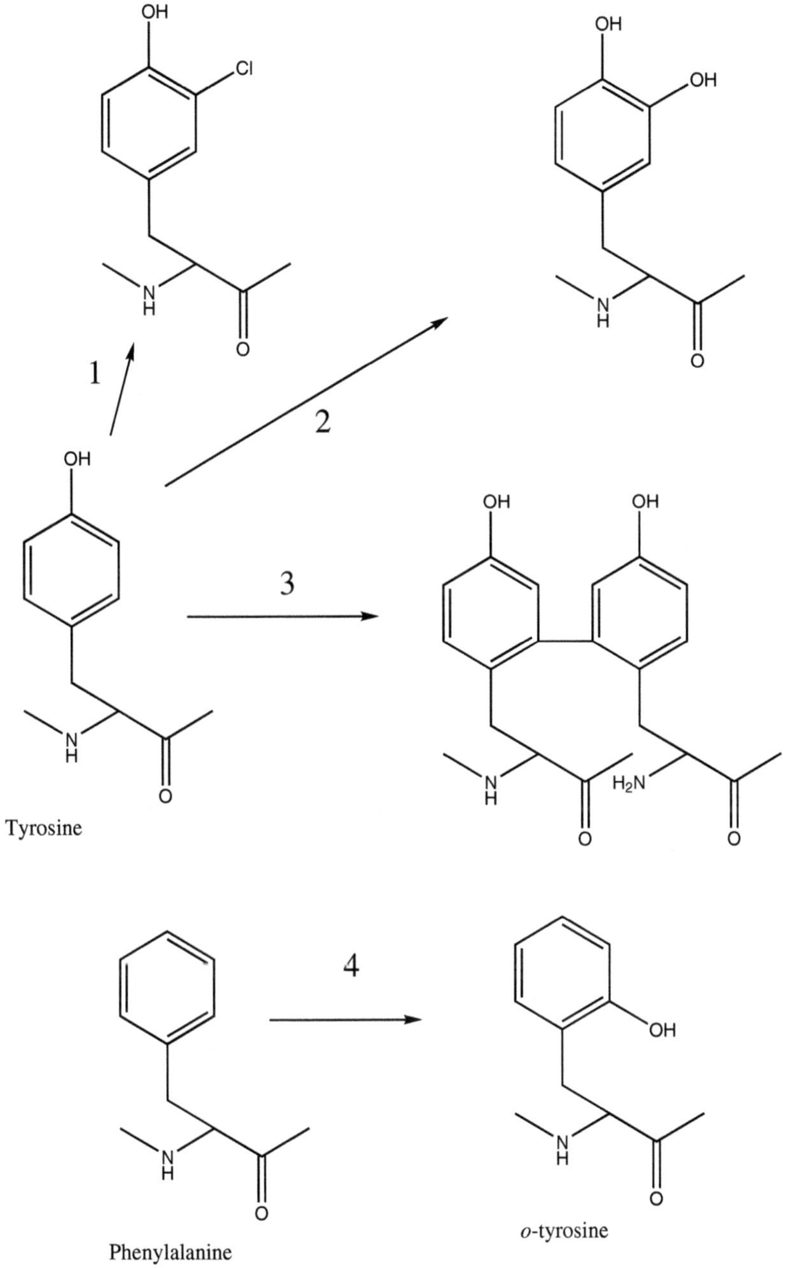

FIGURE 2.38 Multiple modification of tyrosine residues via oxidation reaction. Also included is the oxidation of phenylalanine to product *o*-tyrosine.

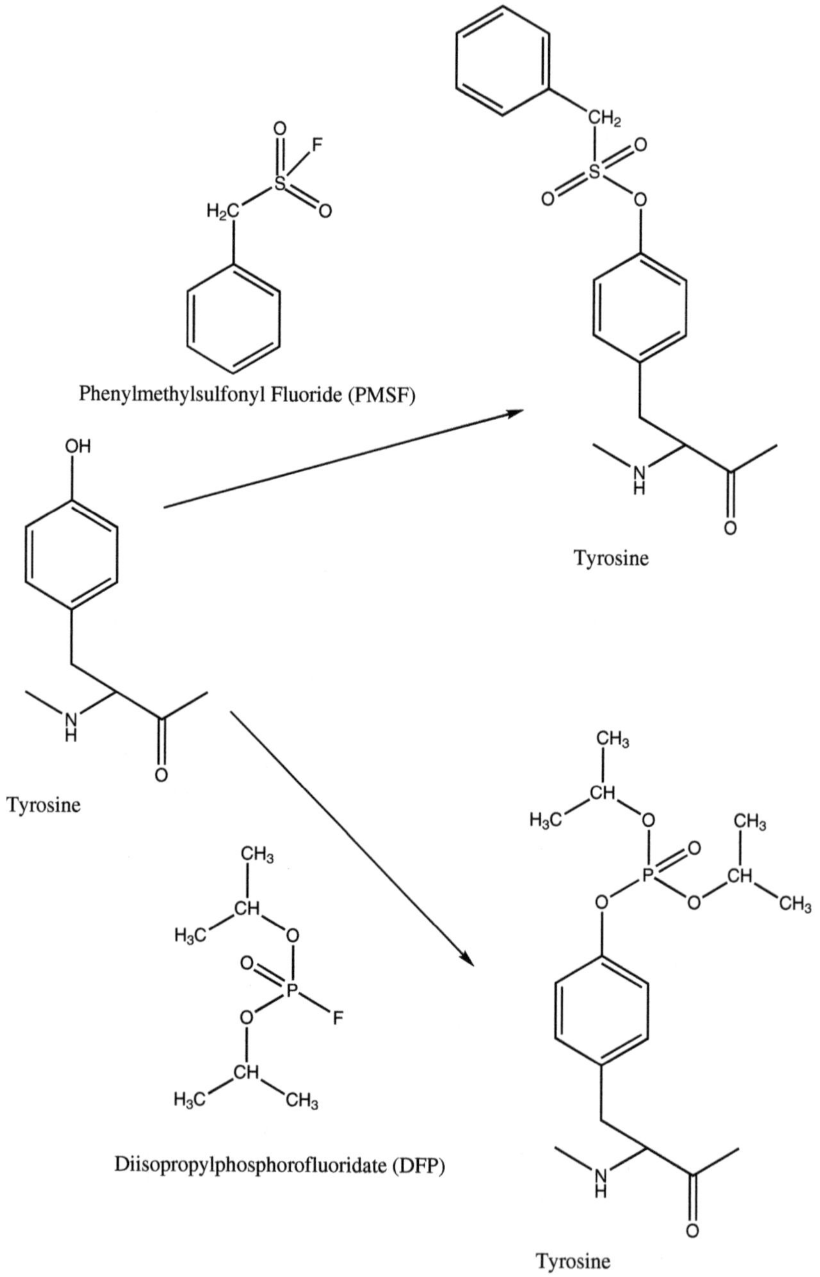

FIGURE 2.39 The modification of tyrosine with phenylmethylsulfonyl fluoride (PMSF) or diisopropylphosphorofluoridate (DFP).

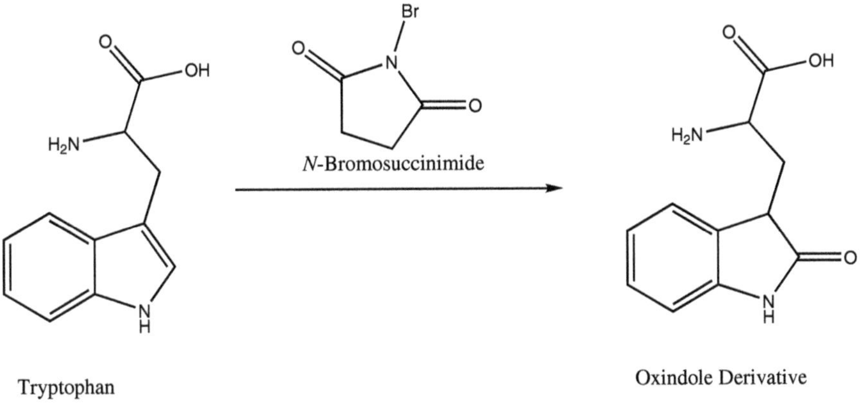

Tryptophan

Oxindole Derivative

FIGURE 2.40 The modification of tryptophan residues in protein with *N*-bromosuccinimide. This irreversible reaction results in the formation of an oxindole derivative with altered spectral properties. Oxidation with *N*-bromosuccinimide has been used for the quantitative determination of tryptophan residues in proteins.

Tryptophan   2-Hydroxy-5-Nitrobenzyl Bromide

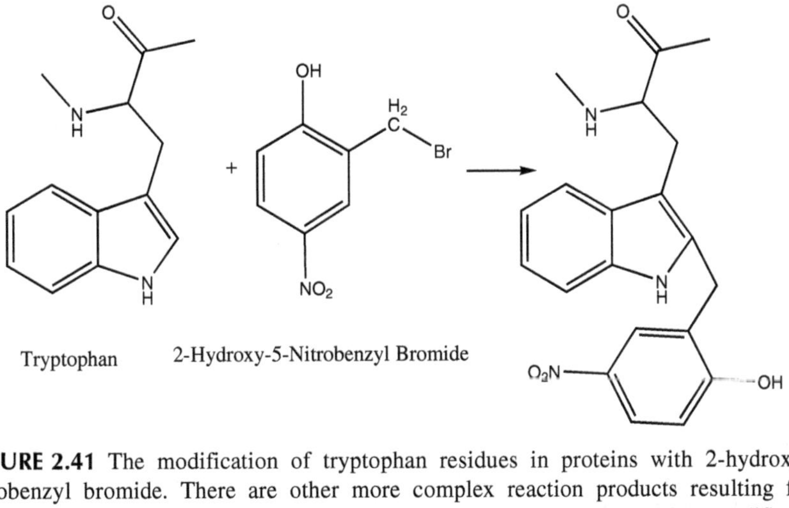

FIGURE 2.41 The modification of tryptophan residues in proteins with 2-hydroxy-5-nitrobenzyl bromide. There are other more complex reaction products resulting from multiple substitutions (Strohalm, M., Kodìk, M. and Pechar, M., Tryptophan modification by 2-hydroxy-5-nitrobenzyl bromide studied by MALDI-TOF mass spectrometry, *Biochem. Biophys. Res. Comm.*, 312, 811–816, 2003).

cleavage. The peptide was partially reduced with phosphine and the resulting cysteine residues immediately cyanylated with 1-cyano-4-(dimethylamino)pyridinium tetrafluoroborate. The cyanylated peptides were isolated by HPLC and cleaved with aqueous ammonia.

Douady and coworkers[117] have used *N*-chlorosuccinimide in acetic acid to cleave peptide bonds in the major polypeptide component of the light-harvesting complex from a brown alga (*Laminaria saccharina*). Droste and colleagues[118] used *N*-chlorosuccinimide for the fragmentation of adenyl cyclase type I in a study on the identification

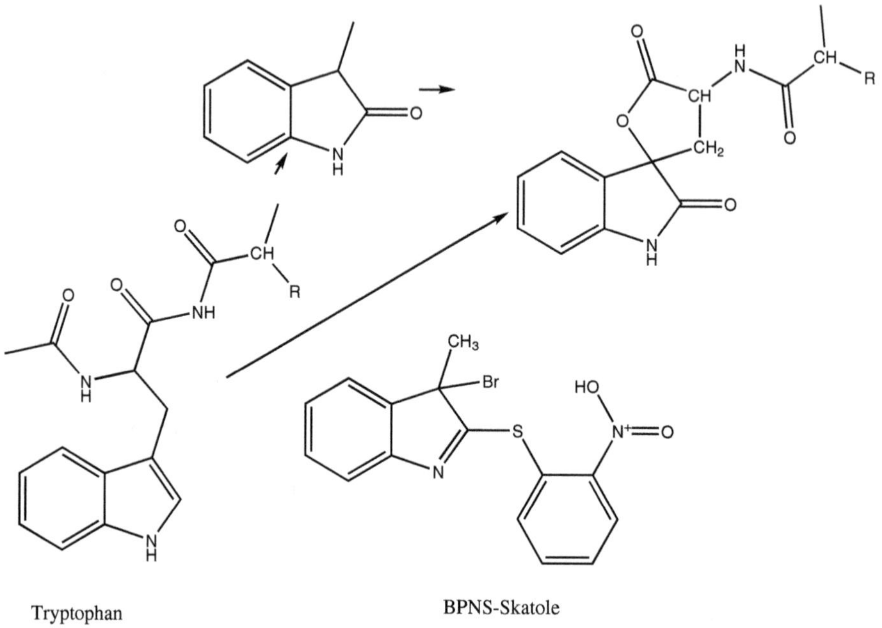

Tryptophan   BPNS-Skatole

FIGURE 2.42 The modification of tryptophan residues in proteins with BPNS-skatole.

of an ATP-binding site. Pliszka and coworkers[119] used *N*-chlorosuccinimide to identify EDC cross-links in subfragment 1 or skeletal muscle myosin. The gel slices were first washed with water for 20 minutes with one change of solvent. The gel slices were then washed with urea/H$_2$O/acetic acid (1g/1 mL/1 mL) for 20 minutes with one change of solvent. Cleavage is accomplished with *N*-chlorosuccinimide (15 mM) in urea/H$_2$O/acetic acid for 30 minutes.

CROSS-LINKING OF PROTEINS

This material is placed last for a good reason: site-specific chemical cross-linking is an application of the various reagents that have been discussed. Covering the many applications of protein cross-linking is simply impossible and the reader is directed to several excellent reviews.[120–128]

Zero-length cross-linking[129] is a procedure which joins peptide chains via existing functional groups such that a "spacer" group is not utilized. Isopeptide bond formation mediated by carbodiimide (Figure 2.45)[130,131] is an extensively used approach. *N*-(3-(dimethylamino)propyl)-N′-ethylcarbodiimide has been used to form an intramolecular cross-link protein between a lysine residue and a glutamic acid residue.[132] This reaction is referred to as zero-length cross-linking and is used frequently to determine areas of protein contact.[133–135] Cyanogen (ethanedinitrile) can be used for zero-length cross-linking of salt bridges such as one between an aspartic acid and arginine in hen egg white lysozyme (Figure 2.46).[136] The cyanogen-mediated cross-linking can occur between glutamic acid or aspartic acid and arginine, lysine or histidine residues.

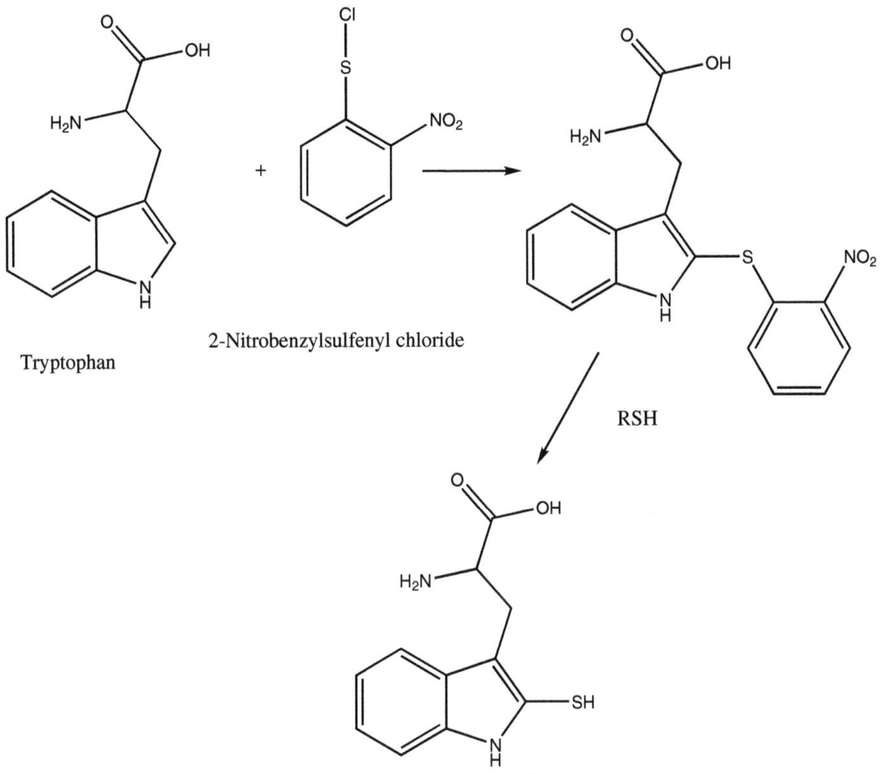

2-Nitrobenzylsulfenyl chloride

Tryptophan

RSH

2-Thiotryptophan

FIGURE 2.43 The modification of tryptophan residues in proteins with 2-nitrophenylsulfonyl chloride.

Zero-length cross-linking between tyrosine residues (3,3′-dityrosine, Figure 2.36) is obtained via reaction with tetranitromethane or oxidation.[137,138] Glutamine, the γ-amido derivative of glutamic acid, is a participant in the transamidation reactions discussed in Chapter 3 that results in the biological cross-linking of proteins.

EDC-mediated zero-length cross-linking has been used to study the interaction between skeletal myosin light chain 1 and F-actin,[139] *Escherichia coli* T-protein and methylenetetrahydropteroyl-tetraglutamate,[140] the calmodulin-melittin complex[141] and a complex between biotin protein ligase and biotin carboxyl carrier protein.[142] In the calmodulin-melittin complex study,[141] zero-length cross-linking was combined with limited proteolysis to develop a model for the complex. The cross-linking experiments were performed in 50 mM MES, 30 μM CaCL$_2$, pH 7.0 at 25°C. Equimolar (15 nmoles) quantities of calmodulin and melittin were incubated in the above solvent for 10 minutes prior to the addition of EDC (tenfold molar excess to calmodulin carboxyl groups). The reaction was allowed to proceed for two hours and then terminated by the addition of trifluoroacetic acid. The cross-linked protein was purified by HPLC (C$_4$ column) and effluent fractions assay with electrospray

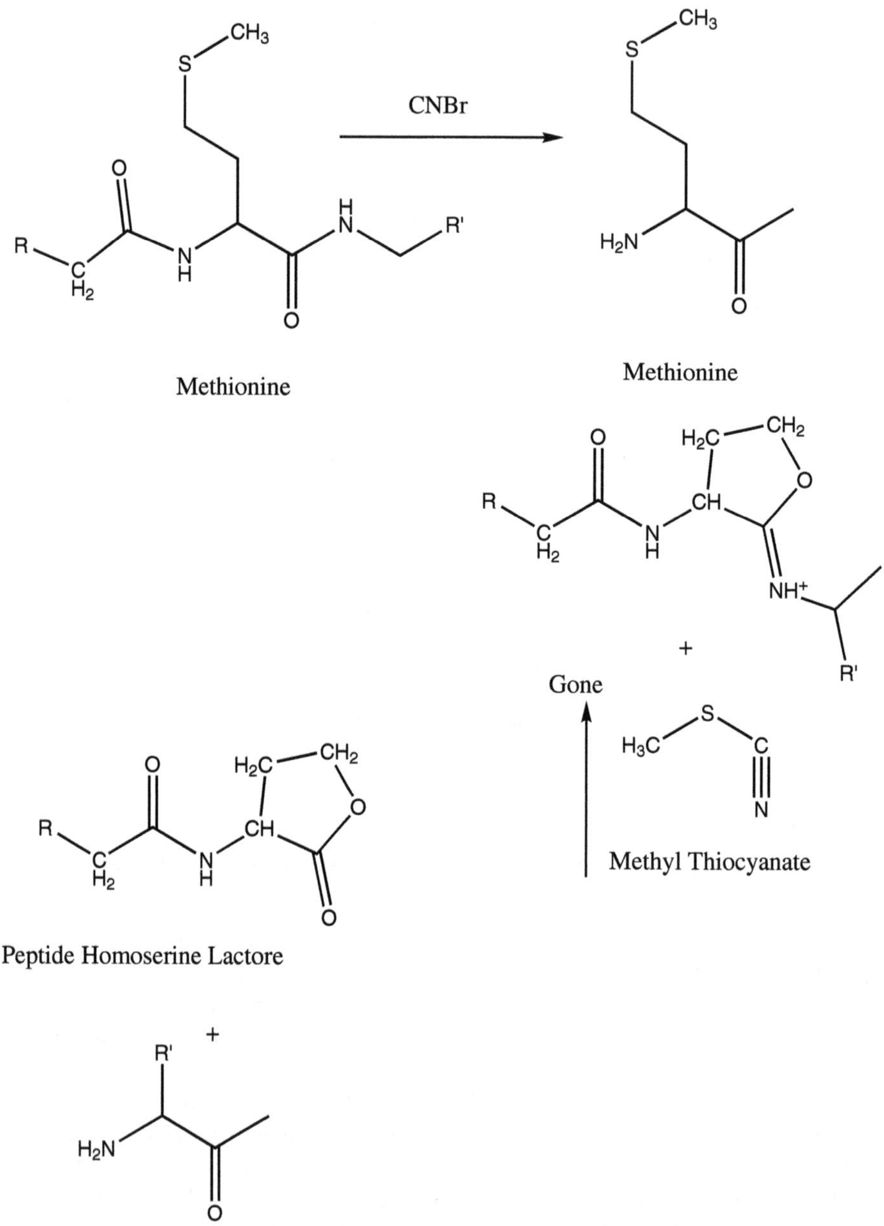

FIGURE 2.44 The reaction of cyanogen bromide with methionine residues in proteins. Note the resulting cleavage of the peptide bond and the release of methyl isothiocyanate.

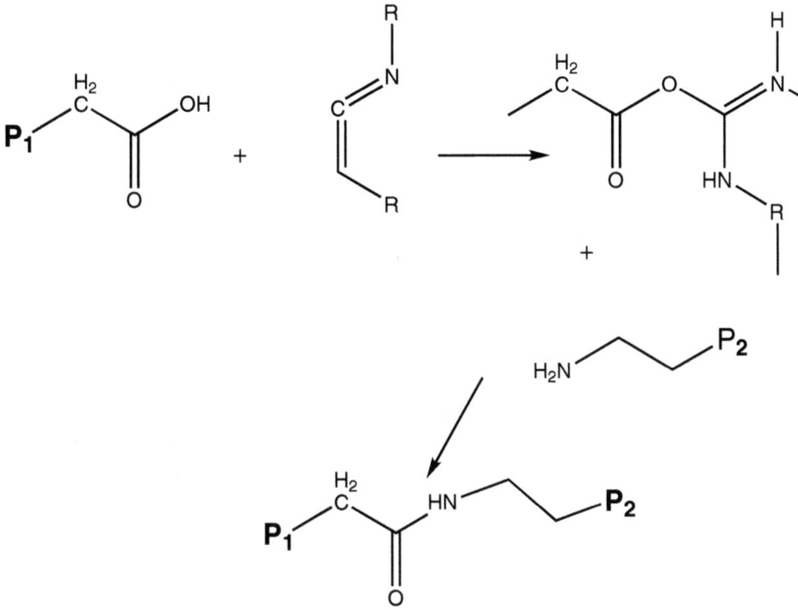

Isopeptide Bond for Zero-Length Cross-Linking

FIGURE 2.45 Zero-length cross-linking in proteins with a carbodiimide.

mass spectrometry. Further analysis by mass spectrometry followed proteolysis of the cross-linked complexes following CNBr cleavage.

Mass spectrometry has been of increasing importance in the study of both intra- and intermolecularly cross-linked proteins. Bennett, Kussman and coworkers[143] have combined thiol-cleavable cross-linking reagents such as 3,3′-dithio-bis(succini-midyl-propionate) (Figure 2.47) with mass spectrometry to determine intermolecular protein contacts in two associating protein systems (ParR, homodimeric DNA pro-tein, CD28-I$_g$G and CD80-F$_{ab}$). Cross-linking was accomplished in phosphate-buffer saline, pH 7.5 at room temperature. In the case of CD28-I$_g$G and CD80-F$_{ab}$, the cross-linking reaction was terminated by the addition of 1.0 M Tris, pH 7.5. The cross-linked proteins were isolated and digested with trypsin (overnight, 37°C) and the digests divided into two equal portions. One portion was taken directly to mass spectrometric analysis while the second portion was reduced with DTT to cleave the cross-link prior to mass spectrometric analysis. Comparison of the "maps" obtained by mass spectrometry allowed the identification of regions of intermolec-ular contact. Several groups have developed isotopically labeled cross-linking reagents to improve the sensitivity of detection of low-abundance cross-linked peptides.[144,45] Homobifunctional imidoesters (Figure 2.48) are highly specific for primary amines. Buffer effects on the reaction have not been extensively investigated, except to specify that the use of potential competing nucleophiles (e.g., Tris, imidazole) should be avoided; ammonium acetate is frequently used to quench

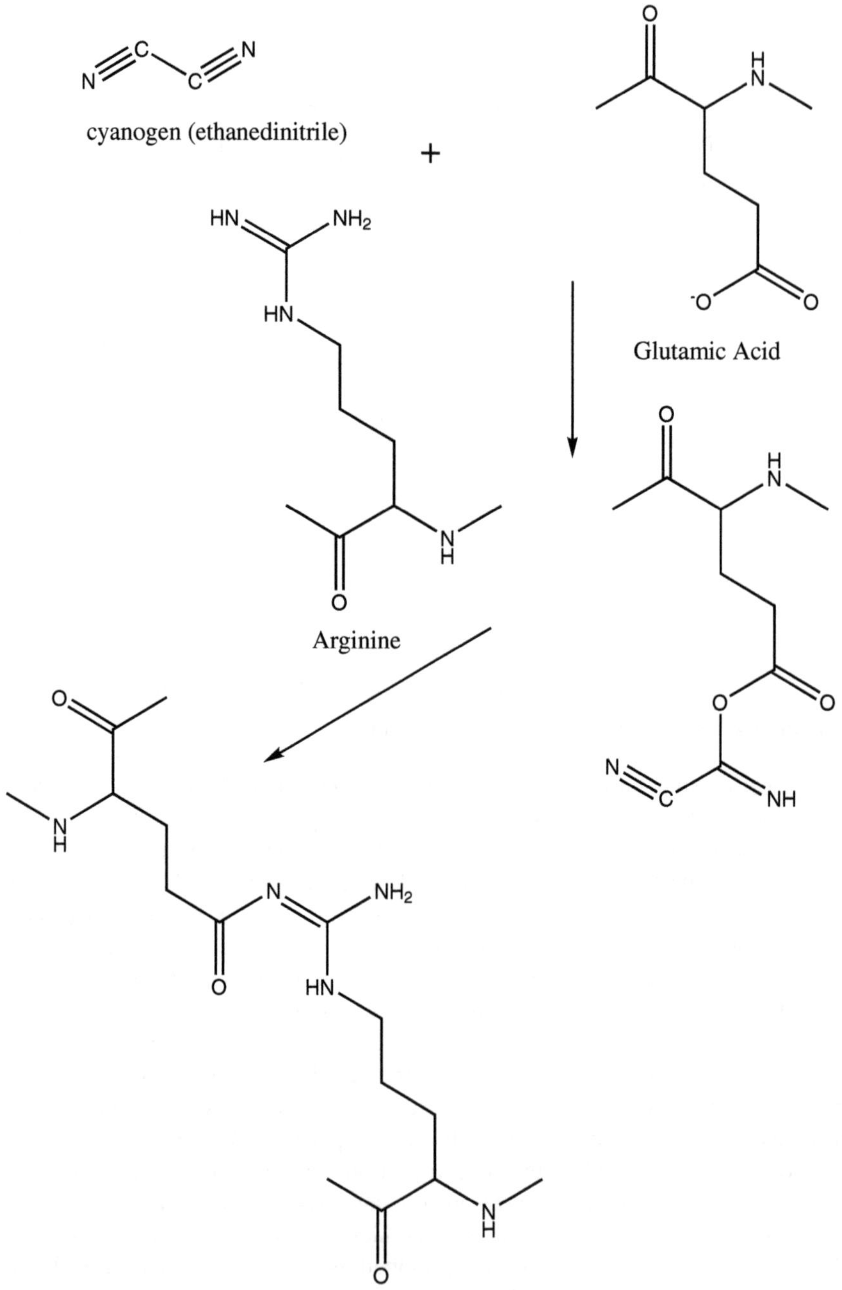

cyanogen (ethanedinitrile)

+

Glutamic Acid

Arginine

FIGURE 2.46 Zero-length cross-linking in proteins with cyanogen.

the reactions. Most studies have used 0.02 to 0.1 M triethanolamine in the range of pH 8.0 to 9.0. The presence of triethanolamine has been suggested to enhance the amidation reaction in studies on the reaction of methyl-4-mercaptobutyrimidate.[146] Photo-activated cross-linking agents (Figure 2.49) can also be useful as well photosensitizing agents.[147–149]

IN VIVO NONENZYMATIC CHEMICAL MODIFICATION

Several post-translational modifications exist that might be considered to occur in a "random" manner in that such modifications are not catalyzed by an enzyme or enzymes and subject to the environment of the proteins. This is a bit of a picky point as the reagents that carry out the chemical modification reactions are frequently produced *in situ* by enzymatic action; examples would be peroxynitrite produced from nitric oxide and peroxide or hypochlorous acid.[150–152] However, the proteins are not necessarily modified in a random manner in that the modification occurs at specific residues or residues of a given class. Reactivity of a given residue can be an indication of solvent exposure (or lack thereof) or intrinsic reactivity of a specific residue. As a result, it is possible to use proteomic analytical techniques to assay for these *in vivo* modifications.[153–156]

OXIDATION

Oxidation of proteins occurs *in vivo* with generally unfavorable consequences. For example, the oxidation of the "active-site" methionine residue in alpha-1-antitrypsin (alpha-1-antiprotease inhibitor) results in pulmonary damage.[157–159] Methionine is quite susceptible to oxidation, first to the sulfoxide, a reversible reaction, and to the sulfone (Figure 2.35).[160,161] A wide spectrum of oxidizing agents exists including free radicals such as hydroxyl radical,[162,163] organic peroxides[164] and hypochlorites (Figure 2.2).[165] Organic peroxides are most notable as lipid peroxides observed in oxidative stress.[166,167] Hypochlorites can be formed by the action of myeloperoxidase and may be responsible for protein oxidation in atherosclerosis and Alzheimer's disease.[168,169] Oxidative covalent modifications of proteins are also suggested to be signals for degradation.[170]

NITRIC OXIDE

Nitric oxide is a potent physiologic agent with diverse systemic effects.[171] Peroxynitrite, formed from nitric oxide by reaction with superoxide,[172–175] is a mediator of some of these physiologic effects of nitric oxide (Figure 2.2). The reaction of proteins with peroxynitrite results primarily in the modification of tyrosine residues to form 3-nitrotyrosine.[176,177] Peroxynitrite can also modify tryptophanyl residues[178] and oxidize methionine residues.[179] Other oxidation pathways exist, resulting in carbonyl formation and dityrosine.[180,181] Carbon dioxide does influence the reaction of peroxynitrite with proteins[182,183] by forming an adduct with peroxynitrite (Figure 2.50).[184] Peroxynitrite can also react with nucleic acids,[185] resulting in chain cleavage.

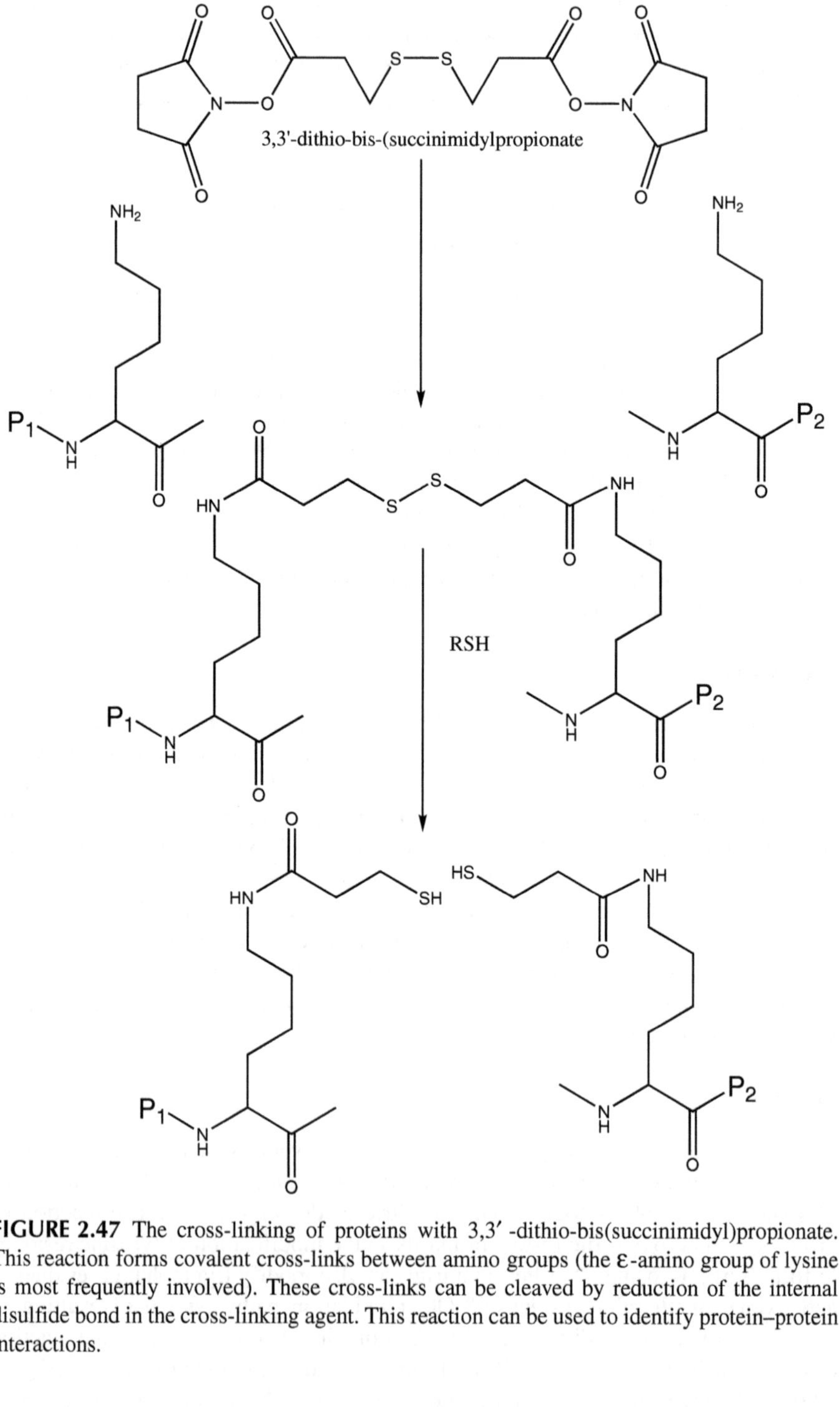

FIGURE 2.47 The cross-linking of proteins with 3,3′ -dithio-bis(succinimidyl)propionate. This reaction forms covalent cross-links between amino groups (the ε-amino group of lysine is most frequently involved). These cross-links can be cleaved by reduction of the internal disulfide bond in the cross-linking agent. This reaction can be used to identify protein–protein interactions.

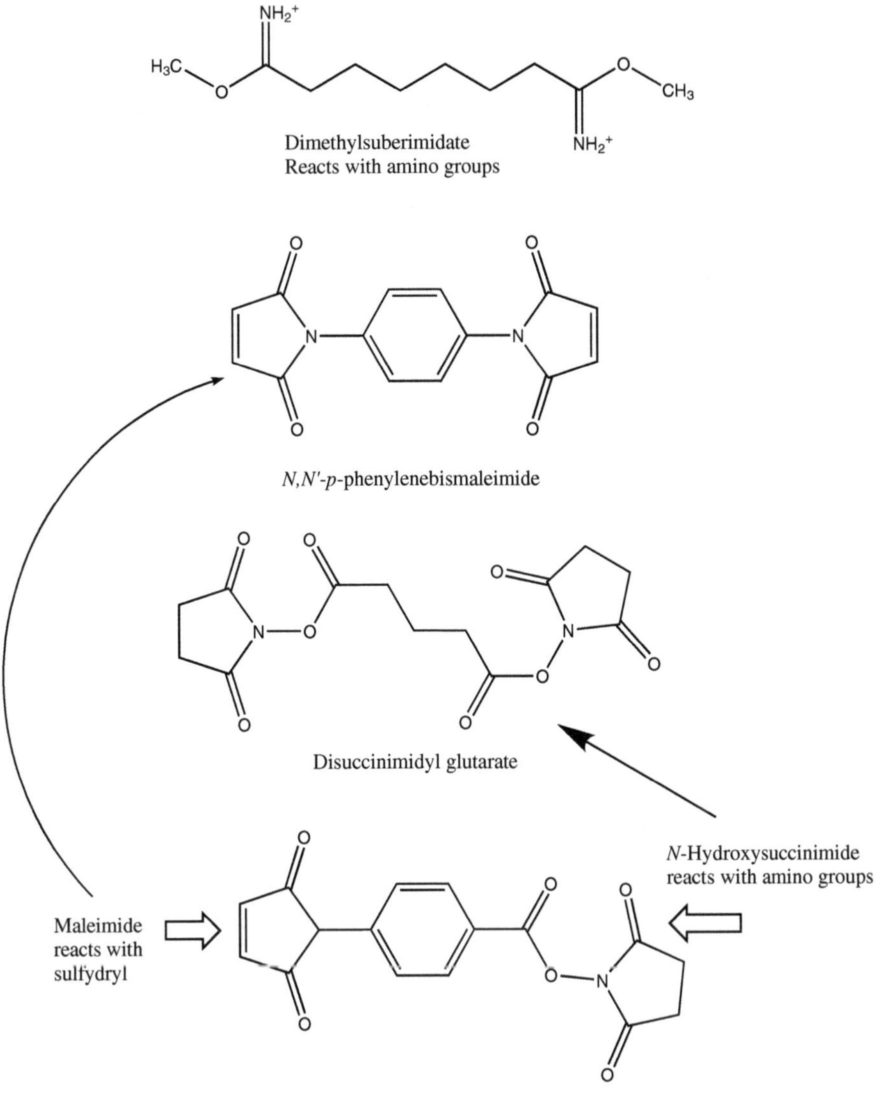

FIGURE 2.48 Some homobifunctional and heterobifunctional cross-linking agents. Homobifunctional reagents react with the same protein functional groups such as sulfydryl groups or amino groups as shown. A heterobifunctional reagent can react with different functional groups — an example is shown with a maleimide function that would react with a sulfydryl group and with a lysine residue in a separate protein; reaction with the same protein is possible as well. Intermolecular and intramolecular cross-linking, in the absence of specific interactions, is a function of protein concentration (Nadeau, O.W. and Carlson, G.M., Chemical cross-linking in studying protein–protein interactions, in *Protein–Protein Interaction: A Molecular Cloning Manual*, E. Golemis, Ed., Cold Spring Harbor Laboratory Press, Cold Spring Harbor, NY, 2002).

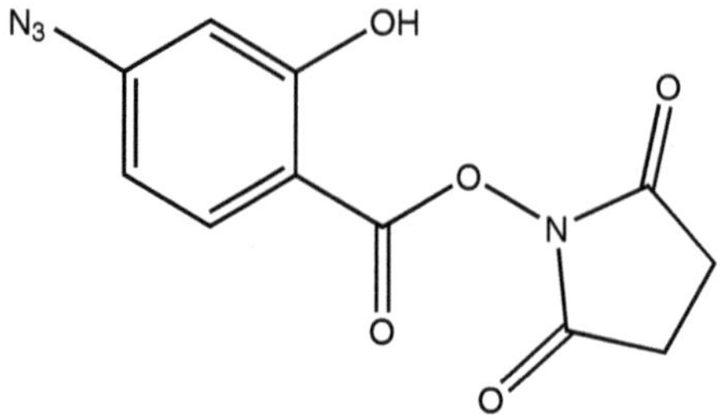

N-Hydroxysuccinimide ester of 4-azido-salicyclic acid

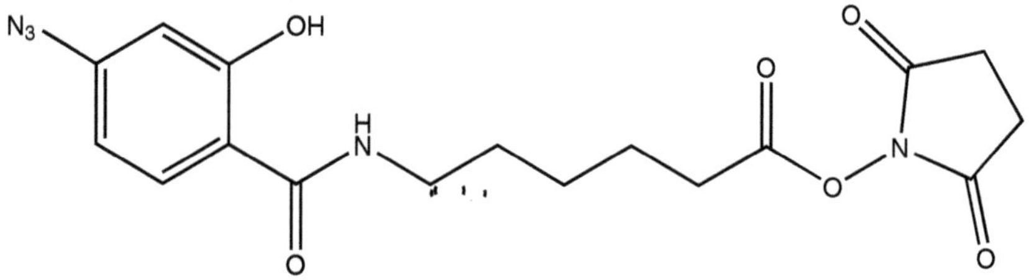

N-Hydroxysuccinimide ester of N-(4-azidosalicyl)-6-aminocaproic acid

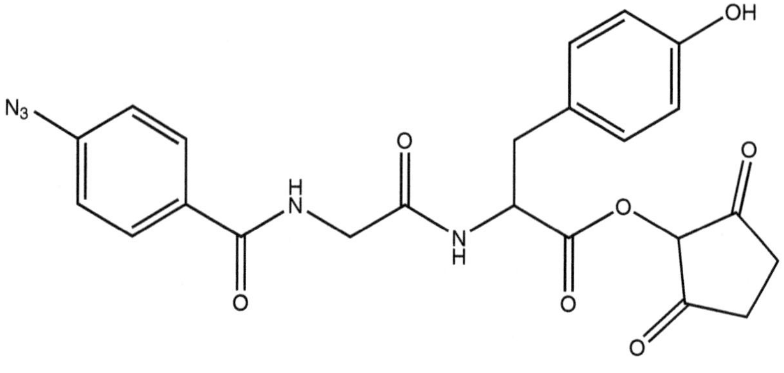

N-Hydroxysuccinimide ester of 4-azidobenzoylglycyltyrosine

FIGURE 2.49 Some photo-activated protein cross-linking reagents. Modification of the protein amino groups via the N-hydroxysuccinimide function is performed at alkaline pH in the dark; following this reaction, photolysis activates the azido function. (Ji, T.H. and Ji, I., Macromolecular photoaffinity labeling with radioactive bifunctional heterobifunctional reagents, *Anal. Biochem.*, 121, 286, 1982).

Nitric oxide can react directly with sulfydryl groups on proteins but the reactions appear to be somewhat complicated, resulting in the formation of an *S*-nitroso derivative of cysteine (Figure 2.51).[186] The *S*-nitroso proteome of *Mycobacterium tuberculosis* has been described[187] using a biotin-switch assay (Figure 2.52) developed by Snyder and coworkers.[188]

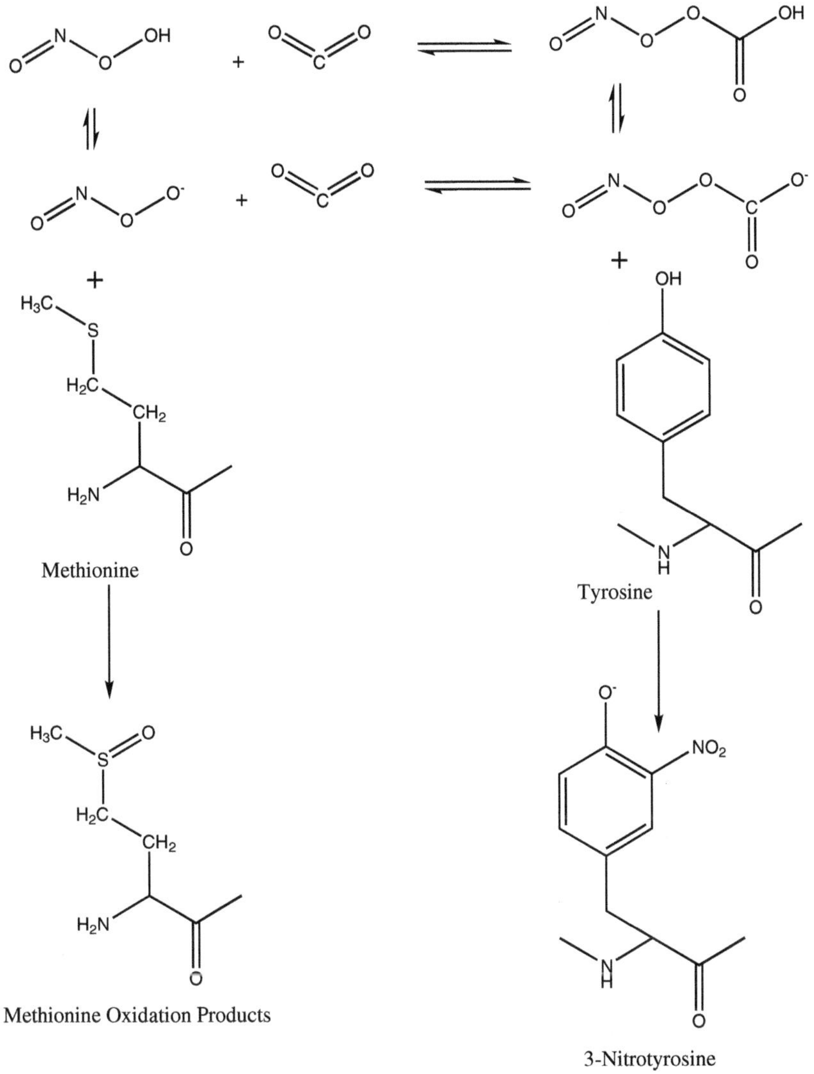

FIGURE 2.50 The reaction of peroxynitrite with methionine and tyrosine: Control by the presence of carbon dioxide.

GLYCATION

Glycation is the term used to identify the reaction of reducing sugars with proteins. This involves the initial formation of a Schiff base followed by rearrangement in the Maillard reaction,[189,190] eventually resulting in advanced glycation end (AGE) products (Figure 2.53).[191] Reaction can occur at lysine and arginine residues with resulting cross-link formation.[192–196] The glycation of proteins by methyl-glyoxal has been of specific interest (Figure 2.54 and Figure 2.55).[197,198]

4

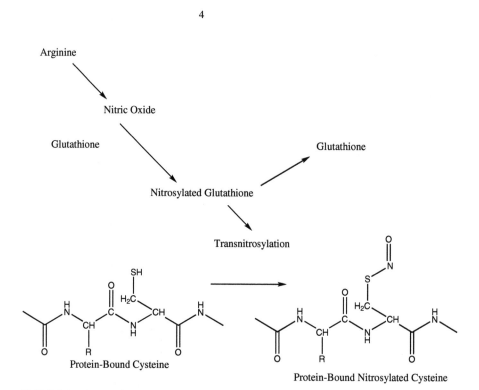

FIGURE 2.51 The nitrosylation of cysteine residues in proteins.

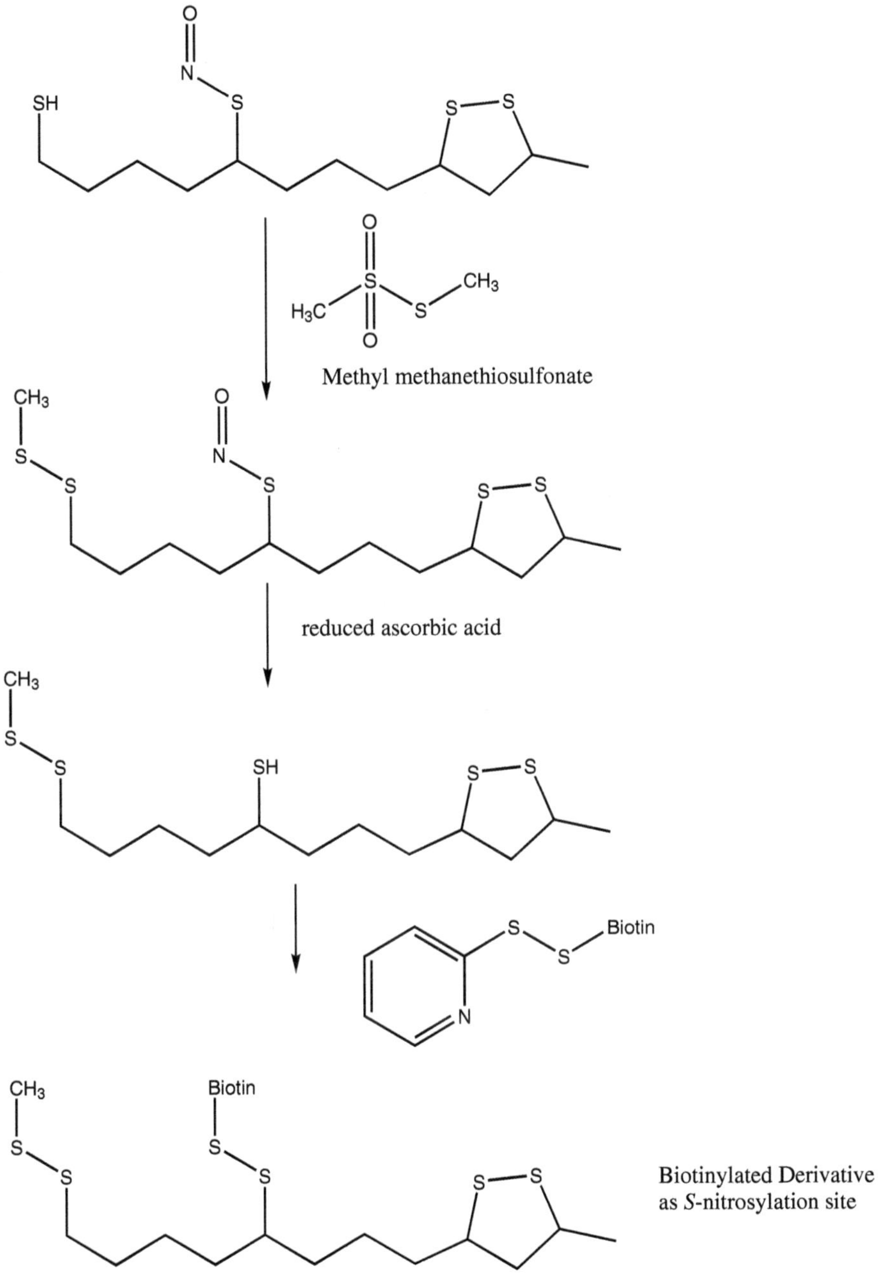

FIGURE 2.52 The use of a "biotin switch" technique to identify the site of cysteine nitrosylation in proteins.

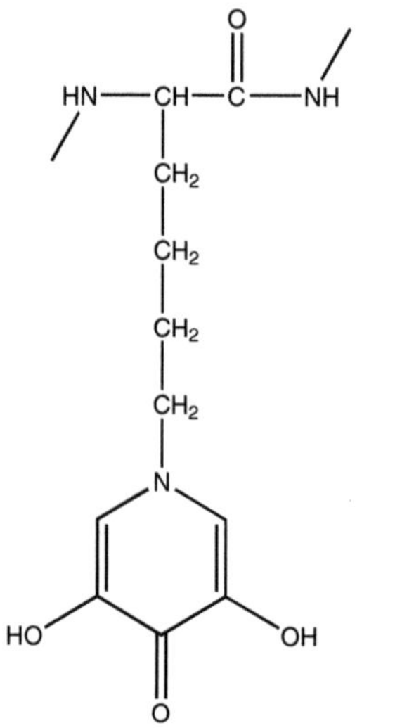

trihydroxy-triosidine, derived from
reaction with glyceraldehyde

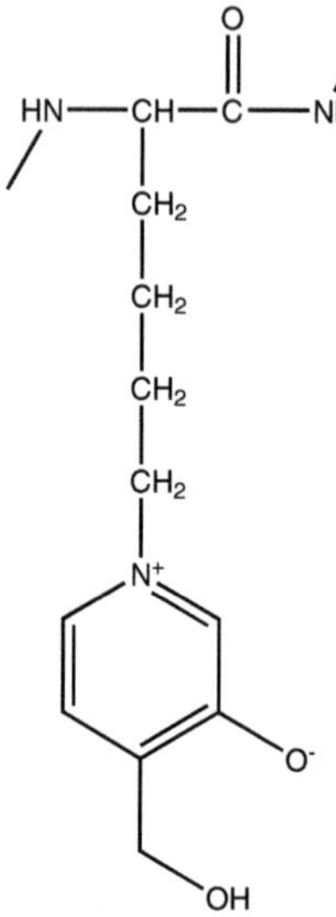

2-ammonium-6-[4-(hydroxymethyl)-3-oxidopyridium-
1-yl-]hexanoate; derived from reaction with
dehydroascorbic acid

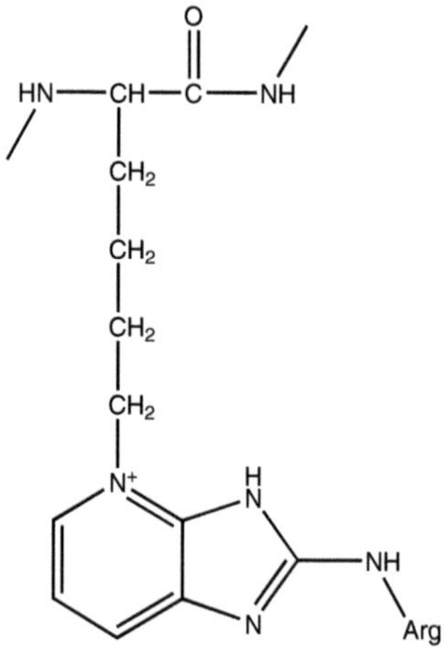

Pentosidine

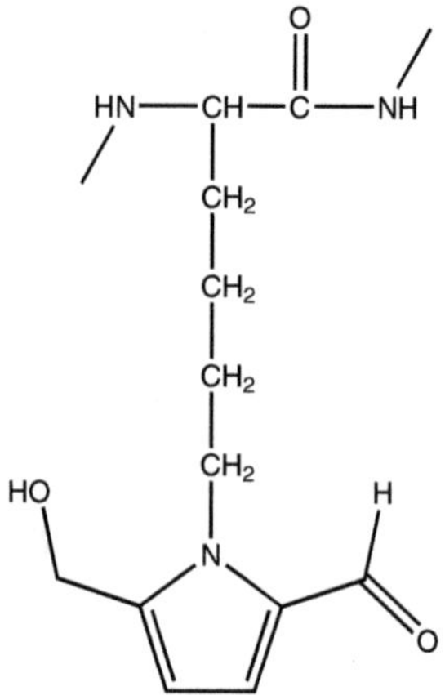

Pyrraline

FIGURE 2.53 The glycation of protein and formation of advanced glycation end-products
(AGE products).

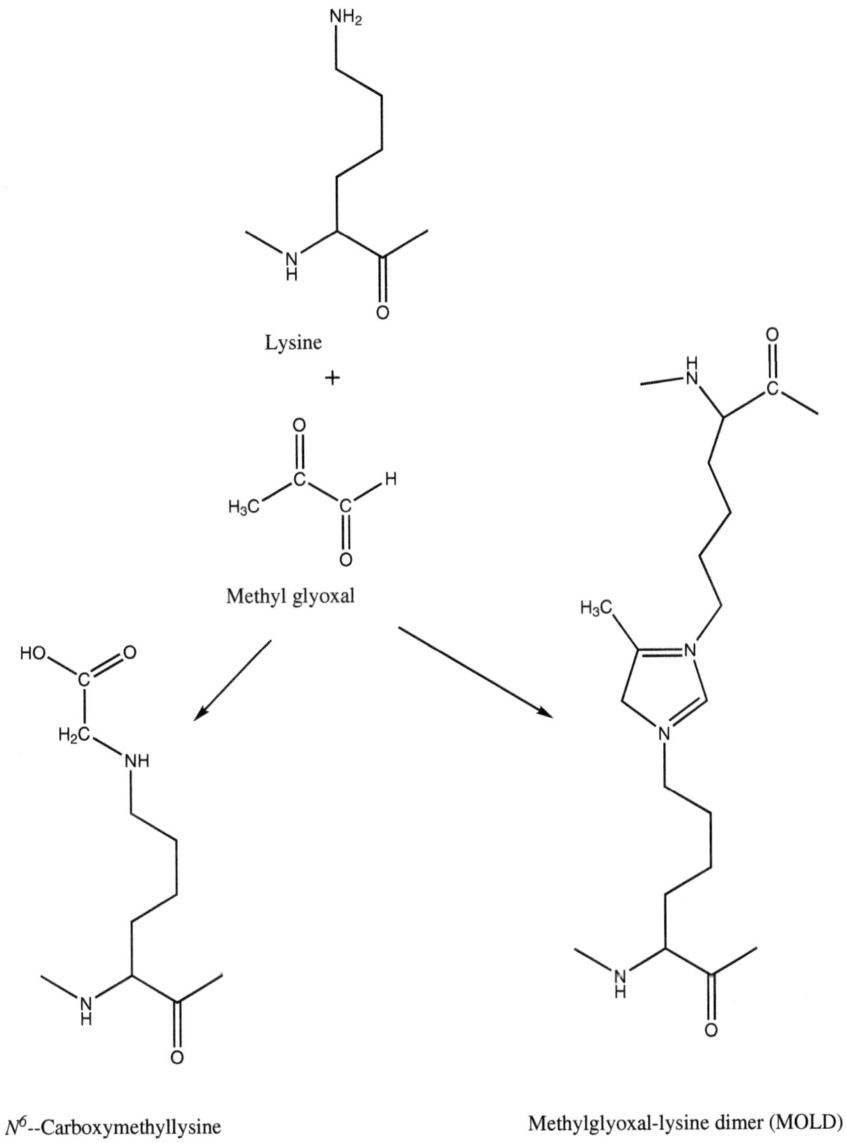

FIGURE 2.54 The complex reaction of methyl glyoxal with lysine.

Lysine

+

Methyl glyoxal

N^6--Carboxymethyllysine   Methylglyoxal-lysine dimer (MOLD)

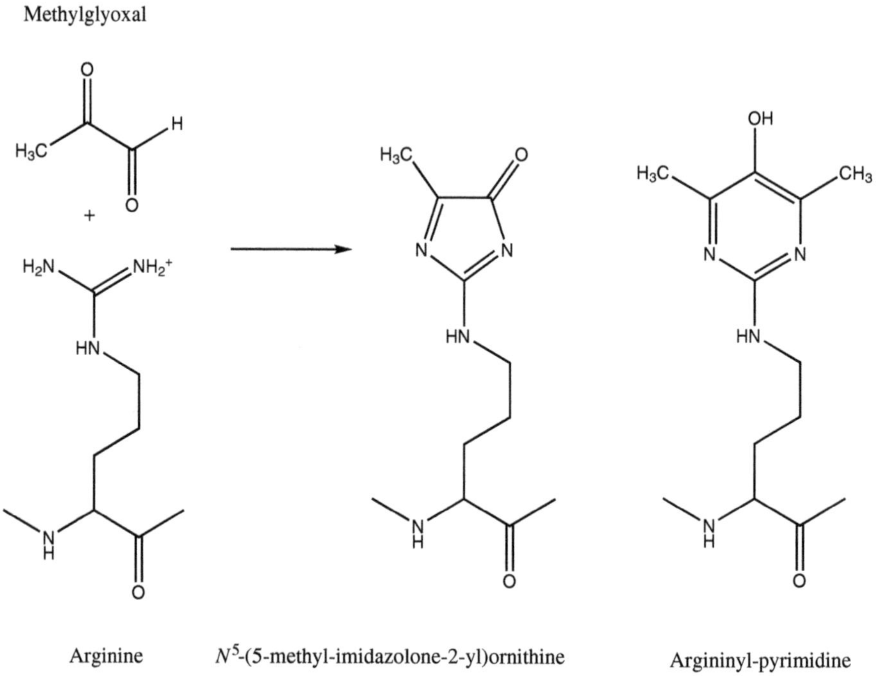

Methylglyoxal

Arginine N^5-(5-methyl-imidazolone-2-yl)ornithine   Argininyl-pyrimidine

FIGURE 2.55 The complex reaction of methyl glyoxal with arginine.

REFERENCES

1. Lundblad, R.L., *Chemical Reagents for Protein Modification*, 3rd Edition, CRC Press, Boca Raton, FL, Ch. 1, 2004.
2. Means, G.E., Zhang, H. and Le, M., The Chemistry of Protein Functional Groups, in *Protein: A Comprehensive Treatise*, G. Allen, Ed., JAI Press, Stamford, CT, Vol. 2, Ch. 2, 23–59, 1999.
3. DeSantis, G. and Jones, J.B., Chemical modification of enzymes for enhanced functionality. *Curr. Opin. Biotech.*, 10, 324, 1999.
4. Isom, A.L., Barnes, S., Wilson, L., Kirk, M., Coward, L. and Darley-Usmar, V., Modification of cytochrome c by 4-hydroxy-2-nonenol: Evidence for histidine, lysine, and arginine-aldehyde adducts, *J. Am. Soc. Mass Spec.*, 15, 1136, 2004.
5. Knott, L. and Bailey, A.J., Collagen cross-links in mineralizing tissues: A review of their chemistry, function, and clinical relevance, *Bone*, 22, 181, 1998.
6. Chaiken, I.M. and Smith, E.L., Reaction of chloroacetamide with the sulfydryl groups of papain, *J. Biol. Chem.*, 244, 5087, 1969.
7. Chaiken, I.M. and Smith, E.L., Reaction of the sulfydryl group of papain with chloroacetic acid, *J. Biol. Chem.*, 244, 5095, 1969.
8. Zhang, G., Mazurkie, A.S., Dunaway-Mariano, D. and Allen, K.N., Kinetic evidence for a substrate-induced fit in phosphonoacetaldehyde hydrolase catalysis, *Biochemistry*, 41, 13370, 2002.
9. Fisher, B.M., Schultz, L.W. and Raines, R.T., Coulombic effects of remote subsites on the active site of ribonuclease A, *Biochemistry*, 37, 17386, 1998.

10. Schmidt, D.E., Jr. and Westheimer, F.H., pK of the lysine amino group at the active site of acetoacetate decarboxylase, *Biochemistry*, 10, 1249, 1971.
11. Plapp, B.V., Application of affinity labeling for studying structure and function in enzymes, *Meth. Enzymology*, 87, 469, 1982.
12. Myers, B.H. and Glazer, A.N., Spectroscopic studies of the exposure of tyrosine residues in proteins with special references to the subtilisins, *J. Biol. Chem.*, 26, 412, 1971.
13. Skov, K., Hofmann, T. and Williams, G.R., The nitration of cytochrome c, *Canadian J. Biochem.*, 47, 750, 1969.
14. Galvani, M., Hamdan, M., Herbert, B. and Righetti, P.G., Alkylation kinetics of proteins in preparation for two-dimensional maps: A matrix assisted laser desorption/ionization mass spectrometry investigation, *Electrophoresis*, 22, 2058, 2001.
15. Hamden, M. and Righetti, P.G., Modern strategies for protein quantification in proteome analysis: Advantages and limitations, *Mass Spec. Rev.*, 21, 287, 2002.
16. Ratner, V., Kahana, E., Eichler, M. and Haas, E., A general strategy for site-specific double labeling of globular proteins for kinetic FRET studies, *Bioconjugate Chem.*, 13, 1163, 2002.
17. Smith, J.J., Conrad, D.W., Cuneo, M.J. and Hellinga, H.W., Orthogonal site-specific protein modification by engineering reversible thiol protection mechanisms, *Protein Sci.*, 14, 64, 2005.
18. Degenhardt, T.P., Thorpe, S.R. and Baynes, J.W., Chemical modification of proteins by methylglyoxal, *Chem. Mol. Biol.*, 44, 1139, 1998.
19. Seidler, N.W. and Kowalewski, C., Methylglyoxal-induced glycation affects protein topography, *Arch. Biochem. Biophys.*, 410, 149, 2003.
20. Odani, H., Sninzato, T., Usami, J. et al., Imidazolium crosslinks derived from reaction of lysine with glyoxal and methylglyoxal are increased in serum proteins of uremic patients: Evidence for increased oxidative stress in uremia, *FEBS Letters*, 427, 381, 1998.
21. Sheng, J. and Preiss, J., Arginine[294] is essential for the inhibition of *Anabaena* PCC 7120 ADP-glucose pyrophosphorylase by phosphate, *Biochemistry*, 36, 13077, 1997.
22. Wu, X., Chen, S.G., Petrash, J.M. and Monnier, V.M., Alteration of substrate selectivity through mutation of two arginine residues in the binding site of Amadoriase II from *Aspergillus* sp., *Biochemistry*, 41, 4453, 2002.
23. Borders, C.L., Jr. and Johansen, J.T., Identification of Arg-143 as the essential arginyl residue in yeast Cu, Zn superoxide dismutase by use of a chromophoric arginine reagent, *Biochem. Biophys. Res. Comm.*, 96, 1071, 1980.
24. Gärtner, E.M., Liebold, K., Legrum, B. et al., Three different actions of phenylglyoxal on band 3 protein-mediated anion transport across the red cell membrane, *Biochimica et Biophysica Acta*, 1323, 208, 1997.
25. Irwin, W.A., Gaspers, L.D. and Thomas, J.A., Inhibition of the mitochondrial permeability transition by aldehydes, *Biochem. Biophys. Res. Comm.*, 291, 215, 2002.
26. Kuera, I., Passive penetration of nitrate through the plasma membrane of *Paracoccus denitrificans* and its potentiation by the lipophilic tetraphenylphosphonium cation, *Biochim. Biophys. Acta*, 1557, 119, 2003.
27. Schepens, I., Ruelland, E., Miginiac-Maslow, M. et al., The role of active site arginines of sorghum NADP-malate dehydrogenase in thioredoxin-dependent activation and activity, *J. Biol. Chem.*, 275, 35792, 2000.
28. Smith, D.J., Maggio, E.T. and Kenyon, G.L., Simple alkanethiol groups for temporary blocking of sulfhydryl groups of enzymes, *Biochemistry*, 14, 766, 1975.
29. Kenyon, G.L. and Bruice, T.W., Novel sulfhydryl reagents, *Meth. Enzymology*, 47, 407, 1977.

30. Roberts, D.D., Lewis, S.D., Ballou, D.P., Olson, S.T. and Shafer, J.A., Reactivity of small thiolate anions and cysteine 25 in papain toward methyl methanethiosulfonate, *Biochemistry*, 25, 5595, 1986.

31. Bednar, R.A., Reactivity and pH dependence of thiol conjugation to *N*-ethylmaleimide detection of a conformational change in chalcone isomerase, *Biochemistry*, 29, 3684, 1990.

32. Lutolf, M.P., Tirelli, N., Cerritelli, S. et al., Systematic modulation of Michael-type reactivity of thiols through the use of charged amino acids, *Bioconjugate Chem.*, 12, 1051, 2001.

33. Herbert, B., Galvani, M., Hamdan, M. et al., Reduction and alkylation of proteins in preparation of two-dimensional map analysis: Why, when, and how? *Electrophoresis*, 22, 2046, 2001.

34. Galvani, M., Hamdan, M., Herbert, B. and Righetti, P.G., Alkylation kinetics of proteins in preparation for two-dimensional maps: A matrix assisted laser desorption/ionization-mass spectrometry investigation, *Electrophoresis*, 22, 2058, 2001.

35. Pihl, A. and Lange, R., The interaction of oxidized glutathione, cystamine monosulfoxide, and tetrathionate with the –SH groups of rabbit muscle D-glyceraldehyde 3-phosphate dehydrogenase, *J. Biol. Chem.*, 237, 1356, 1962.

36. Karlin, A. and Akabas, M.H., Substituted-cysteine accessibility method, *Meth. Enzymology*, 293, 123, 1998.

37. Britto, P.J., Knipling, L. and Wolff, J., The local electrostatic environment determines cysteine reactivity of tubulin, *J. Biol. Chem.*, 277, 29018, 2002.

38. Friedman, M., Application of the *S*-pyridylethylation reaction to the elucidation of the structures and functions of proteins, *J. Protein Chem.*, 20, 431, 2001.

39. Lindorff-Larsen, K. and Winther, J.K., Thiol alkylation below neutral pH, *Anal. Biochem.*, 286, 308, 2000.

40. Ren, D., Julka, S., Inerowicz, H.D. and Regnier, F.E., Enrichment of cysteine-containing peptides from tryptic digests using a quaternary amine tag, *Anal. Chem.*, 76, 4522, 2004.

41. Thevis, M., Loo, R.R.O. and Loo, J.A., In-gel derivatization of proteins for cysteine-specific cleavages and their analysis by mass spectrometry, *J. Proteome Res.*, 2, 162, 2003.

42. Crestfield, A.M., Moore, S. and Stein, W.H., The preparation and enzymatic hydrolysis of reduced and *S*-carboxymethylated proteins, *J. Biol. Chem.*, 238, 622, 1963.

43. Hirs, C.H.W., Reduction and *S*-carboxymethylation of proteins, *Meth. Enzymology*, 11, 199, 1967.

44. Lundblad, R.L., *The Modification of Cystine: Techniques in Protein Modification*, CRC Press, Boca Raton, FL, Ch. 6, 193–204, 2004.

45. Yano, H., Kuroda, S. and Buchanan, B.B., Disulfide proteome in the analysis of protein function and structure, *Proteomics*, 2, 1090, 2002.

46. Permyakov, E.A., Permyakov, S.E., Deikus, G.Y., Morozova-Roche, L.A., Grishchenko, V.M., Kalinichenko, L.P. and Uversky, V.N., Ultraviolet illumination-induced reduction of α-lactalbumin disulfide bridges, *Proteins Struct. Func. Gen.*, 51, 498, 2003.

47. Rüegg, U.T. and Rudinger, J., Reductive cleavage of cystine disulfides with tributylphosphine, *Meth. Enzymology*, 47, 111, 1977.

48. Gray, W.R., Disulfide structures of highly bridged peptides: A new strategy for analysis, *Protein Sci,.* 2, 1732, 1993.

49. Wu, H. and Watson, J.T., A novel methodology for assignment of disulfide bond pairings in proteins, *Protein Sci.*, 6, 391, 1997.

50. Breton, J., Avanzi, N., Valsasina, B. et al., Detection of traces of a trisulphide derivative in the preparation of a recombinant truncated interleukin-6 mutein, *J. Chrom. A*, 709, 135, 1995.

51. Iciek, M. and Wlodek, L., Biosynthesis and biological properties of compounds containing highly reactive, reduced sulfane sulfur, *Polish J. Pharm.*, 53, 215, 2001.
52. Righetti, P.G., Bossi, A., Wenisch, E. et al., Protein purification in multicompartment electrolyzers with isoelectric membranes, *J. Chrom. B*, 699, 105, 1997.
53. George, A.L. and Borders, C.L., Jr., Essential carboxyl groups in yeast enolase, *Biochem. Biophys. Res. Comm.*, 87, 59, 1977.
54. van Ballmoos, C., Appoldt, Y., Brunner, J. et al., Membrane topology of the coupling ion binding site in Na^+-translocating F_1F_0 ATP synthase, *J. Biol. Chem.*, 277, 3504, 2002.
55. Salhany, J.M., Sloan, R.L. and Cordes, K.S., The carboxyl side chain of glutamate 681 interacts with a chloride binding modifier site that allosterically modulates the dimeric conformational state of band 3 (AE1). Implications for the mechanism of anion/proton contransport, *Biochemistry*, 42, 1589, 2003.
56. Liu, P. and Regnier, F.E., An isotope coding strategy for proteomics involving both amine and carboxyl group labeling, *J. Proteome Res.*, 1, 443, 2002.
57. Dage, J.L., Sun, H. and Halsall, H.B., Determination of diethylpyrocarbonate-modified amino residues in α_1-acid glycoprotein by high-performance liquid chromatography electrospray ionization-mass spectrometry and matrix-assisted laser desorption/ionization time-of-flight-mass spectrometry, *Anal. Biochem.*, 257, 176, 1998.
58. Willard, B.B. and Kinter, M., Effects of the position of internal histidine residues on the collision-induced fragmentation of triply protonated tryptic peptides, *J. Am. Soc. Mass Spec.*, 12, 1262, 2001.
59. Qin, K., Yang, Y., Mastrangelo, P. and Westaway, D., Mapping Cu(II) binding sites in prion proteins by diethylpyrocarbonate modification and matrix-assisted laser desorption ionization time-of-flight (MALDI-TOF) mass spectrometric footprinting, *J. Biol. Chem.*, 277, 1981, 2002.
60. Alderton, A.L., Faustman, C., Liebler, D.C. and Hill, D.W., Induction of redox instability of bovine myoglobin by adduction with 4-hydroxy-2-nonenal, *Biochemistry*, 42, 4398, 2003.
61. Reubsaet, J.L., Beijnen, J.H., Bult, A. et al., Analytical techniques used to study the degradation of proteins and peptides: Chemical instability, *J. Pharm. Biomed. Anal.*, 17, 955, 1998.
62. Davies, M.J., Singlet oxygen-mediated damage to proteins and its consequences, *Biochem. Biophys. Res. Comm.*, 305, 761, 2003.
63. Gurd, F.R.N., Carboxymethylation, *Meth. Enzymology*, 11, 532, 1967.
64. Crestfield, A.M., Stein, W.H. and Moore, S., Alkylation and identification of the histidine residues at the active site of ribonuclease, *J. Biol. Chem.*, 238, 2413, 1963.
65. Wylie, D.C., Voloch, M., Lee, S. et al., Carboxylated histidine is a pH-dependent product of pegylation with SC-PEG, *Pharm. Res.*, 18, 1534, 2001.
66. Marcus, J.P. and Dekker, E.E., Identification of a second active site residue in *Escherichia coli* L-threonine dehydrogenase: Methylation of histidine-90 with methyl-*p*-nitrobenzenesulfonate, *Arch. Biochem. Biophys.*, 316, 413–420, 1995.
67. Kusnezow, W., Jacob, A., Walijew, A. et al., Antibody microarrays: An evaluation of production parameters, *Proteomics*, 3, 254, 2003.
68. Haugland, R.P., *Handbook of Fluorescent Probes and Research Products*, 9th Edition, Molecular Probes, Eugene, OR, 2002.
69. Muller, E.C., Lapko, A., Otto, A. et al., Covalently crosslinked complexes of bovine adrenodoxin with adrenodoxin reductase and cytochrome P450scc. Mass spectrometry and Edman degradation of complexes of the steroidogenic hydroxylase system, *Eur. J. Biochem.*, 268, 1837, 2001.

70. Duncan, R.J.S., Weston, P.D. and Wrigglesworth, R., A new reagent which may be used to incorporate sulfydryl groups into proteins and its use in the preparation of conjugates for immunoassay, *Anal. Biochem.*, 132, 68, 1983.

71. Jue, R., Lambert, J.M., Pierce, L.R. and Traut, R.P., Addition of sulfydryl groups to *Escherichia coli* ribosomes by protein modification with 2-iminothiolane(methyl-4-mercaptobutyrimidate), *Biochemistry*, 17, 5399, 1978.

72. Kadlík, V., Strohalm, M. and Kodíek, M., Citraconylation — A simple method for high protein coverage in MALDI-TOF mass spectrometry, *Biochem. Biophys. Res. Comm.*, 305, 1091, 2003.

73. Naminatsu, S., Ghazizadeh, M. and Sugisaki, Y., Reversing the effects of formalin fixation with citraconic anhydride and heat: A universal antigen retrieval method, *J. Histochem. Cytochem.*, 53, 3, 2005.

74. Kimmel, J., Guanidation of proteins, *Meth. Enzymology*, 11, 584, 1967.

75. Beardsley, R.L., Karty, J.A. and Reilly, J.P., Enhancing the intensities of lysine-terminated tryptic peptide ions in matrix-assisted laser desorption/ionization mass spectrometry, *Rapid Comm. Mass Spec.*, 14, 2147–2153, 2000.

76. Banks, T.E. and Shafer, J.A., Isoureas as alkylating agents. Inactivation of papain by *S*-methylation of its cysteinyl residue with *O*-methylisourea, *Biochemistry*, 11, 110, 1972.

77. Hundle, B.S. and Richards, W.R., Use of a new membrane-impermeable guanidi-nating reagent, 2-*S*-[^{14}C]thiuroniumethanesulfonate, for the labeling of intracyto-plasmic membrane proteins in *Rhodobacter sphaeroides*, *Biochemistry*, 26, 4505, 1987.

78. Gundlach, H.G., Moore, S. and Stein, W.H., The reaction of iodoacetate with methion-ine, *J. Biol. Chem.*, 234, 1761, 1959.

79. Gurd, F.R.N., Carboxymethylation, *Meth. Enzymology*, 11, 532, 1967.

80. Gurd, F.R.N., Carboxymethylation, *Meth. Enzymology*, 25, 424, 1972.

81. Hopkins, C.E., O'Connor, P.B., Allen, K.N. et al., Chemical-modification rescue assessed by mass spectrometry demonstrates that gamma-thia-lysine yields the same activity as lysine in aldolase, *Protein Sci.*, 11, 1591, 2002.

82. Schroeder, W.A., Shelton, J.R. and Robberson, B., Modification of methionyl residues during aminoethylation, *Biochim. Biophys. Acta*, 147, 590, 1967.

83. Crim, J.W., Garczynski, S.F. and Brown, M.R., Approaches to radioiodination of insect neuropeptides, *Peptides*, 23, 2045, 2002.

84. Beckingham, J.A., Nousden, N.G., Muir, N.M. et al., Studies on a single immuno-globulin-binding domain of protein L from *Peptostreptococcus magnus*: The role of tyrosine-53 in the reaction with human IgG, *Biochem. J.*, 353, 395, 2001.

85. Leite, J.F. and Cascio, M., Probing the topology of the glycine receptor by chemical modification coupled to mass spectrometry, *Biochemistry*, 41, 6140, 2002.

86. Korn, C., Scholz, S.R., Gimadutdinow, O. et al., Involvement of conserved histidine, lysine and tyrosine residues in the mechanism of DNA cleavage by the capase-3 activated DNase CAD, *Nucleic Acids Res.*, 30, 1325, 2002.

87. Haas, J.A., Frederick, M.A. and Fox, B.G., Chemical and posttranslational modifi-cation of *Escherichia coli* acyl carrier protein for preparation of dansyl-acyl carrier proteins, *Prot. Express. Purific.*, 20, 274, 2000.

88. Kato, Y., Kitamoto, N., Kawai, Y. and Osawa, T., The hydrogen peroxide/copper ion system, but not other metal-catalzyed oxidation systems, produces protein-bound dityrosine, *Free Rad. Biol. Med.*, 31, 624, 2001.

89. Sardana, V., Carlson, J.D., Breslow, E. and Peyton, D., Chemical modification and cross-linking of neurophysin tyrosine-49, *Biochemistry*, 26, 995, 1987.

90. De Vendittis, E., Ursby, T., Rullo, R. et al., Phenylmethylsulfonyl fluoride inactivates an archaeal superoxide dismutase by chemical modification of a specific tyrosine residue. Cloning, sequencing and expression of the gene coding for *Sulfolobus solfataricus* superoxide dismutase, *Eur. J. Biochem.*, 268, 1794, 2001.

91. Means, G.E. and Wu, H.L., The reactive tyrosine residues of human serum albumin: Characterization of its reaction with diisopropylphosphorofluoridate, *Arch. Biochem. Biophys.*, 194, 525, 1979.

92. Frey, J., Kordel, W. and Schneider, F., The reaction of aminoacylase with chloromethylketone analogues of amino acids, *Zeit. Naturforsch.*, 32, 769, 1977.

93. Abate, A., Oberle, S. and Schröder, H., Lipopolysaccharide-induced expression of cyclooxygenase-2 in mouse macrophages is inhibited by chloromethylketones and a direct inhibitor of NF-κB translocation, *Prostaglandins other Lipid Mediators*, 56, 277, 1998.

94. Lee, M.J. and Goldsworthy, G.J., Chloromethyl ketones are insulin-like stimulators of lipid synthesis in locust fat body, *Arch. Insect Biochem. Phys.*, 39, 9, 1998.

95. McDermott, L., Moore, J., Brass, A. et al., Mutagenic and chemical modification of the ABA-1 allergen of the nematode *Ascaris*: Consequences for structure and lipid binding properties, *Biochemistry*, 40, 9918, 2001.

96. Strohalm, M., Kodíek, M. and Pecher, M., Tryptophan modification by 2-hydroxy-5-nitrobenzyl bromide studies by MALDI-TOF mass spectrometry, *Biochem. Biophys. Res. Comm.*, 312, 811, 2003.

97. Celestina, F. and Suryanarayana, T., Biochemical characterization and helix stabilizing properties of HSNP-C' from the thermophilic archeon, *Sulfolobus acidocaldarius*, *Biochem. Biophys. Res. Comm.*, 267, 614, 2000.

98. Fontana, A. and Scoffone, E., Sulfenyl halides as modifying reagents for polypeptides and proteins, *Meth. Enzymology*, 25, 482, 1972.

99. Cui, H., Leon, J., Reusaet, E. and Bult, A., Selective determination of peptides containing specific amino acid residues by high-performance liquid chromatography and capillary electrophoresis, *J. Chrom. A*, 704, 27, 1995.

100. Hassani, O., Mansuelle, P., Cestele, S. et al., Role of lysine and tryptophan residues in the biological activity of toxin VII (Ts gamma) from the scorpion, *Tityus serrulatus*, *Eur. J. Biochem.*, 260, 76, 1999.

101. Wilchek, M. and Miron, T., The conversion of tryptophan to 2-thioltryptophan in peptides and proteins, *Biochem. Biophys. Res. Comm.*, 47, 1015, 1972.

102. Spande, T.F. et al., Selective cleavage and modification of peptides and proteins, *Adv. Prot. Chem.*, 24, 97, 1970.

103. Smith, B.J., Chemical cleavage of proteins, *Meth. Mol. Biol.*, 32, 197, 1994.

104. Gross, E., The cyanogen bromide reaction, *Meth. Enzymol.*, 11, 238, 1967.

105. Spande, T.F. and Witkop, B., CNBr cleavage of peptides and proteins, *CRC Handbook of Biochemistry-Selected Data for Molecular Biology*, 2nd Edition, H.A. Sober and R.A. Harte, Eds., CRC Press, Boca Raton, FL, C-137, 1970.

106. Tristram, G.R. and Smith, R.H., Amino acid composition of certain proteins, in *The Proteins*, 2nd Edition, H. Neurath, Ed., Academic Press, New York, NY, 1963, 45.

107. Kelleher, N.L. et al., Top down versus bottom up protein characterization by tandem high-resolution mass spectrometry, *J. Am. Chem. Soc.*, 121, 806, 1999.

108. Morrison, J.R., Fidge, N.N. and Grego, H., Studies on the formation, separation, and characterization of CNBr fragments of human A1 apolipoprotein, *Anal. Biochem.*, 186, 145, 1990.

109. Liao, T.-H., Salnikow, J., Moore, S. and Stein, W.H., Bovine pancreatic deoxyribonuclease A. Isolation of CNBr peptides; complete covalent structure of the polypeptide chain, *J. Biol. Chem.*, 248, 1489, 1973.

110. Sokolov, B.P., Sher, B.M. and Kalinin, V.N., Modified method for peptide mapping of collagen chains using CNBr-cleavage of protein within polyacrylamide gels, *Anal. Biochem.*, 176, 365, 1989.

111. Jahnen, W., Ward, L.D., Reid, G.E., Moritz, R.L. and Simpson, R.J., Internal amino acid sequencing of proteins by *in situ* CNBr cleavage in polyacrylamide gels, *Biochem. Biophys. Res. Comm.*, 166, 139, 1990.

112. van Montfort, B.A. et al., Improved in-gel approaches to generate peptide maps of integral membrane proteins with matrix-assisted laser desorption/ionization time-of-flight mass spectrometry, *J. Mass. Spec.*, 37, 322, 2002.

113. Hulmes, J.D. and Pan, Y.-C.E., Selective cleavage of polypeptides with trifluoroacetic acid: Applications for microsequencing, *Anal. Biochem.*, 197, 368, 1991.

114. Ye, J.M. et al., Degradation of antiflammin 2 under acidic conditions, *J. Pharm. Sci.*, 85, 695, 1996.

115. Lu, H.S. and Gracy, R.W., Specific cleavage of glucosephosphate isomerase at cysteinyl residues using 2-nitro-5-thiocyanobenzoic acid: Analyses of peptides eluted from polyacrylamide gels and localization of active site histidyl and lysyl residues, *Arch. Biochem. Biophys.*, 212, 347, 1981.

116. Qi, J. et al., Determination of the disulfide structure of sillucin, a highly knotted, cysteine-rich peptide, by cyanylation/cleavage mass mapping, *Biochemistry*, 40, 4531, 2001.

117. Douady, D., Rousseau, B. and Caron, L., Fucoxanthin-chlorophyll a/c light-harvesting complexes of *Laminaria saccharina* — Partial amino acid sequences and arrangement in the thylakoid membranes, *Biochemistry*, 33, 3165, 1994.

118. Droste, M., Mollner, S. and Pfeuffer, T., Localisation of an ATP-binding site on adenyl cyclase type I after chemical and enzymatic fragmentation, *FEBS Letters*, 391, 208, 1996.

119. Pliszka, B., Karczewska, E. and Wawro, B., Nucleotide-induced movements in the myosin head near the converter region, *Biochim. Biophys. Acta*, 1481, 55, 2000.

120. Wong, S.S., *Chemistry of Protein Conjugation and Cross-Linking*, CRC Press, Boca Raton, FL, 1991.

121. Brinkley, M., A brief survey of methods for preparing protein conjugates with dyes, haptens, and cross-linking reagents, *Bioconjugate Chem.*, 3, 2, 1992.

122. Nadeau, O.W. and Carlson, G.M., Chemical cross-linking in studying protein–protein interactions, in *Protein–Protein Interactions: A Cloning Manual*, E. Golemis, Ed., Cold Spring Harbor Laboratory Press, Cold Spring Harbor, NY, Ch. 6, 2002.

123. Meisenheimer, K.M. and Koch, T.H., Photocross-linking of nucleic acids to associated proteins, *Crit. Rev. Biochem. Mol. Biol.*, 32, 101, 1997.

124. Fágáin, C.O., Understanding and increasing protein stability, *Biochim. Biophys. Acta*, 1252, 1, 1995.

125. Desantis, G. and Jones, J.B., Chemical modification of enzymes for enhanced functionality, *Curr. Opin. Biotech.*, 10, 324, 1999.

126. Löster, K. and Josi, D., Analysis of protein aggregates by combination of cross-linking reactions and chromatographic separations, *J. Chrom. B*, 699, 439, 1997.

127. Fancy, D.A., Elucidation of protein–protein interactions using chemical cross-linking or label transfer techniques, *Curr. Opin. Chem. Biol.*, 4, 28, 2000.

128. Sinz, A., Chemical cross-linking and mass spectrometry for mapping three-dimensional structures of proteins and protein complexes, *J. Mass Spec.*, 38, 1225, 2003.

129. Kunkel, G.R., Mehrabian, M. and Martinson, H.G., Contact-site cross-linking agents, *Mol. Cell. Biochem.*, 34, 3, 1981.

130. Sheehan, J.C. and Hlavka, J.J., The use of water-soluble and basic carbodiimides in peptide synthesis, *J. Org. Chem.*, 21, 439, 1956.
131. Sheehan, J.C. and Hlavka, J.J., The cross-linking of gelatin using a water-soluble carbodiimide, *J. Am. Chem. Soc.*, 79, 4528, 1957.
132. Schmid, B., Einsle, O., Chiu, H.J. et al., Biochemical and structural characterization of the cross-linked complex of nitrogenase: Comparison to the ADP-A1F4(-)-stabilized structure. *Biochemistry*, 41, 15557, 2002.
133. Scaloni, A., Miraglia, N., Orrù, S. et al., Topology of the calmodulin-melittin complex., *J. Mol. Biol.*, 277, 945, 1998.
134. Nyman, T., Page, R., Schutt, C.E. et al., A cross-linked profilin-actin heterodimer interferes with elongation at the fast-growing end of F-actin, *J. Biol. Chem.*, 277, 15828, 2002.
135. Clarke, D.J., Coulson, J., Baillie, R. and Campopiano, D.J., Biotinylation in the hyperthermophile *Aquifex aeolicus*. Isolation of a cross-linked BPL:BCCP Complex, *Eur. J. Biochem.*, 270, 1277, 2003.
136. Winters, M.S. and Day, R.A., Identification of amino acid residues participating in intermolecular salt bridges between self-associating proteins, *Anal. Biochem.*, 309, 48, 2002.
137. Heinecke, J.W., Tyrosyl radical production by myeloperoxidase: A phagocyte pathway for lipid peroxidation and dityrosine cross-linking of proteins, *Toxicology*, 177, 11, 2002.
138. Ghezzi, P. and Bonetto, V., Redox proteomics: Identification of oxidatively modified proteins, *Proteomics*, 3, 1145, 2003.
139. Andreev, O.A. et al., Interaction of the *N*-terminus of chicken skeletal essential light chain 1 with F-actin. *Biochemistry*, 38, 2480, 1999.
140. Okamura-Ikeda, K., Fujiwara, K. and Motokawa, Y., Identification of the folate binding sites on *Escherichia coli* T-protein of the glycine cleavage system, *J. Biol. Chem.*, 274, 17471, 1999.
141. Scaloni, A. et al., Topology of the calmodulin-melittin complex. *J. Mol. Biol.*, 277, 945, 1998.
142. Clarke, D.J. et al., Biotinylation in the hyperthermophile *Aquivex aeolicus*. Isolation of a cross-linked BPL:BCCP complex, *Eur. J. Biochem.*, 270, 1277, 2003.
143. Bennett, K.L. et al., Chemical cross-linking with thiol-cleavable reagents, combined with differential mass spectrometric peptide mapping — A novel approach to assess intermolecular protein contacts, *Protein Sci.*, E9, 1503, 2000.
144. Pearson, K.M., Pannell, L.K. and Fales, H.M., Intramolecular cross-linking experiments on cytochrome C and ribonuclease A using an isotope multiplet method, *Rapid Comm. Mass Spec.*, 16, 149, 2002.
145. Collins, C.J. et al., Isotopically labeled crosslinking reagents: Resolution of mass degeneracy in the identification of crosslinked peptides, *Bioorg. Med. Chem. Letters*, 13, 4023, 2003.
146. Birnbaumer, M.F., Schrader, W.T. and O'Malley, B.W., Chemical cross-linking of chick oviduct progesterone-receptor subunits by using a reversible bifunctional cross-linking agent, *Biochem. J.*, 181, 201, 1979.
147. Hagenstein, M.C., Mussgnug, J.H., Lotte, K. et al., Affinity-based tagging of protein families with reversible inhibitors: A concept for functional proteomics, *Angew. Chem. Int. Ed. Engl.*, 42, 5635, 2003.
148. Khodyreva, S.N. and Lavrik, O.I., Photoaffinity labeling technique for studying DNA replication and DNA repair, *Curr. Med. Chem.*, 12, 641, 2005.

149. Basu, S., Rodionov, V., Terasaki, M. and Campagnola, P.J., Multiphoton-excited microfabrication in live cells via Rose Bengal cross-linking of cytoplasmic proteins, *Opt. Letters*, 30, 159, 2005.

150. Radi, R., Nitric oxide, oxidants, and protein tyrosine nitration, *Proc. Natl. Acad. Sci. USA*, 101, 4003, 2004.

151. Hazell, L.J., van den Berg, J.J.M. and Stocker, R., Oxidation of low-density lipoprotein by hypochorite causes aggregation that is mediated by modification of lysine residues rather than lipid peroxidation, *Biochem. J.*, 302, 297, 1994.

152. Akagawa, M., Sasaki, T. and Suyama, K., Oxidative deamination of lysine residue in plasma protein of diabetic rats. Novel mechanism via the Mallard reaction, *Eur. J. Biochem.*, 269, 5451, 2002.

153. Merrick, B.A. and Tomer, K.B., Toxicoproteomics: A parallel approach to identifying biomarkers, *Env. Health Pers.*, 111, A578, 2003.

154. LoPachin, R.M., Jones, R.C., Patterson, T.A., Slikker, W., Jr. and Barber, D.S., Application of proteomics to the study of molecular mechanisms in neurotoxicology, *Neurotoxicology*, 24, 761, 2004.

155. Wetmore, B.A. and Merrick, B.A., Toxicoproteomics: Proteomics applied to toxicology and pathology, *Tox. Path.*, 32, 619, 2004.

156. Witzmann, F.A., Bai, F., Hong, S.M. et al., Gels and more gels: Probing toxicity, *Curr. Opin. Mol. Ther.*, 6, 608, 2004.

157. Johnson, D. and Travis, J., The oxidative inactivation of human alpha-1-proteinase inhibitor. Further evidence for methionine at the reactive center, *J. Biol. Chem.*, 254, 4022, 1979.

158. Matheson, N.R. and Travis, J., Differential effects of oxidizing agents on human plasma alpha 1 proteinase inhibitor and human neutrophils myeloperoxidase, *Biochemistry*, 24, 1941, 1985.

159. Ueda, M., Mashiba, S. and Uchida, K., Evaluation of oxidized alpha-1-antitrypsin in blood as an oxidative stress marker using anti-oxidative alpha1-AT monoclonal antibody, *Clin. Chiem. Acta*, 317, 125, 2002.

160. Vogt, W., Oxidation of methionyl residues in proteins: Tools, targets, and reversal, *Free Rad. Biol. Med.*, 18, 93, 1995.

161. Maleknia, S.D., Brenowitz, M. and Chance, M.R., Millisecond radiolytic modification of peptides by synchrotron x-rays identified by mass spectrometry, *Anal. Chem.*, 71, 3965, 1999.

162. Heyduk, E. and Heyduk, T., Mapping protein domains involved in macromolecular interactions: A novel protein footprinting approach, *Biochemistry*, 33, 9643, 1994.

163. Rashidzadeh, H., Khrapunov, S., Chance, M.R. and Brenowitz, M., Solution structure and interdomain interactions of the *Saccharomyces cerevisiae* "TATA binding protein" (TBP) probed by radiolytic protein footprinting, *Biochemistry*, 42, 3655, 2003.

164. Gay, C.A. and Gebicki, J.M., Measurement of protein and lipid hydroperoxides in biological systems by the ferric-xylenol orange method, *Anal. Biochem.*, 315, 29, 2003.

165. Hawkins, C.L. and Davies, M.J., Hypochlorite-induced oxidation of proteins in plasma: Formation of chloramines and nitrogen-centered radicals and their role in protein fragmentation, *Biochem. J.*, 340, 539, 1999.

166. Refsgaard, H.H., Tsai, L. and Stadtman, E.R., Modifications of proteins by polyunsaturated fatty acid peroxidation products, *Proc. Natl. Acad. Sci. USA*, 97, 611, 2000.

167. Hidalgo, F.J., Alaiz, M. and Zamora, R., A spectrophotometric method for the determination of proteins damaged by oxidized lipids, *Anal. Biochem.*, 262, 129, 1998.

168. Woods, A.A., Linton, S.M. and Davies, M.J., Detection of HOCl-mediated protein oxidation products in the extracellular matrix of human atherosclerotic plaques, *Biochem. J.*, 370, 729, 2003.

169. Dong, J., Atwood, C.S., Anderson, V.E., Siedlak, S.L., Smith, M.A., Perry, G. and Carey, P.R., Metal binding and oxidation of amyloid within isolated senile plaque cores: Raman microscopic evidence, *Biochemistry*, 42, 2768, 2003.

170. Stadtman, E.R., Covalent modification reactions are marking steps in protein turnover, *Biochemistry*, 29, 6323, 1990.

171. Schwentker, A. and Billiar, T.R., Nitric oxide and wound repair, *Surg. Clin. N. Am.*, 83, 521, 2003.

172. Mamas, M.A., Reynard, J.M. and Brading, A.F., Nitric oxide and the lower urinary tract: Current concepts, future prospects, *Urology*, 61, 1079, 2003.

173. von Haehling, S., Ander, S.D. and Bassenge, E., Statins and the role of nitric oxide in chronic heart failure, *Heart Failure Rev.*, 8, 99, 2003.

174. Ricciardolo, F.L., Multiple roles of nitric oxide in the airways, *Thorax*, 58, 175, 2003.

175. Freels, J.L., Nelson, D.K., Hoyt, J.C. et al., Enhanced activity of human IL-10 after nitration in reducing human IL-1 production by stimulated peripheral blood mononuclear cells, *J. Immunol.*, 169, 4568, 2002.

176. Galiñanes, M. and Matata, B.M., Protein nitration is predominantly mediated by a peroxynitrite-dependent pathway in cultured human leukocytes, *Biochem. J.*, 367, 467, 2002.

177. Jiao, K., Mandapati, Skipper, P.L. et al., Site-selective nitration of tyrosine in human serum albumin by peroxynitrite, *Anal. Biochem.*, 293, 43, 2003.

178. Ischiropoulos, H., Biological selectivity and functional aspects of protein tyrosine nitration, *Biochem. Biophys. Res. Comm.*, 305, 776, 2003.

179. Pryor, W.A., Jin, X. and Squadrito, G.L., One- and two-electron oxidations of methionine by peroxynitrite, *Proc. Natl. Acad. Sci. USA*, 91, 11173, 1994.

180. Beckman, J.S., Chen, J., Ichiropoulos, H. and Crow, J.P., Oxidative chemistry of peroxynitrite, *Meth. Enzymology*, 233, 229, 1994.

181. Norris, E.H., Giasson, B.I., Ishciropoulos, H. and Lee, V.M.-Y., Effects of oxidative and nitrative challenges on α-synuclein fibrillogenesis involve distinct mechanisms of protein modification, *J. Biol. Chem.*, 278, 27230, 2003.

182. Berlett, B.S., Levine, R.L. and Stadtman, E.R., Carbon dioxide stimulates peroxynitrite-mediated nitration of tyrosine residues and inhibits oxidation of methionine residues of glutamine synthetase: Both modifications mimic effects of adenylation, *Proc. Natl. Acad. Sci. USA*, 95, 2784, 1998.

183. Tien, M., Berlett, B.S., Levine, R.L. et al., Peroxynitrite-mediated modification of proteins at physiological carbon dioxide concentration: pH dependence of carbonyl formation, tyrosine nitration, and methionine oxidation, *Proc. Natl. Acad. Sci. USA*, 96, 7809, 1999.

184. Lymar, S.V. and Hurst, K., Rapid reaction between peroxynitrite ion and carbon dioxide: Implications for biological activity, *J. Am. Chem. Soc.*, 117, 8867, 1995.

185. Saigo, M.G., Stone, K., Squadrito, G.L., Battista, J.R. and Pryor, W.A., Peroxynitrite causes DNA nicks in plasmid pBR232, *Biochem. Biophys. Res. Comm.*, 210, 1025, 1995.

186. Tao, L. and English, A.M., Mechanism of S-nitrosation of recombinant human brain calbindin D_{28K}, *Biochemistry*, 42, 3326, 2003.

187. Rhee, K.Y., Erdjument-Bromage, H., Tempst, P. and Nathan, C.F., *S*-Nitroso proteome of *Mycobacterium tuberculosis*: Enzymes of intermediary metabolism and antioxidant defense, *Proc. Natl. Acad. Sci. USA*, 102, 467, 2005.

188. Jaffrey, S.R., Erdjument-Bromage, H., Ferris, C.D., Tempst, P. and Snyder, S.H., Protein *S*-nitrosylation: A physiological signal for neuronal nitric oxide, *Nat. Cell Biol.*, 3, 193, 2001.

189. Maillard, L.C., Action des acides amines sur les sucres: Formation des melanoidines par voie methodique, *C.R. Hebd. Seances Acad. Sci.*, 154, 66, 1912.

190. Ulrich, P. and Cerami, A., Protein glycation, diabetes, and aging, *Recent Prog. Horm. Res.*, 56, 1, 2001.

191. Biemel, K.M. and Lederer, M.O., Site-specific quantitative evaluation of the protein glycation product N(6) O-2,3 -dihydroxy-5,6-dioxohexyl-1-lysinate by LC-(ESI) MS peptide mapping: Evidence for its key role in AGE formation, *Bioconjugate Chem.*, 14, 619, 2003.

192. Tessier, F.J., Monnier, V.M., Sayre, L.M. and Kornfield, J.A., Triosidines: Novel Maillard reaction products and cross-links from the reaction of triose sugars with lysine and arginine residues, *Biochem. J.*, 369, 705, 2003.

193. Thornally, P.J., Battah, S., Ahmed, N. et al., Quantitative screening of advanced glycation endproducts in cellular and extracellular proteins by tandem mass spectrometry, *Biochem. J.*, 375, 581, 2003.

194. Voziyan, P.A., Khalifah, R.G. and Thibaudeau, C., Modification of proteins *in vitro* by physiological levels of glucose. Pyridoxamine inhibits conversion of amadori intermediate to advanced glycation end-products through binding of redox metal ions, *J. Biol. Chem.*, 278, 46616, 2003.

195. Argirov, O.K., Lin, B., Olesen, P. and Ortwerth, B.J., Isolation and characterization of a new advanced glycation endproduct of dehydroascorbic acid and lysine, *Biochim. Biophys. Acta*, 1620, 235, 2003.

196. Miller, A.G., Meade, S.J. and Gerrard, J.A., New insights into protein crosslinking via the Maillard reaction: Structural requirements, the effects on enzyme function, and predicted efficacy of crosslinking inhibitors as anti-aging therapeutics, *Bioorg. Med. Chem.*, 11, 843, 2003.

197. Degenhardt, T.P., Thorpe, S.R. and Baynes, J.W., Chemical modification of proteins by methylglyoxal, *Cell. Mol. Biol.*, 44, 1139, 1998.

198. Seidler, N.W. and Kowalewski, C., Methylglyoxal-induced glycation affects protein topography, *Arch. Biochem. Biophys.*, 410, 149, 2003.

3 The Application of Site-Specific Chemical Modification to Proteomics: Chemical Proteomics

CONTENTS

INTRODUCTION

The chemical modification of proteins has extensive use in proteomics in several different categories. Chemical proteomics could be described as the application of "classical" solution chemistry, such as the use of alkylating agents, isotope-coded affinity tags and fluorescent probes, in quantitative proteomics.[1–5] These reagents are described later in this chapter. The term chemical proteomics is used more frequently to describe chemical probes used to modify intact proteins[6–9] for analysis by proteomic technologies. Chemiproteomics — which would appear to be a closely related concept — has been defined as "the study of interactions between proteins and small molecules" as part of the process of the identification of drug candidates. Figeys[10] described chemiproteomics as an approach using small molecules as affinity materials for the discovery of small-molecule binding proteins. This approach is best characterized by the use of affinity chromatography.[11–15] As of this writing, neither "chemical proteomics" nor "chemiproteomics" was recognized as a search term by PUBMED ("chemical" and "proteomics" yielded 368 citations). While it can be argued that all proteomics are chemical, we will attempt to keep the current discussion within the general area of functional proteomics as defined by Lefkowitz ("description of changes in protein expression during differentiation of cells").[16] We will be a little broader in the definition in that we suggest functional proteomics concerns changes in protein expression measured either by mass or function. Consistent with this approach, Yanagida[17] has defined functional proteomics as an "emerging field of systematic protein analyses focusing on protein function, its interactions, and biological phenomena" and includes analysis of functional protein complexes, affinity purification, protein chips and quantitative proteomics including differential staining (differential gel electrophoresis, DIGE) and isotope-coded affinity tags (ICAT). Quantitative proteomics refers to the use of the aforementioned alkylating agents (ICAT) and fluorescent dyes for the labeling of proteins and peptides prior to analysis.[18] Quantitative proteomics has also been described as comparative proteomics. Quantitative proteomics uses reagents that can measure differences in protein expression in paired experimental samples[19,20] and permitted the analysis of changes in global protein expression in response to system perturbation.

This chapter will focus on the use of specific chemical modification as it is useful within proteomics. The following areas will be discussed in detail; it is recognized that this list is somewhat parochial but represents a useful coverage of investigations where there is combination of experimental data and detail:

- Activity-based proteomics
- Site-specific chemical modification of proteins and peptides for proteomic analysis
- Selective enzymatic modification of targets for proteomic analysis
- Prelabeling of proteins prior to fractionation
- Chemistry of protein microarrays

ACTIVITY-BASED PROTEOMICS

Activity-based proteomics is a term that has been used to describe methods for the identification and measurement of enzymes *in situ* in cell and biological fluids.[21–32] Activity-based proteomics appears to use the chemistry that was developed for the affinity labeling of enzymes[32–37] and suicide enzyme inhibitors.[38–42] Modification of proteins at specific binding sites is mostly driven by selective binding as opposed to functional group reactivity; photoaffinity reagents (Figure 3.1) have been quite useful in the study of such sites in proteins.[43–49] Such modification is critical in considering the work in tissue microarrays.[50] Consideration of the early work in these areas is useful, as is the immense literature in enzyme histocytochemistry.[51–53]

The reactivity of functional groups in proteins is, in general, the product of individual site reactivity (nucleophilicity) and steric accessibility, as discussed in Chapter 2. Obtaining site-specific or residue-specific modification is relatively

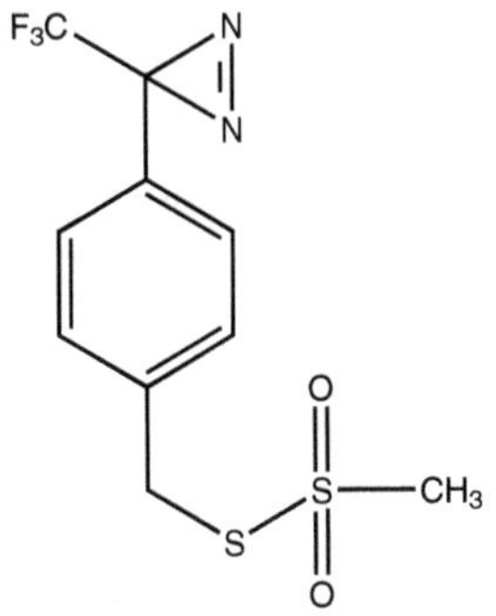

Diazirine methylthiosulfate derivative (1)

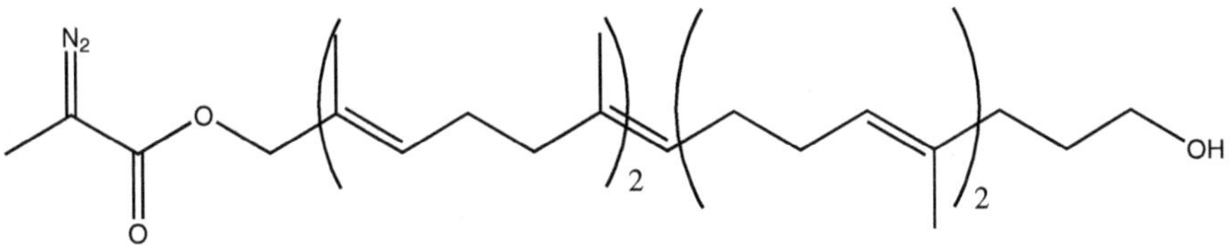

Trinorsqualine alcohol, diazo derivative (2)

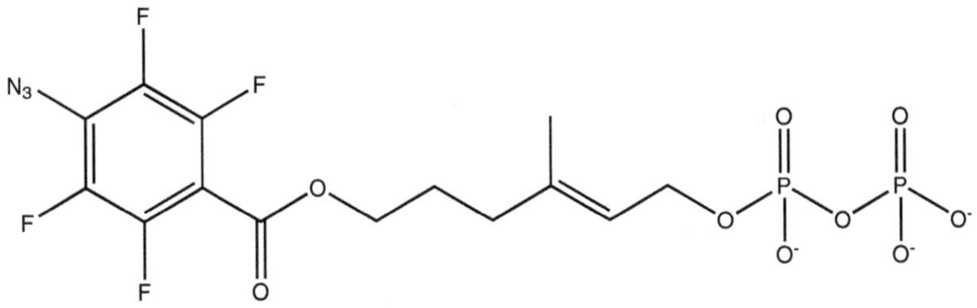

6-(4-azido-tetrafluorobenzylester)-3-methyl-2-hexane pyrophosphate;
photoaffinity label based on farnesyl pyrophosphate (3)

FIGURE 3.1 Some photoaffinity reagents. (1) An affinity label which was used to label glyceraldehyde-3-phosphate dehydrogenase by modification of a sulfydryl group with the alkylmethylthiosulfonate derivative. This enabled the identification of some specific ligand (Kaneda, M., Sadakane, Y. and Hatanaka, Y., A novel approach for affinity-based screening of target specific ligands: Application of photoreactive D-glyceraldehyde-3-phosphate dehydrogenase, *Bioconjugate Chem.*, 14, 849–852, 2003). (2) An alkyl diazo derivative for labeling the substrate binding site of squalene epoxidase (Lee, H.-K., Zheng, Y.F., Ziao, X.-Y. et al., Photoaffinity labeling identifies the substrate-binding site of mammalian squalene epoxidase, *Biochem. Biophys. Res. Comm.*, 315, 1–9, 2004). (3) A photoaffinity analogue of farnesyl pyrophosphate (Chehade, K.A.H., Kiegiel, K., Isaacs, R.J. et al., Photoaffinity analogues of farnesyl pyrophosphate transferable by protein farnesyl transferase, *J. Am. Chem. Soc.*, 124, 8206–8219, 2002).

difficult and this difficulty should be considered in the interpretation of data. The chemical modification of functional groups in enzymes is a combination of specific binding and the chemical reactivity of functional groups at the enzyme active site. Thus, active-site directed reagents (affinity labels) are frequently modeled on the structure of substrates. Figure 3.2 shows the structure of a substrate for chymotrypsin and the peptide chloromethyl ketone designed to modify chymotrypsin. As seen below, considerable information can be obtained with simple chemical reagents without having to go to the effort of designing a specific reagent;[32] for example, can reaction with a reagent such as bromoacetamido or iodoacetamide provide meaningful information? Isotope-coded affinity tags[1,54] are chemically similar reagents used for a different purpose but do provide valuable information about differential expression of proteins.[55] In this situation, obtaining saturation labeling is desirable rather than selective modification of a functional group at the enzyme active site.

Several approaches have been developed for the study of protease activity expression in cellular systems and subsequently for proteomics. One approach has promoted derivatives of diisopropylphosphorofluoridate (DFP), one of the earliest described inhibitors of serine proteases.[56–58] Alkylphosphoro-fluoridates react with the serine residues at the active site of enzymes such as trypsin and chymotrypsin resulting in phosphorylation and inactivation of the enzyme (Figure 3.3). In the process of inhibition, the formation of the phosphorylated serine is analogous to the formation of an acyl-enzyme intermediate.[59–62] Some mechanistic similarities exist to active-site titration[63–64] and the reaction of proteases with proteases inhibitors.[65] Reaction also occurs with enzymes other than proteases, including esterases such as acetylcholine esterase[66] and an N-acylphosphatidylethanolamine.[67] Reactivity of the serine residue at the active site of these enzymes is driven by enhanced nucleophilicity, a consequence of the charge-relay system where an aspartic acid residue polarizes a histidine residue that in turn partially abstracts the hydrogen from the serine hydroxyl group.

ALKYLFLUOROPHOSPHONATE DERIVATIVES

Several groups have used alkylphosphorofluoridate derivatives in proteomic research. While chemical characterization data is largely missing, these derivatives most likely act by modifying serine residues at the enzyme active site as shown in Figure 3.3. Cravatt and colleagues have prepared several derivatives (Figure 3.4). One derivative — fluorophosphonyl-biotin, 10-(fluoroethoxyphosphinyl)-N-(biotinamidopentyl) decanamide — has been used to isolate modified proteins while another derivative, labeling with fluorescein instead of biotin, has been used for intracellular localization of serine proteases.[68] The FP-biotin derivative [10-(fluoroethoxyphosphinyl)-N-(biotinamidopentyl) decanamide] was reacted with either purified proteins or tissue homogenates in 50 mM Tris, 0.32 M sucrose, pH 8.0. Samples were subjected to electrophoresis and detection of the modified proteins accomplished by a Western blot technique using avidin-horse radish peroxidase with chemiluminescence reagents. Nonspecific binding was determined with heat-treated samples. Under these conditions, the fluorophosphorate derivatives reacting with serine residues at enzyme active sites is a reasonable assumption. Examination

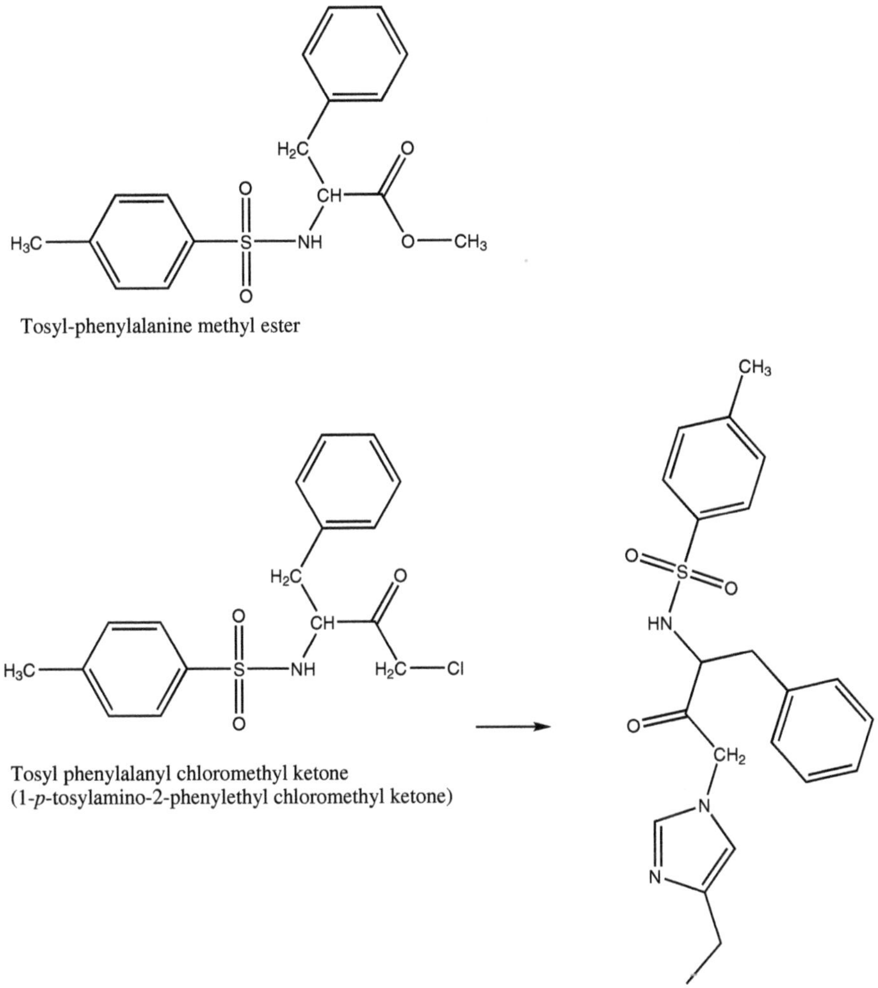

FIGURE 3.2 The design of affinity labels for enzymes. Shown is the basis for the design of tosyl-phenylalanine chloromethyl ketone, an affinity label designed to modify the enzyme active site of chymotrypsin. Shown is the structure of an ester substrate for chymotrypsin, tosyl-phenylalanine methyl ester; also shown is the structure of tosylphenylalanine chloromethylketone (TPCK) and the reaction with the active site histidine.

of different tissue suggested that this technique could evaluate the tissue-dependent expression of serine proteases. In a subsequent study,[69] this group developed a FP-biotin derivative containing a more hydrophilic linker region (FP-Peg-Biotin, Figure 3.5). A comparison of the two derivatives was obtained with rat testis and a similar pattern of reaction was observed; however, some differences existed in the rates of reaction with the two derivatives and the multiplexing with the two FP derivatives is suggested as the most useful approach. By the combined use of the two reagents, identification of a number of serine peptidases, esterase and lipases including fatty acid amide hydrolase was possible. Reaction with a fluorophosphate

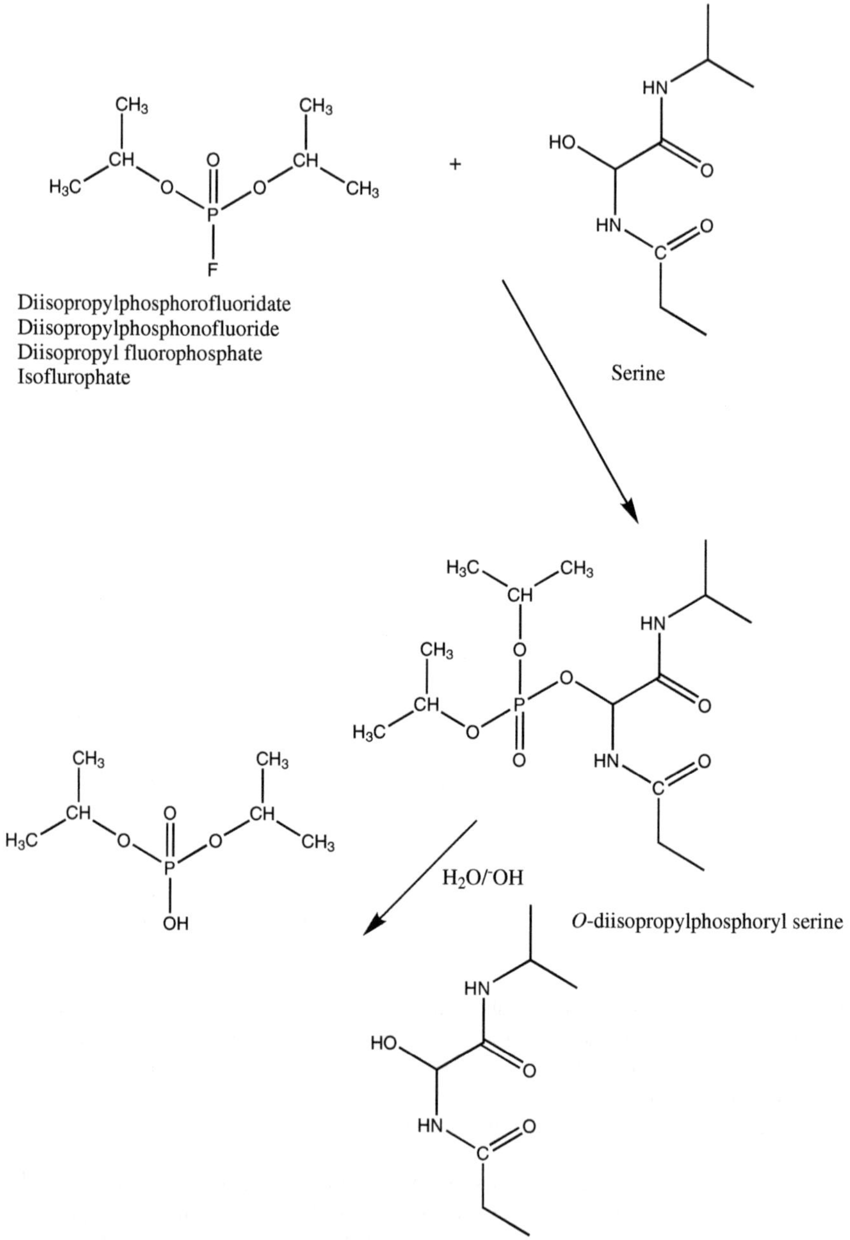

Diisopropylphosphorofluoridate
Diisopropylphosphonofluoride
Diisopropyl fluorophosphate
Isoflurophate

Serine

O-diisopropylphosphoryl serine

FIGURE 3.3 Reaction of diisopropylphosphorofluoridate (diisopropylfluorophosphate, DFP) with serine residues in enzymes. Also shown is the dephosphorylation reaction that occurs more rapidly at alkaline pH; the presence of a nucleophile such as hydroxylamine will accelerate the dephosphorylation reaction.

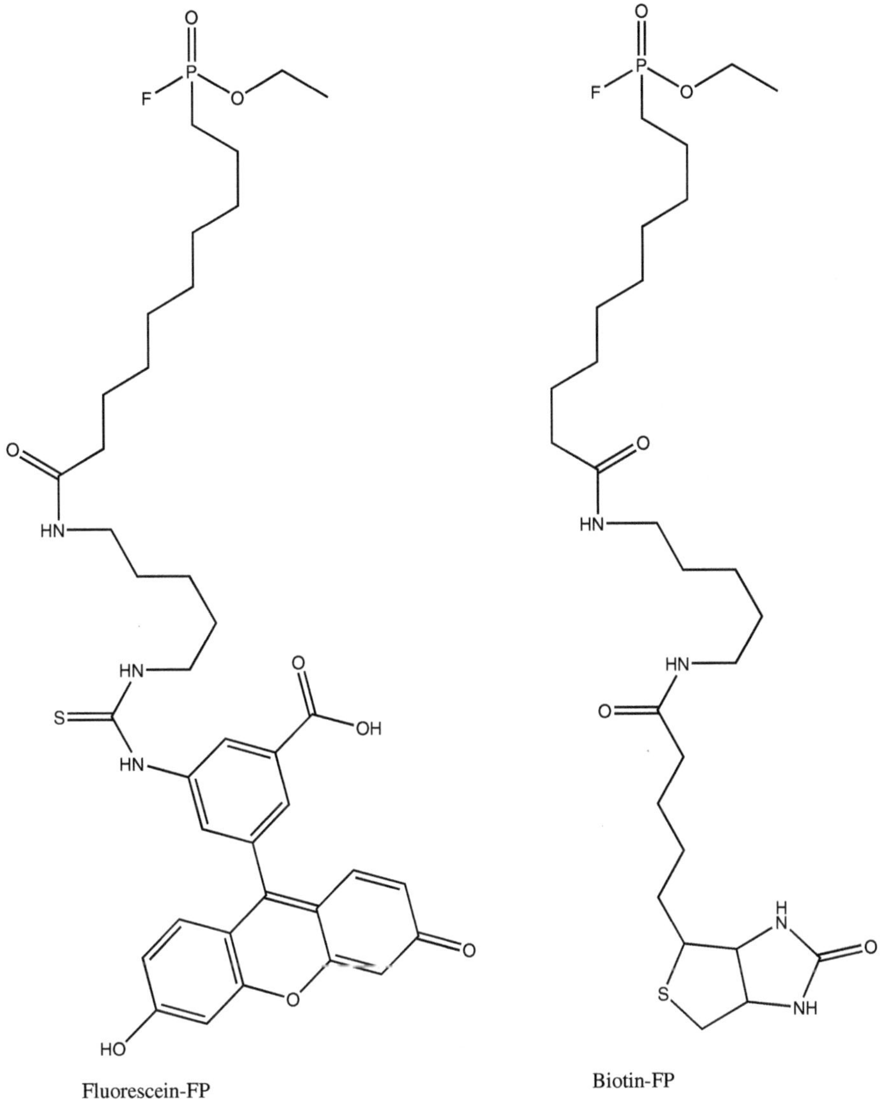

Fluorescein-FP

Biotin-FP

FIGURE 3.4 Some examples of activity-based affinity probes for use in proteomics. One derivative contains a "tag" — biotin in this example — that can be used for the isolation of enzymes. The other derivative contains a flurorescent label. These derivatives can be used for the identification of a class of serine enzymes, most notably proteases but also esterases.

probe, fluorophosphonate-poly(ethylene)glycol-(6-carboxyltetramethylrhodamine), was used to demonstrate that Orlistat® was an inhibitor of fatty acid synthase and had anti-tumor activity (Figure 3.6).[70] Orlistat is a β-lactone developed as an anti-obesity drug. This study reported that Orlistat blocked the reaction of the fluorophosphate probe with fatty synthase in prostate tumor cell extracts. Subsequent studies with a

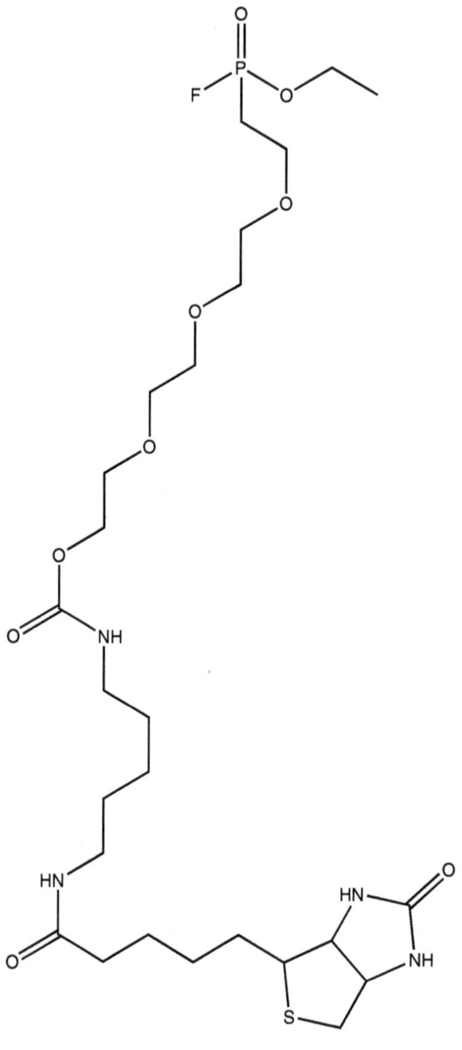

FP-PEG-Biotin

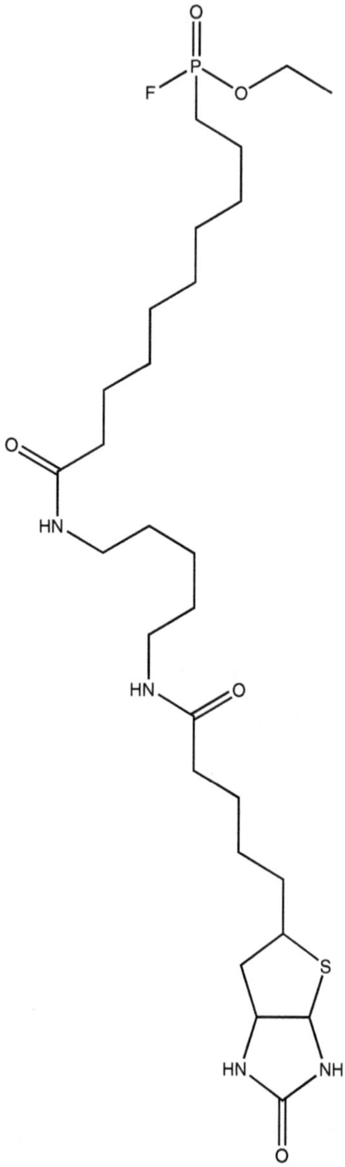

FP-Biotin

FIGURE 3.5 Some examples of activity-based affinity probes for labeling serine enzymes. The poly(ethylene) glycol derivative is designed to have a more hydrophilic quality than the parent DFP-based affinity probe shown on the right.

cell permeable fluorophosphate demonstrated that Orlistat also blocked the reaction with fatty acid synthase in intact cells.

A word of caution: While alkylphorphorofluoridate derivatives are reasonably specific for the modification of serine at enzyme active sites, reaction at other

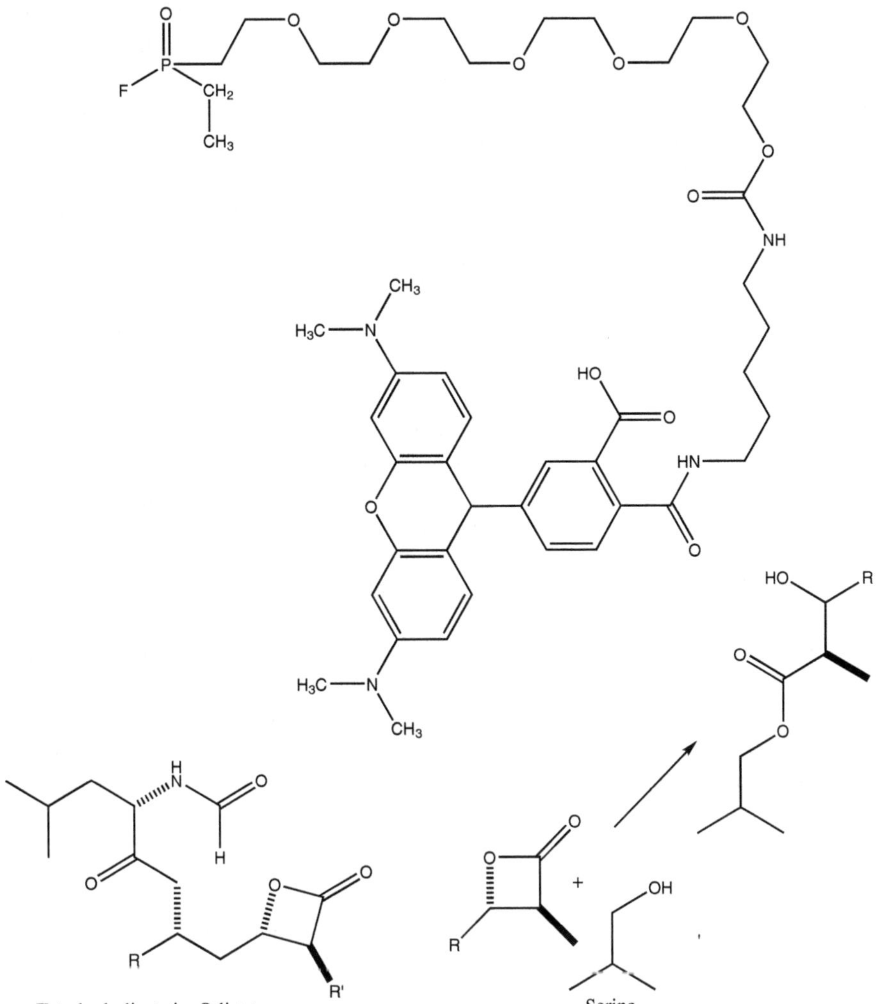

Tetrahydrolipstatin, Orlistat
R=nC$_{11}$H$_{23}$; R'=nC$_6$H$_{13}$

Serine

FIGURE 3.6 A fluorescent fluorophosphate probe. Fluorophosphonate-poly(ethylene)glycol-(6-carboxyltetramethylrhodamine) was demonstrated to show that Orlistat® was a covalent inhibitor of fatty acid synthase reacting via a lactone mechanism.

residues has been reported.[71] Previous studies also exist on the use of radiolabeled DFP for the identification of serine esterases in tissue homogenates.[72] The derivative formed is reasonably stable but, as it is analogous to an acyl-enzyme intermediate,[73–76] the inactivated enzymes can undergo a slow reactivation with loss of the label (Figure 3.3).[73,74] The reactivation occurs more rapidly at acid pH (phthalate buffer, pH 5.1) than at basic pH (borate, pH 8.7) but is still on the order of days. Reactivation does occur more rapidly with the addition of nucleophiles[73–77] such as oximes and hydroxamic acids (Figure 3.3). Similar reactivation occurs with choline esterases inactivated with organophosphorous compounds.[77] Those investigators who have worked with DFP may remember having 2-PAM (2-pyridinealdoxime) close by in case of contact with the reagent. More complex reactions exist involving rearrangement of the alkylphosphoryl label[78] that tend to retard reactivation of the enzyme. These "aging" reactions are thought to involve the hydrolytic removal of one of the O-alkyl substituent groups.[79] Reactivation of human neuropathy target esterase with N,N'-diisopropyl-phosphorodiamidic fluoride (Mipafox, Pestox, Figure 3.7) occurs by the alkyl substituent loss as well as another pathway involving deprotonation.[80] The potential complexity of the labeling of cellular contents with "specific" reagents is illustrated by the studies the neuropathy target esterase (neurotoxic esterase).[80–82] Several studies exist on the modification of this protein in tissue extracts and the ability of pretreatment of tissues with Paraoxon (diethyl-p-nitrophenyl phosphate) to block the reaction of DFP and related compounds with cholinesterase permitted more specific modification and identification of the neuropathy esterase.[83–86] IgM serine protease activity has been modified by reaction with a hapten amidino phosphodiester.[87] A representative structure for these "covalently reactive analogues" is shown in Figure 8 and is designated as a hapten covalently reactive analogue. More complex derivatives incorporate peptide segments to mimic EGF antigen and GP120 antigen.[88] The synthesis and early applications of these reagents to reaction with proteolytic antibody fragments was described earlier.[89] The use of a terminal biotin allows purification of the modified proteins by affinity chromatography on an avidin matrix. Enzymatically active fragments could be obtained by elution with 2-pyridinealdoxamine. These reagents are quite similar to those described by Cravatt and coworkers but application to proteomics has not been described.

Peptide Halomethyl Ketones

Peptide halomethyl ketones, most notably tosyl-phenylalanylchloromethyl ketone (TPCK) and tosyl-lysylchloromethyl ketone (TLCK), react with classical serine protease by alkylation of a histidine residue (Figure 3.2).[90] TPCK reacts preferentially with chymotrypsin-like enzymes while TLCK reacts preferentially with trypsin-like enzymes. The chloro function was selected because of the low reactivity of chloroacetate/chloroacetamide with histidine residues in proteins as compared to the bromo- or iodo-derivatives. 5-(6-Carboxyfluoresceinyl)-L-phenylalanyl-chloromethyl ketone has been used as a specific inhibitor for chymotrypsin-like enzymes (Figure 3.9).[91] This compound and a Texas Red derivative of TPCK[92] are available from Imgenex under the name of Serpas™. Other interesting derivatives of peptide chloromethyl ketones exist that might be useful. N-Terminal biotin-labeled peptide chloromethyl ketones (Figure 3.10) have been developed by Williams, Mann and coworkers[93,94] and have been used to measure various activated coagulation factors in complex mixtures.[95,96] Biotin-labeled[97] and fluorescein-labeled[98,99] peptide chloromethyl ketones have been used to identify capases in complex mixtures. As a precautionary note, peptide chloromethyl ketones do react with sulfydryl groups in a variety of proteins.[100–106] Shaw and coworkers[107] observed that peptidyl fluoromethyl

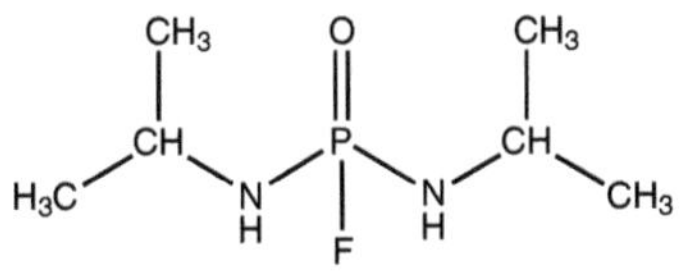

N,N'-Diisopropylphosphoramidic fluoride(Mipafox)

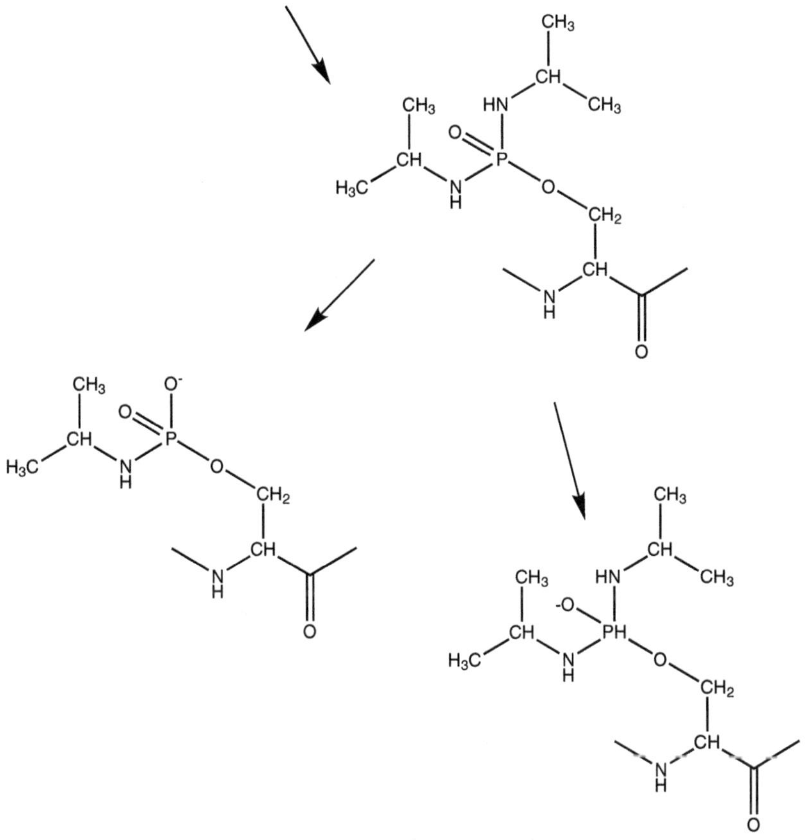

FIGURE 3.7 The reaction of serine enzymes with N,N'-diisopropylphosphorodiamidic fluoride (Mipafox, Pestox). Shown are two pathways for the decomposition of the reaction product. These chemical changes are described as an aging response of the reaction product (Clothier, B. and Johnson, M.K., Rapid aging of neurotoxic esterase after inhibition by diisopropyl phosphorofluoridate, *Biochem. J.*, 177, 549–559, 1979).

ketones are less reactive that the chloromethyl derivatives and can be useful for the study of thiol proteases. Peptide fluoromethyl ketones have been used as relatively specific inhibitors of capases activity.[108] Bergen and coworkers[109] have observed "non-specific" modification of protein with the components of protease inhibitor "cocktails."

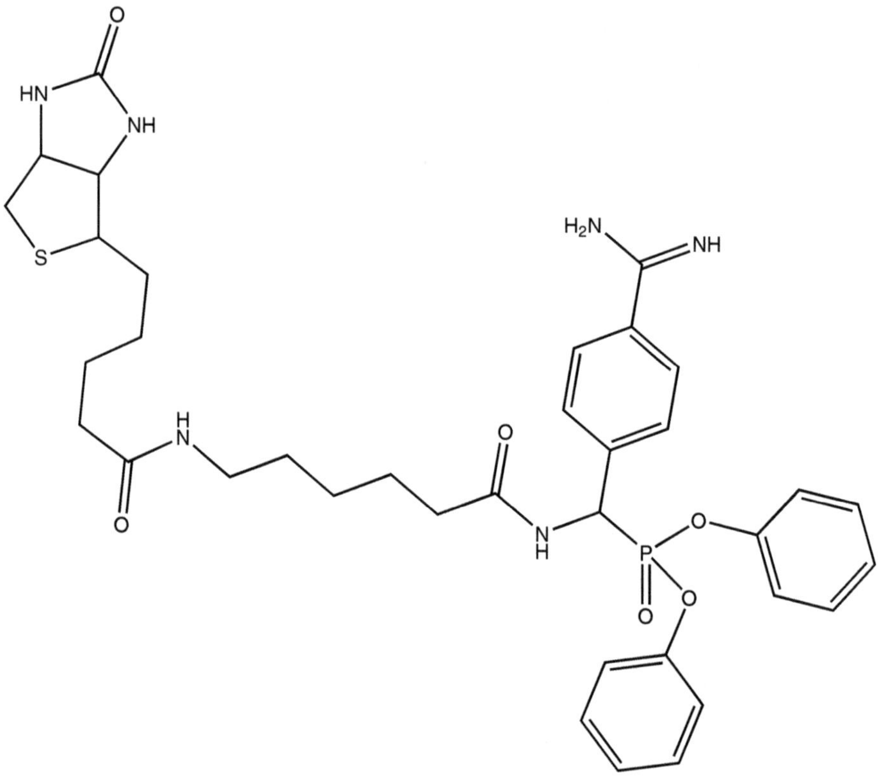

FIGURE 3.8 Phosphonate ester probes for proteolytic antibodies (Hapten Covalently Reactive Analogues). See Paul, S. *et al.*, Phosphonate ester probes for proteolytic antibodies, *J. Biol. Chem.*, 276, 28314–28320, 2001.

OTHER FUNCTIONAL DERIVATIVES

Other approaches exist to the use of active-site-directed reagents (affinity reagents, activity-based affinity probes).[110–113] Zhang and coworkers developed affinity probes for the modification of the active site of protein tyrosine phosphatases (Figure 3.11). These inhibitors are based on earlier work developing α-bromobenzylphosphonate[114] and demonstrated high specificity in the modification of protein tyrosine phosphatases in *Escherichia coli* cell lysates.

Another approach has been developed by Ovaa and colleagues[115] to detect ubiquitin proteases with vinyl sulfone derivatives and related compounds. These are fairly complex reagents (Figure 3.12) that react with cysteine proteases to form stable derivatives that can be detected. Bogyo and coworkers[116] have developed epoxide probes (Figure 3.13) for the modification of cysteine proteases. Epoxides inactivate cysteine proteases by alkylation of the active site cysteine residue via a mechanism-based reaction.[117] The reader is recommended to an excellent article by

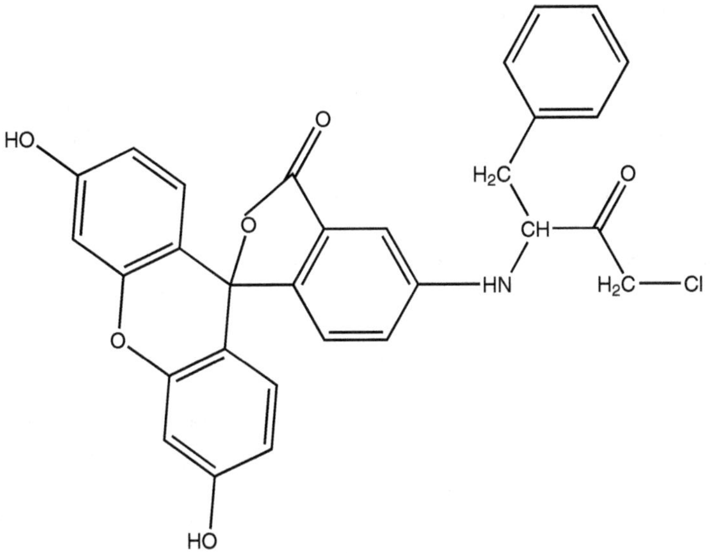

Carboxyfluoresceinyl-phenylalanine chloromethyl ketone

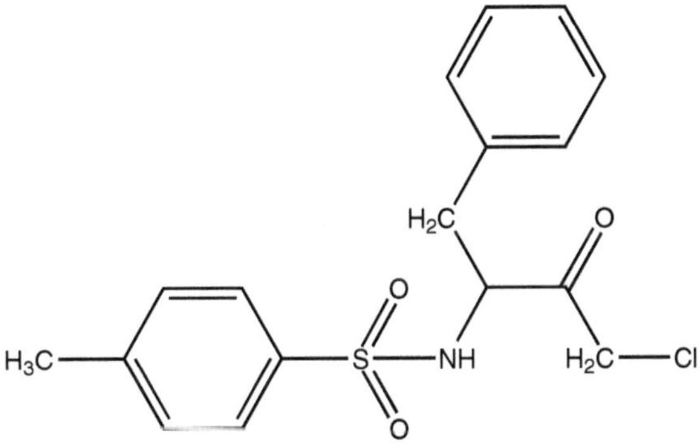

Tosylphenylalanine chloromethyl ketone

FIGURE 3.9 Carboxyfluoresceinyl phenylalanine chloromethyl ketone, a fluorescent analogue of TPCK that can be used to modify a class of proteolytic enzymes.

Powers and coworkers[118] on the development of aza-peptide epoxides as selective inhibitors of capases.

A novel approach developed by Dupont and colleagues[119] uses an antibody directed at the cleavage site region in a specific substrate; after cleavage of the substrate protein, the antibody no longer reacts with the protein. Another proteomic approach to the study of intracellular proteases used two-dimensional differential gel electrophoresis (2D DIGE)[120] to identify substrates for granzyme A and granzyme B in YAC-1 (mouse lymphoma) cell lysates.

SPECIFIC MODIFICATION OF PROTEINS AND PEPTIDES FOR PROTEOMIC ANALYSIS

The above reagents are intended to be active-site directed either on the basis of functional group reactivity at the active site or affinity binding.[32] Some work exists on reagents that are not designed to be active-site directed[121] but react with functional groups on proteins which can then be captured for proteomic analysis. One example is provided by sulfonate esters (Figure 3.14). Not much research exists on the reaction of these compounds with proteins. While here is extensive information on reaction with nucleic acids, a limited amount of information is found on the reaction of simple alkyl sulfonate esters such as methyl methane sulfonate with proteins.[122,123] In principle, these reagents have the potential to react with a variety of amino functional groups in proteins such as the amino-terminal α-amino group, the ε-amino group of lysine, imidazole nitrogens of histidine, the mercaptide function of cysteine and the hydroxyl group of tyrosine. In fact, the potential reactive sites are probably similar to those subject to alkylation by α-halo acid and would then also include glutamic acid. Reaction at a given residue should be a function of nucleophilicity of the protein bound functional group, the electronegativity of the alkyl/aryl sulfenyl leaving group and steric accessibility.

Interesting and perhaps relevant literature is found on the use of methyl alkyl sulfonates as skin sensitizers.[124–127] This work was driven by the need to develop alternatives to animal testing. The various methyl alkyl sulfonates were thought to

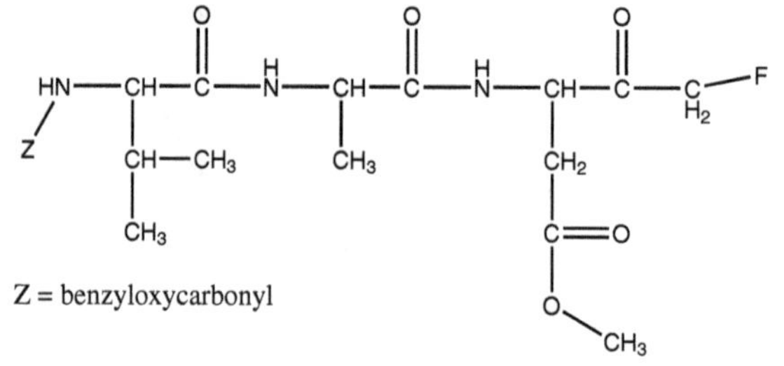

Z = benzyloxycarbonyl

Benzyloxcarbonyl-valylalanylasparate(beta methyl ester) fluoromethyl ketone (1)

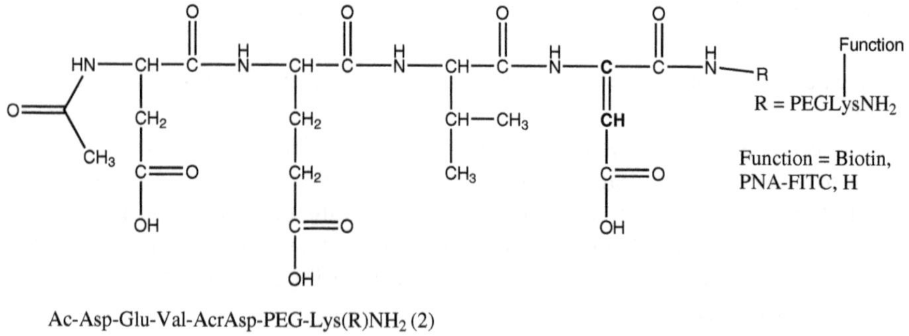

Function

R = PEGLysNH$_2$

Function = Biotin, PNA-FITC, H

Ac-Asp-Glu-Val-AcrAsp-PEG-Lys(R)NH$_2$ (2)

(Continued)

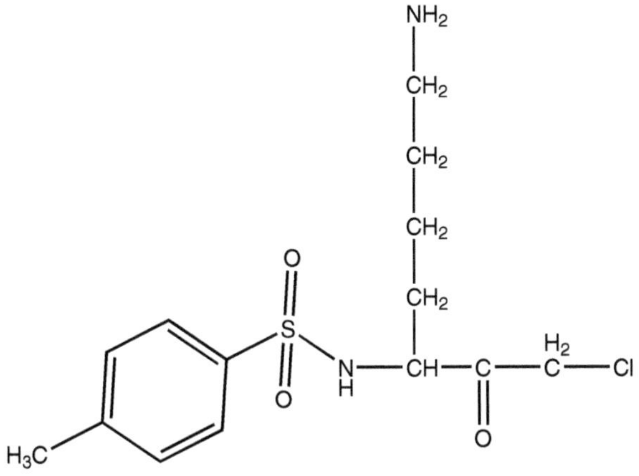

Tosyl-lysine-chloromethylketone, TLCK (1)

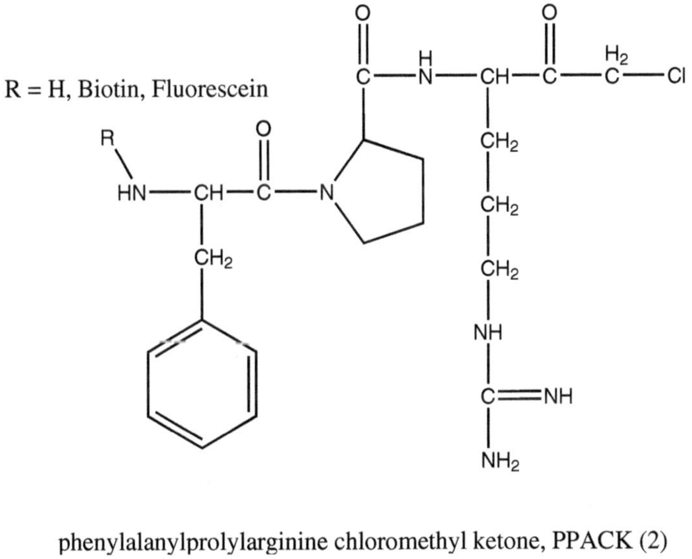

phenylalanylprolylarginine chloromethyl ketone, PPACK (2)

FIGURE 3.10 Peptide chloromethyl ketones. Shown is a simple peptide chloromethyl ketone, tosyl-lysine chloromethyl ketone, and a more complex peptide chloromethyl ketone, phenylalanine arginine chloromethyl ketone. Figure 10a shows the more complex fluoromethyl ketone which can be used as a capase inhibitor.

be methyl proteins and their reactivity is a product of skin permeability and the ability to methylate proteins. The relative effectiveness of the reagents was described by a relative alkylation index (RAI). No differences were found in chemical reactivity and differences in activity were due to partitioning that could be evaluated by a

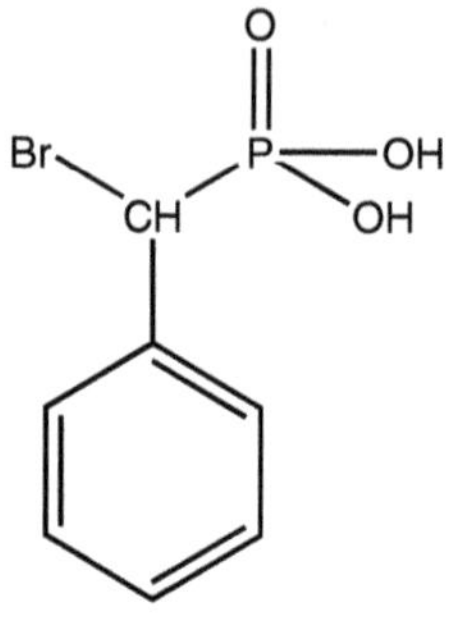

alpha-bromobenzylphosphonate[1]

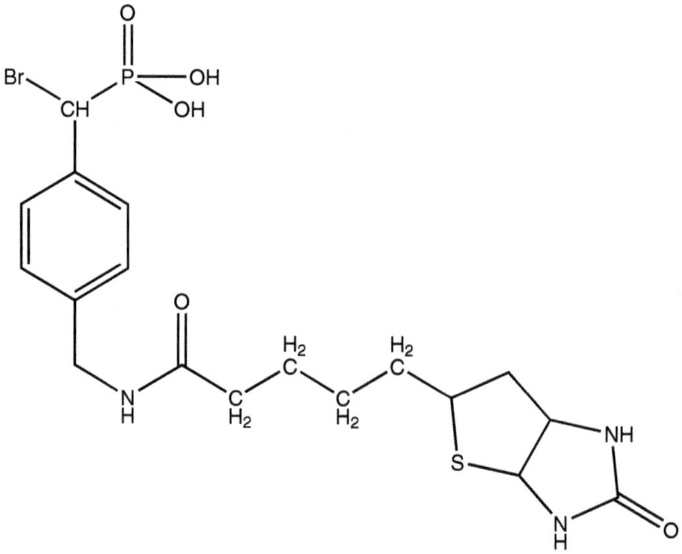

Protein Tyrosine Phosphatase Activity-Based Probe[2]

FIGURE 3.11 Protein tyrosine phosphatase acitivity-based probes for use in proteomic research. (1) Taylor, W.P., Zhang, Z.-Y. and Widlanski, T.S., Quantitative affinity inactivtors of protein tyrosine phosphatases, *Bioorg. Med. Chem.*, 4, 1515, 1996; (2) Kuman, S. et al., Activity-based probes for protein tyrosine phosphatases, *Proc. Natl. Acad. Sci. USA*, 101, 7943, 2004.

methanol/hexane partitioning coefficient.[124] Most of the work was performed on methyl dodecane sulfonate, methyl hexadecane sulfonate and methyl trans-hexadecane-3-ene sulfonate. This group did perform a limited study[128] on the rate of reaction of these and some other alkylating agents with *N*-methylbutylamine and *N*-methyl-analine. Reaction was more rapid with *N*-methylbutylamine than with *N*-methyl-analine; the difference was far more pronounced with methyl dodecylsulfonate than with dodecyl methylsulfonate.

The reaction of chymotrypsin with the toluenesulfonoxy ester and methanesulfon-oxy derivatives of *N*-carbobenzoxyamino-4-phenyl-2-butanone was much faster than

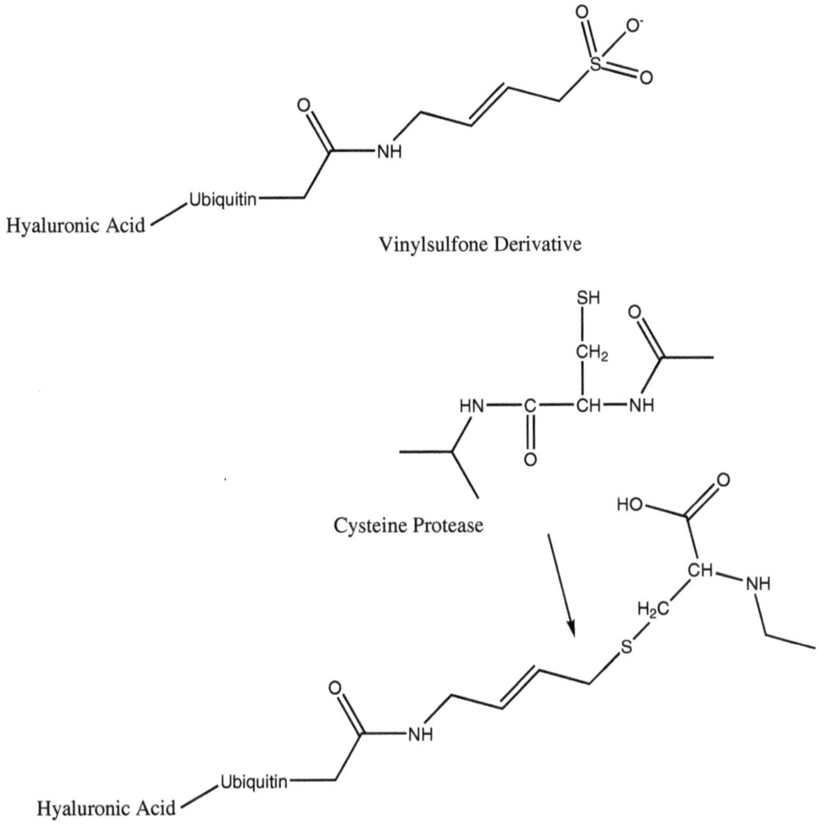

FIGURE 3.12 The reaction of a complex vinyl sulfone derivative with a cysteine proteinase.

reaction with the corresponding halo derivatives.[129] The analogues of TPCK have not been subjected to further study. Nagagawa and Bender[130] studied the methylation of histidine residues in chymotrypsin. Methylation occurred at the N3 nitrogen on the imidazole ring. This reagent has been used by other investigators for the methylation of histidine in proteins.[131–134] Heinrikson[135] has described the reaction of methyl-*p*-nitrobenzene sulfonate with cysteine residues in peptides and proteins. Plapp has examined the reaction of ribonuclease A with tosylglycolate (carboxymethyl *p*-toluenesulfonate)[136] as an analogue of the well known haloacetate inhibitors (Figure 3.15). This reagent alkylated the active site histidine residues; histidine 119 was preferentially alkylated at the N1 position. The kinetics are consistent with affinity labeling and this reagent reacts more slowly than either iodoacetate or bromoacetate. Reaction with 4-(*p*-nitrobenzyl) pyridine is approximately five times slower than that observed with bromoacetate; reaction is also observed with cysteine residues. Oshima and Takahashi[137] reported the inactivation of ribonuclease T$_1$ with tosylglycolate in a reaction involving the modification of glutamic acid residue as observed with iodoacetate.

Alkylsulfonates are S$_N$2 alkylating agents[123] as is methyl iodide (Figure 3.14). Edman and coworkers[138] have shown that methyl iodide will react with cysteine to

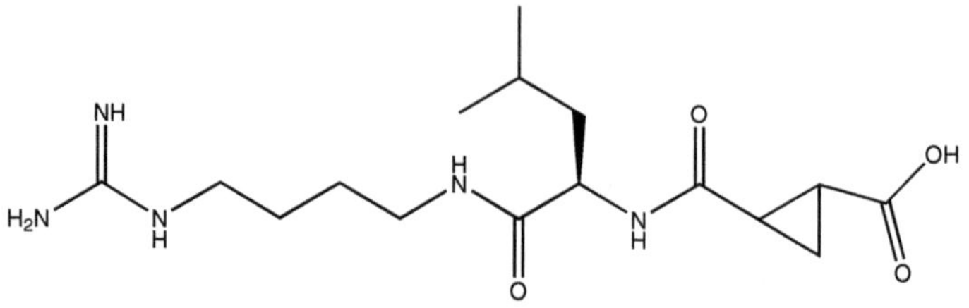

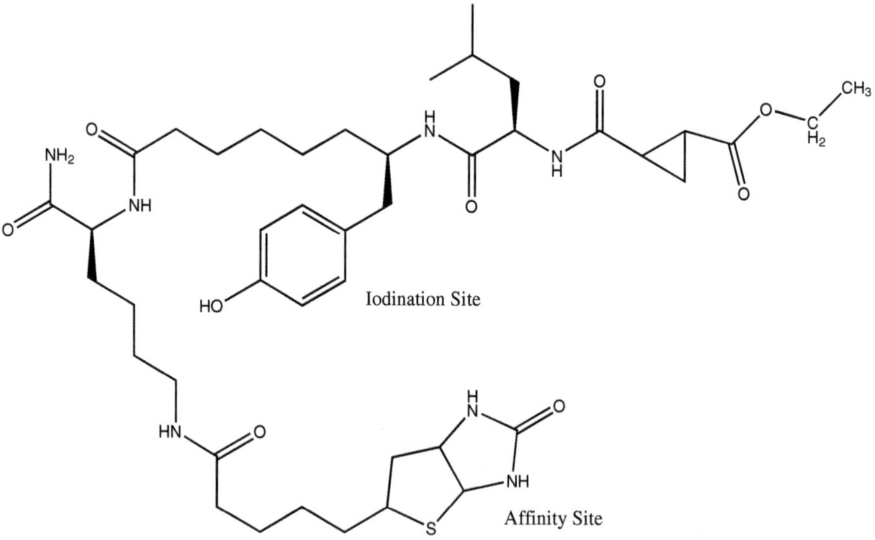

FIGURE 3.13 Epoxide-based inhibitors of cysteine proteinases. Shown is E-64, trans-epoxysuccinyl-leucylamido(4-guanidino)butane (Barrett, A.J., Kembhavi, A.A., Brown, M.A. et al., L-trans-epoxysuccinyl-leucylamido(4-guanidino)butane)) and a more complex derivative (Greenbaum, D., Medzihradszky, K.F., Burlingame, A.L. et al., Epoxide electrophiles as activity-dependent cysteine protease profiling and discovery tools, *Chem. & Biol.,* 7, 569–581, 2000). The presence of a phenolic functional group (tyrosine analogue) permits the preparation of a radiolabeled form of the inhibitor.

form S-methylcysteine; no reaction was observed with lysine, histidine or tryptophan. Thus, the modification reactions observed with the above reagents seem most likely to occur at cysteine residues in proteins if the reaction is driven entirely by the protein functional group nucleophilicity. If the modification reaction is allowed to proceed for longer periods of time, the yield of S-methyl cysteine is reduced, presumably due to the formation of the dimethylsulfonium salt. In a related study, Sjöquist and associates[139] reported the specific methylation of cysteine with dimethyl sulfate.

Cravatt and his colleagues published an extremely interesting series of papers on the use of alkyl sulfonate esters for the modification of proteins. In 2001, Adam and coworkers[121] described the development of a series of biotinylated alkyl-sulfonate esters (Figure 3.16). These various derivatives were used to screen rat testis homogenate to identify reactive proteins. Reaction was assessed after

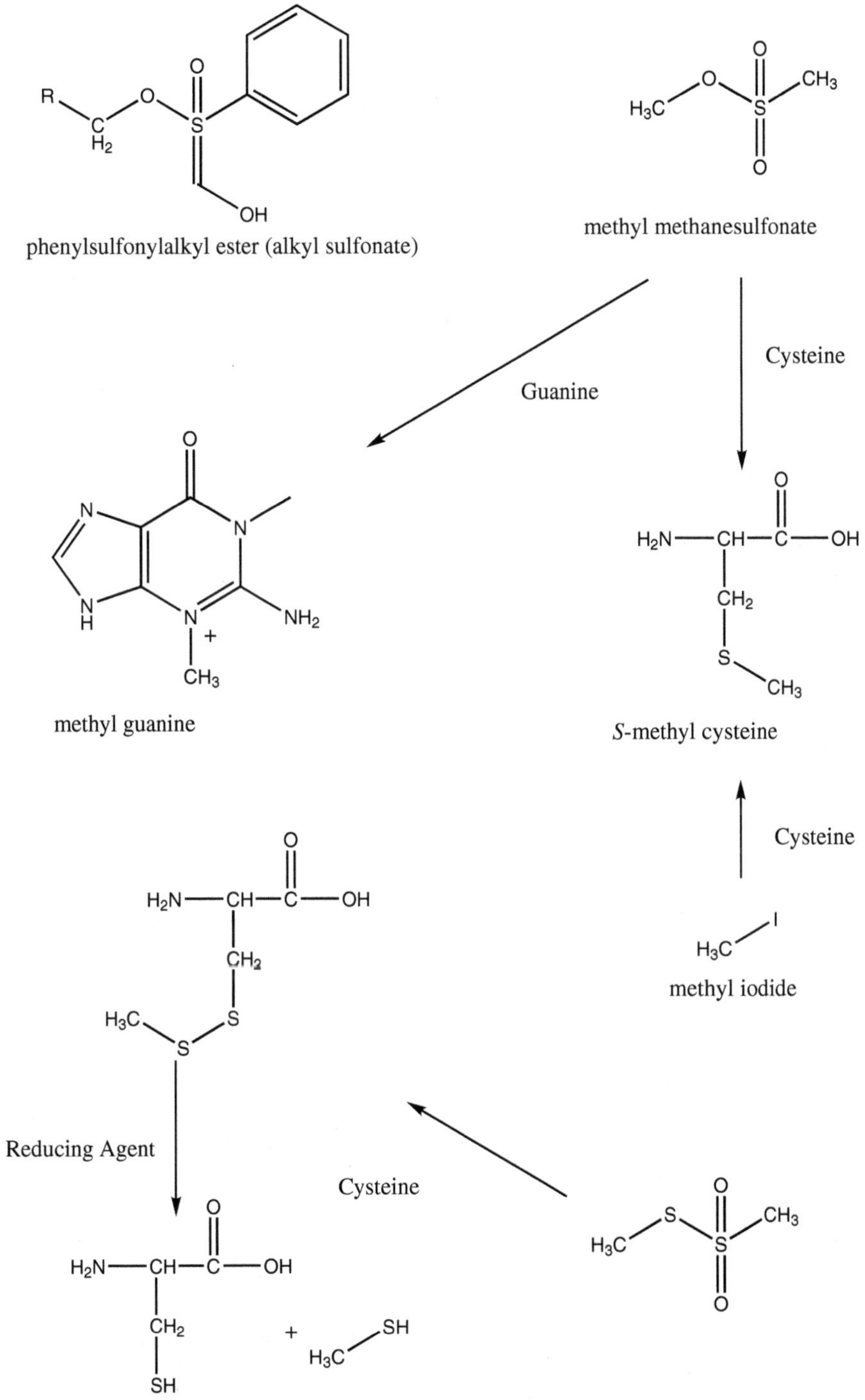

FIGURE 3.14 Structure and reactions of some methylating agents with proteins and nucleic acids.

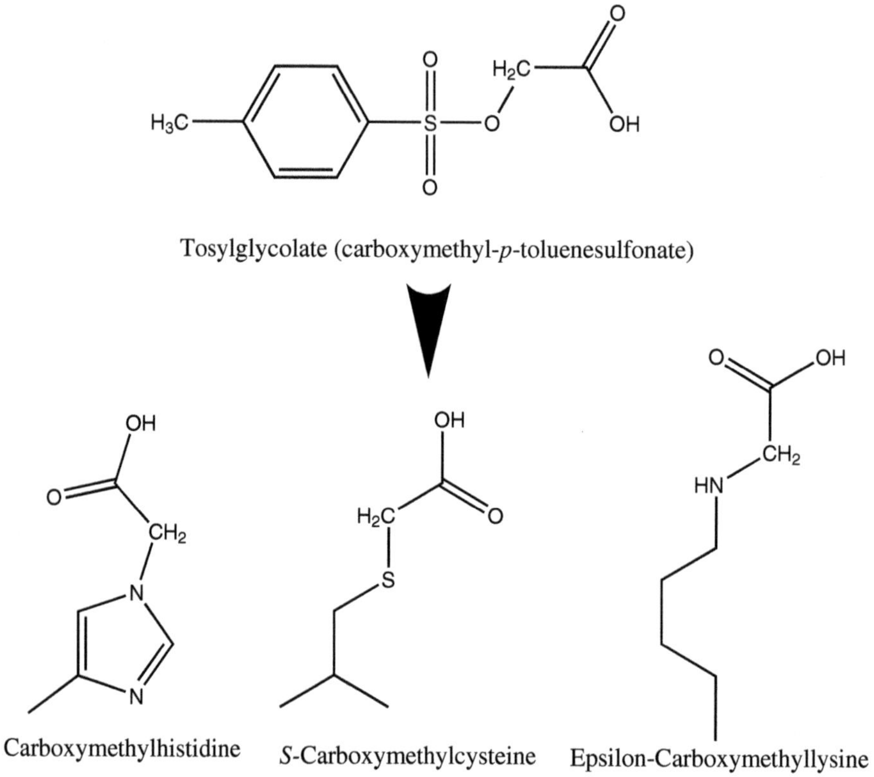

Tosylglycolate (carboxymethyl-*p*-toluenesulfonate)

Carboxymethylhistidine   *S*-Carboxymethylcysteine   Epsilon-Carboxymethyllysine

FIGURE 3.15 The reaction of tosylglycolate with some protein functional groups.

electrophoretic separation on the basis of reaction with an avidin-based detection system. Specificity for reaction with native proteins was demonstrated by the lack of reaction with heat-treated (80°C, 5 minutes) preparations. An aldehyde dehydrogenase was preferentially modified with the pyridyl derivative. In subsequent work,[140] a combined chemical modification approach was used. In order to improve the sensitivity, a rhodamine-labeled probe was used in initial studies and a biotin-labeled probe used to isolate the modify proteins. Activity-based protein labeling occurred with protein extracts that had not been heat-treated. A variety of probes (Figure 3.17) were prepared with different alkyl or aryl sulfenyl leaving groups. The highest extent of labeling of tissue homogenates was obtained with the phenyl derivative and the least with the octyl derivative. The phenyl derivative identified a variety of different enzymes including aldehyde dehydrogenase, acetyl CoA acetyltransferase, an epoxide hydrolase and an NAD/NADP-dependent oxidoreductase. This same group also developed a sulfonate ester probe that contained both a rhodamine function and a biotin function (Figure 3.18).[141] In their most recent work,[142] this group further characterized the reaction of the phenyl sulfonate derivative with enzymes. In this study, a rhodamine tag was inserted via the alkylation reaction. The proteins were subjected to proteolytic digestion and the rhodamine-labeled peptides isolated by immunoaffinity chromatography on a rhodamine monoclonal affinity column. Mass spectrometric analysis combined with oligo-nucleotide-directed mutagenesis established the sites of modification in various enzymes. Reaction occurred at a cysteine residue in a glutathione S-transferase, at aspartic acid residues in enoyl CoA transferase and at a glutamic residue in aldehyde dehydrogenase-1. More recent studies[143,144] have used alkyl sulfonate esters with terminal azide or alkyne functions for reaction with either alkyne or azide functions respectively (see Figure 3.19).[145,146]

OTHER CLEVER REAGENTS

Reaction with 2,4-dinitrophenylhydrazine, fluorescein hydrazide and thiosemicarbi-zide is used for the detection of carbonyl products occurring as results of oxida-tion.[147,148] Selective modification of 3-nitrotyrosine residues has been described by Tannebaum and coworkers.[149] These investigators reduced the 3-nitrotyrosine residues to the corresponding 3-aminotyrosine with sodium dithionite. The 3-amino-tyrosine was modified with sulfosuccinimidyl-2-(biotinamido)ethyl-1,3-dithiopro-pionate (Figure 3.20). Selective alkylation of the 3-aminotyrosine is obtained at pH 5.0. After tryptic cleavage, the biotin-containing peptides are obtained by adsorption to a streptavidin matrix. Reduction of the disulfide linkage releases the peptide for subsequent analysis by mass spectrometry. Two somewhat more complex approaches to the identification of phosphorylated peptides have been developed.[150–152] One approach is based on the β-elimination of phosphoserine or phosphothreonine under alkaline conditions and elevated temperature followed by Michael addition of a dithiol and subsequent modification of the thiol group with a biotin-containing alkylating reagents (see Figure 3.21). The other approach[153] involves a series of chemical reactions resulting in the selective modification of the phosphoryl group and subsequent affinity isolation (Figure 3.22). As a word of caution, β-elimination has been observed for non-phosphorylated residues under the above reaction conditions.[154] Thaler and colleagues have extended some of this earlier work on β-elimination.[155] First, solvent conditions for the β-elimination reaction were optimized by increasing the amount of dimethyl sulfoxide. Second, an alterative approach to the isolation of the thiol derivative was developed using disulfide exchange on a dithiopyridine matrix. Affinity labeling of integral membrane proteins has been obtained by using (+)biotinyl-iodoacetamidyl-3,6-dioxaoctanediamine (Figure 3.23) as an alkylating agent for cysteine residues.[156]

STABLE ISOTOPE LABELING — QUANTITATIVE PROTEOMICS

The use of stable isotope labeling for quantitative proteomics was introduced by Aebersold and coworkers.[157] This method is a clever application of the concept of isotope dilution, a well-accepted technique in analytical chemistry.[158,159] Specifically, Aebersold and coworkers developed a somewhat complex reagent composed of a "tag" that bound to a affinity matrix facilitating purification of a labeled peptide or protein, a "linker" region that can be labeled with a "heavy" isotope such as deuterium

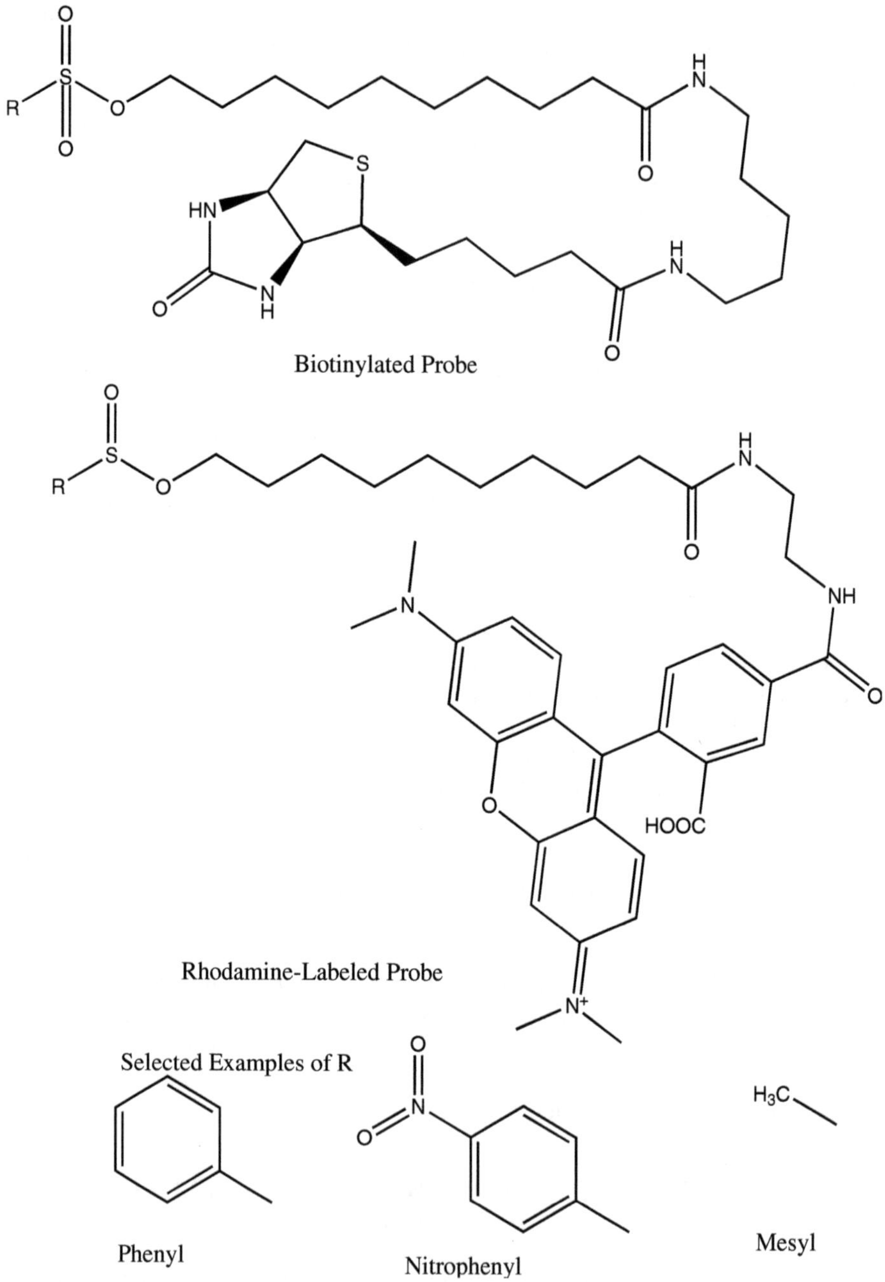

FIGURE 3.16 Some complex alkyl sulfonate esters used for proteomic profiling.

in place of hydrogen, and a reactive probe such as an α-ketohalo function analogous to iodoacetamide (Figure 3.24). These reagents, described as isotope-coded affinity tags (ICAT), were developed by Aebersold and colleagues.[160–162] ICAT reagents enable the relatively specific introduction of a deuterium-labeled moiety on the

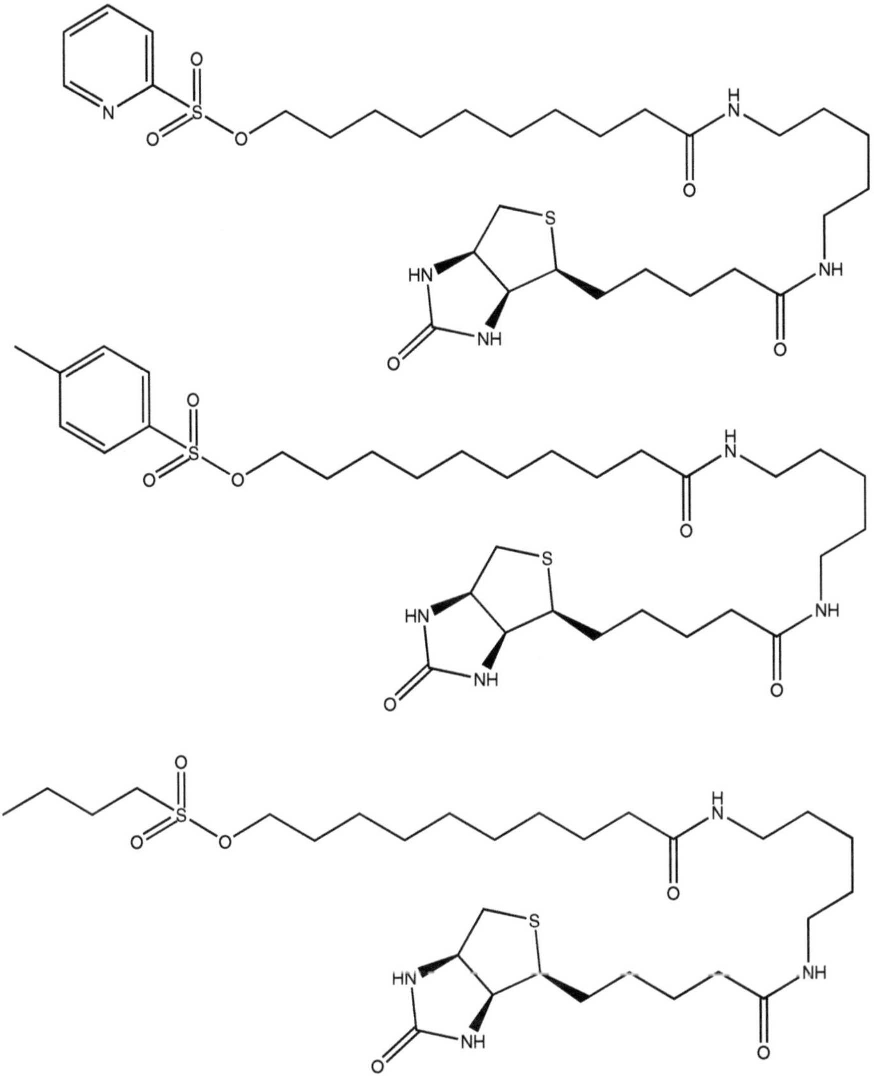

FIGURE 3.17 Alkyl and aryl sulfonate esters used for non-activity directed probes in proteomic research.

sulfydryl groups in a protein mixture. The use of a chemically identical modifying reagent not containing deuterium on a different protein allows the comparison of protein expression.[163] Since the derived peptides were presumed chemically identical — with the exception of the isotope — a difference in molecular weight on mass spectrometric analysis would differentiate between the peptides in the two samples. Consider an experiment where two cell cultures, one of which is challenged, are compared by labeling the experimental sample with a deuterated reagent. The cell culture is reduced and modified with the labeled or unlabled ICAT reagent. Proteolysis of the alkylated protein yields peptides that have been modified at sulfydryl

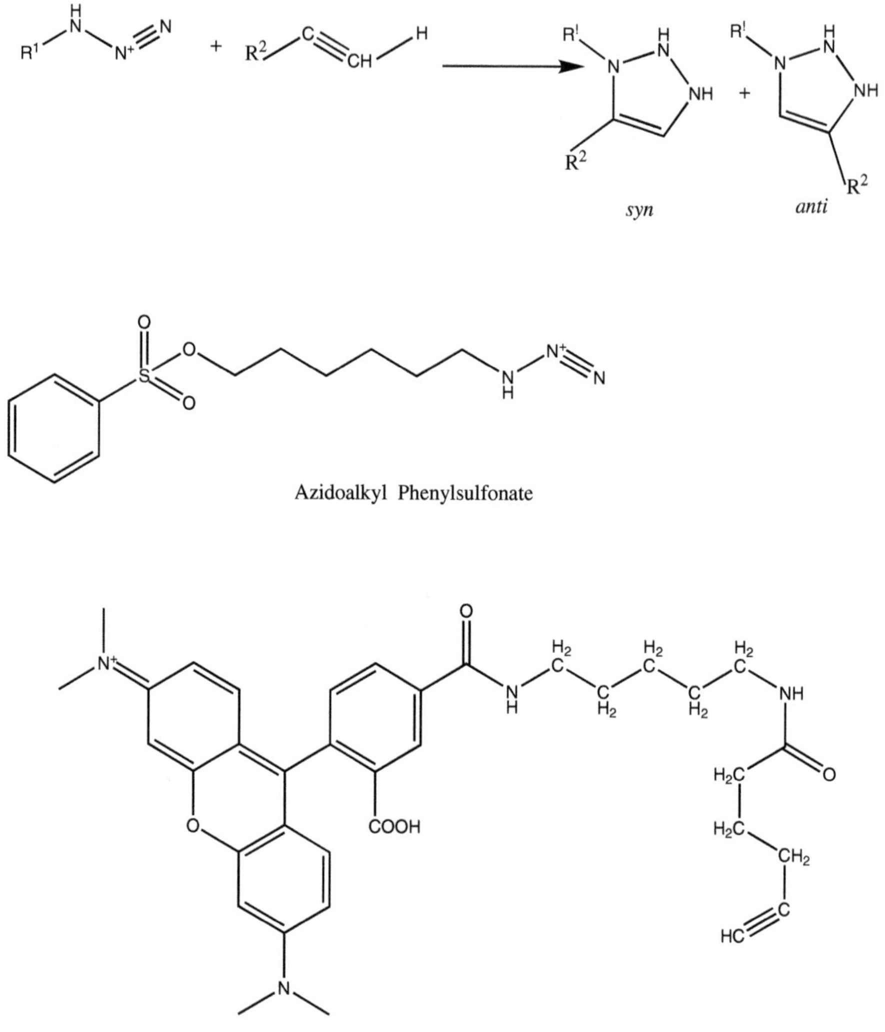

Azidoalkyl Phenylsulfonate

Alkyne Rhodamine Derivatives

FIGURE 3.18 Azide- and alkyne-based chemical probes for proteomic profiling. Note that these reagents can be bifunctional and useful for the study of protein–protein interaction (Breinbauer, R. and Kohn, M., Azide-alkyne coupling: A powerful reaction for bioconjugate chemistry, *ChemBioChem*, 4, 1147–1149, 2003).

groups with biotinylated reagents which can then be isolated by affinity chromatography on streptavidin/avidin matrices. The isolated peptides are then analyzed by mass spectrometry. The ratio of deuterated peptide to unlabeled peptide is an indication of a change induced by the change. Subsequent work has refined this technique[164,165] and reagents that target residues other than cysteine have been developed.[166–168] The development of a reagent with an acid labile link to a resin permits the facile purification of peptides.[169] More recently, visible ICAT reagents (VICAT

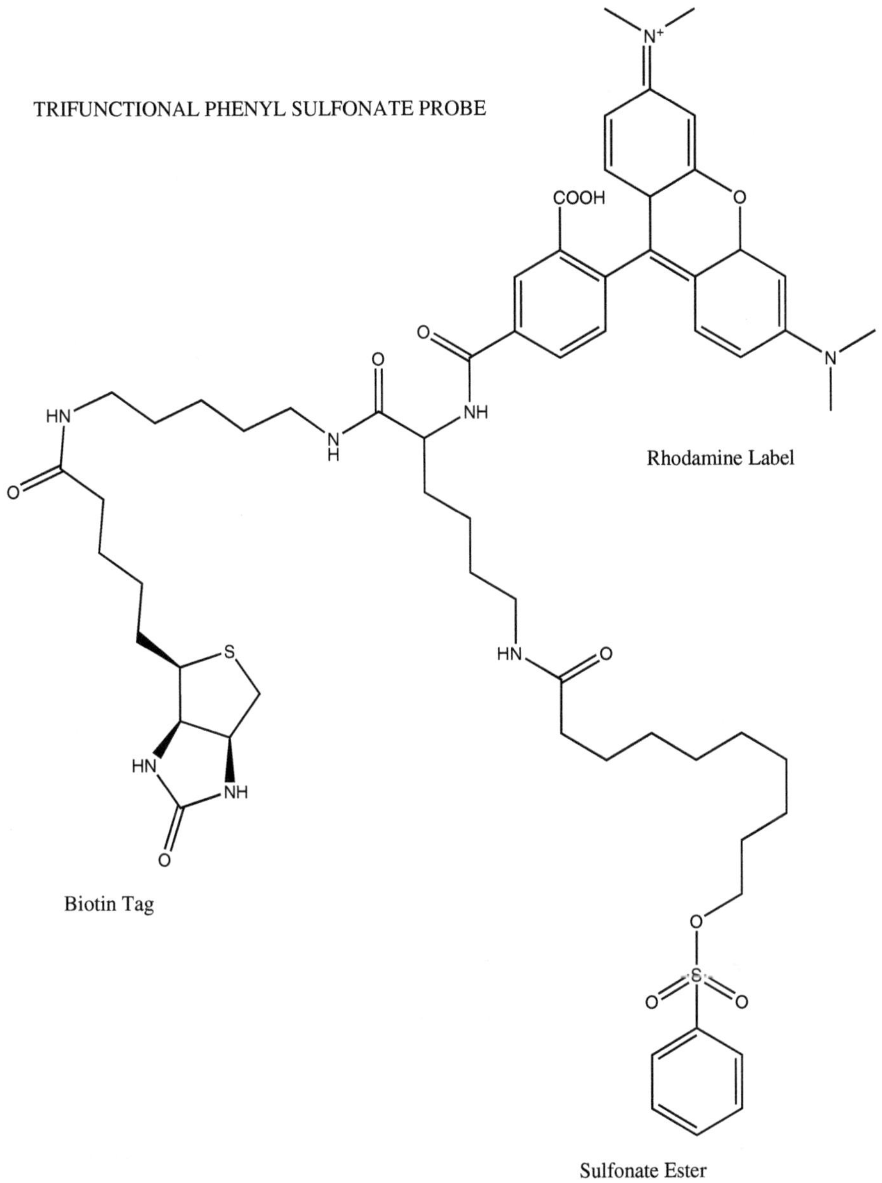

TRIFUNCTIONAL PHENYL SULFONATE PROBE

COOH

Rhodamine Label

Biotin Tag

Sulfonate Ester

FIGURE 3.19 A trifunctional phenyl sulfonate probes for the study of macromolecular inter-actions in proteomic research.

reagents) have been developed (Figure 3.25).[170] A related approach uses the incor-poration of stable isotope-labeled amino acids in cell culture (SILAC).[171,172]

A bit of a problem arose with the deuterium-labeled reagents, as Regnier and coworkers demonstrated in that the derivatives were, in fact, chemically different and separated on HPLC.[173] Regnier and colleagues then introduced a different type of ICAT reagents (Figure 3.26) that did not have this problem in that the reagents

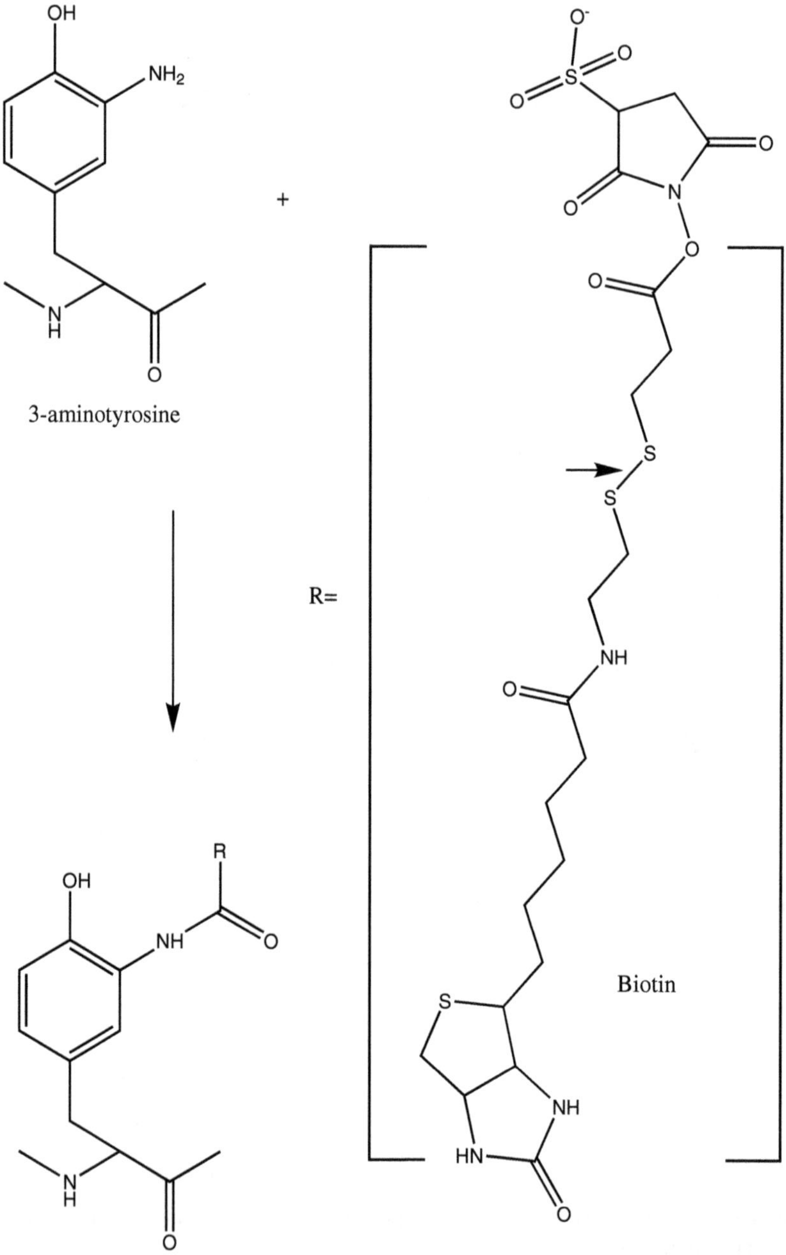

FIGURE 3.20 The reduction of nitrotyrosine to aminotyrosine and subsequent modification with an affinity label, showing a method for the isolation of nitrotyrosine peptides. Nitrotyrosine, either endogenous as a result of reaction with peroxynitrite or exogenous by modification with tetranitromethane, is reduced to aminotyrosine. As a result of the lower pKa, this amino group can be selectively modified with the N-hydroxysuccinimide derivative containing a biotin tag. The biotin tag permits the isolation of the modified protein or peptide on a streptavidin matrix; the disulfide bond can be reduced with the release of the isolated material.

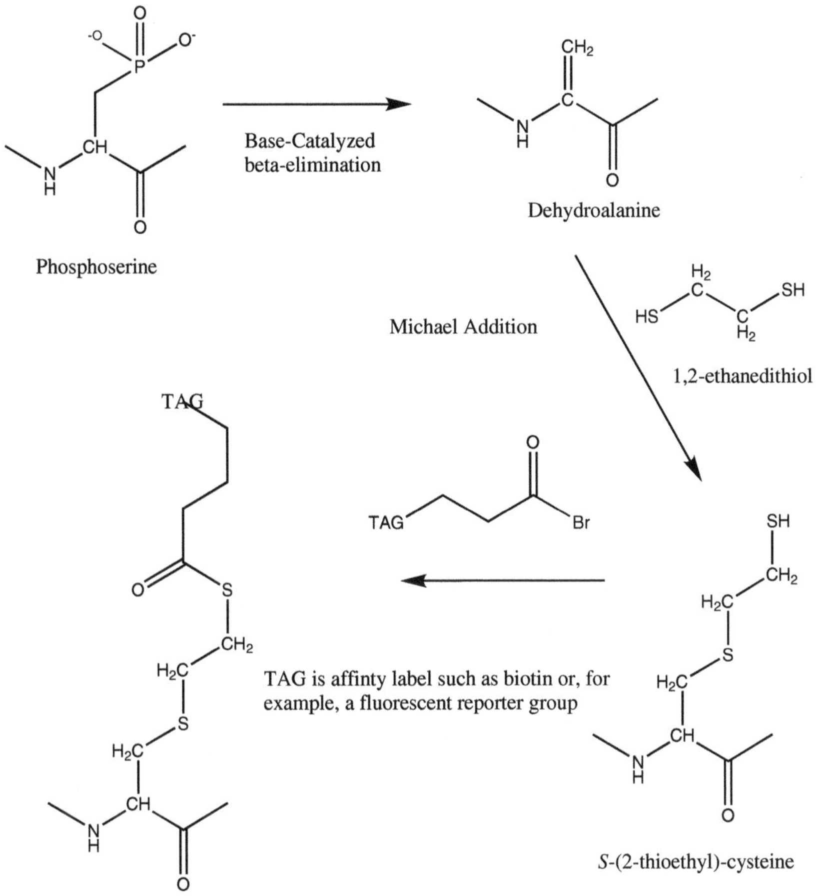

FIGURE 3.21 Modification and isolation of phosphoserine peptides via beta-elimination followed by a Michael addition.

were labeled with ^{13}C.[174] This reagent modified amino groups with a ^{13}C-labeled acetyl group via an N-hydroxysuccinimide reagent and is referred to as Global Internal Standard Technology (GIST). Other reagents are also shown in Figure 3.26.

A variety of applications exist that stem directly from earlier research in protein chemistry as discussed in Chapter 2. One particular example is provided from the work of Vanderkerckhove and coworkers.[175] These investigators used the change in chromatographic mobility of methionine-containing peptides after oxidation with hydrogen peroxide (H_2O_2) to form the sulfoxide.[176,177] This approach is useful and is based on earlier work by Hartley and colleagues on diagonal electrophoresis.[178] Another approach to the selective isolation of methionine-containing peptides is derived from the selectivity of alkylation at low pH.[179] In this approach, methionine-containing peptides are isolated by reaction with solid phase bromoacetyl derivative[180,181] and subsequently eluted with a mild reducing agent such as 2-mercaptoethanol. This approach has been used to differentiate methionine peptides from methionine sulfoxide peptides.[182]

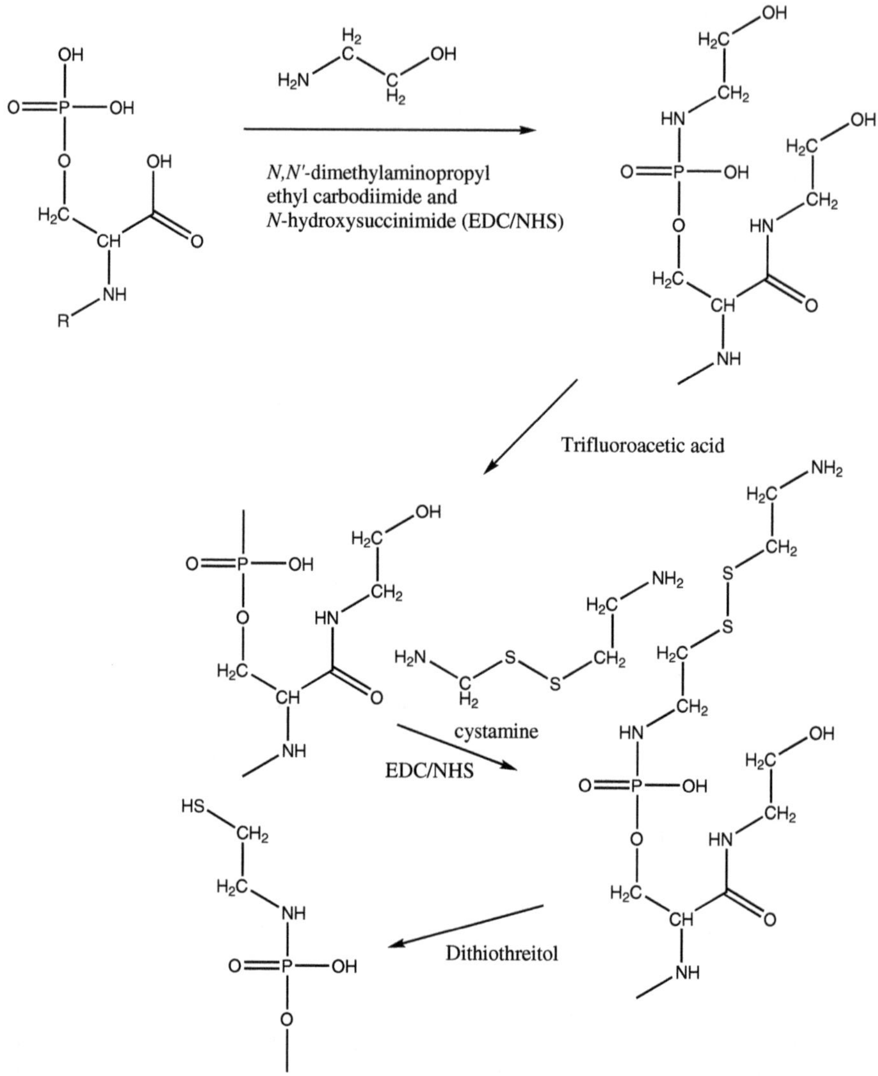

FIGURE 3.22 The differential modification of phosphoserine with carbodiimide in phosphoproteins.

A number of opportunities for the application of "classical" solution protein chemistry in the emerging discipline of proteomics should be obvious from the above information. Unlike approaches where information is obtained regarding the relationship between structure and function, application to proteomics does not necessarily require the rigorous identification of the sites of modification, although this would be useful information. The primary requirement in the application to proteomics is that the reaction be reproducible. Careful attention to experimental conditions is incumbent upon the investigator in order to assure reproducibility. Also quite useful is that the investigator understand the chemistry of the reagents. All too

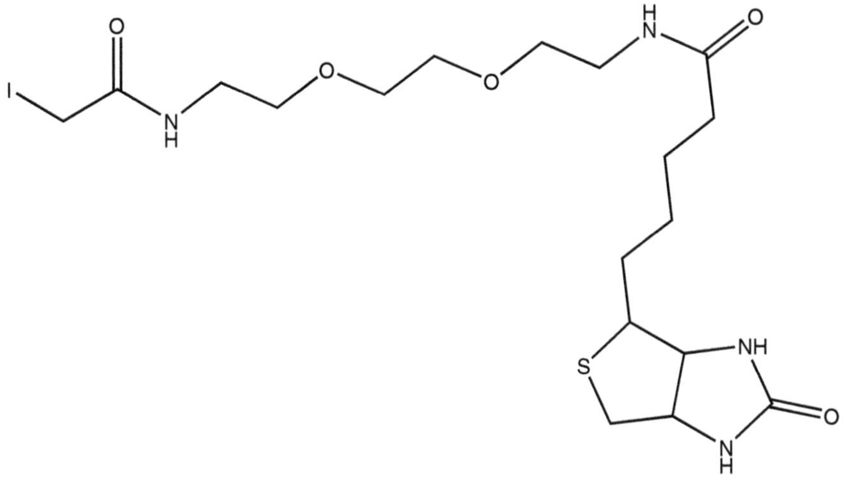

1-iodoacetamidyl,8-biotinyl, 3,5-dioxaoctanediamine (idoacetyl-PEO-biot in

FIGURE 3.23 A complex probe for the modification and isolation of cysteine-containing membrane proteins.

frequently today, reagents and kits are available with little information regarding the chemistry of their contents.

OTHER CLEVER APPROACHES TO THE MODIFICATION OF PROTEINS FOR PROTEOMIC ANALYSIS

Hsieh-Wilson and colleagues have reported an extremely novel approach to the proteomic analysis of O-linked β-N-acetylglucosamine-modified proteins.[183–185] An engineered galactosyltransferase is used to label O-linked β-N-acetylglucosamine proteins with a ketone-biotin tag, N-(aminooxyacetyl)-N′-(D-biotinoyl) hydrazine (Figure 3.27). A related approach to the identification of O-linked β-N-acetyl-glucosamine has been developed by Bertozzi and colleagues.[186,187] In this approach, N-azidoacetyl-glucosamine is incorporated into proteins and derivatized at the azido function via Staudinger ligation. This same group has used this chemistry for the incorporation of a fluorogenic dye (Figure 3.28).[188]

Several chemical approaches to the identification of sites of phosphorylation of proteins are described above. Also possible is the combination of an enzymatic approach and a chemical approach to the identification of these sites. A clever approach to this problem has been developed by Zhao and coworkers.[189] The use of adenosine 5′-O-(thiotriphosphate) resulted in the labeling of sites of phosphorylation with a thiophosphoryl function. The difference in the acid dissociation constant for the thiophosphoryl group and protein sulfydryl groups permitted the selective alkylation of the thiophosphoryl function with a tagged reagent allowing isolation

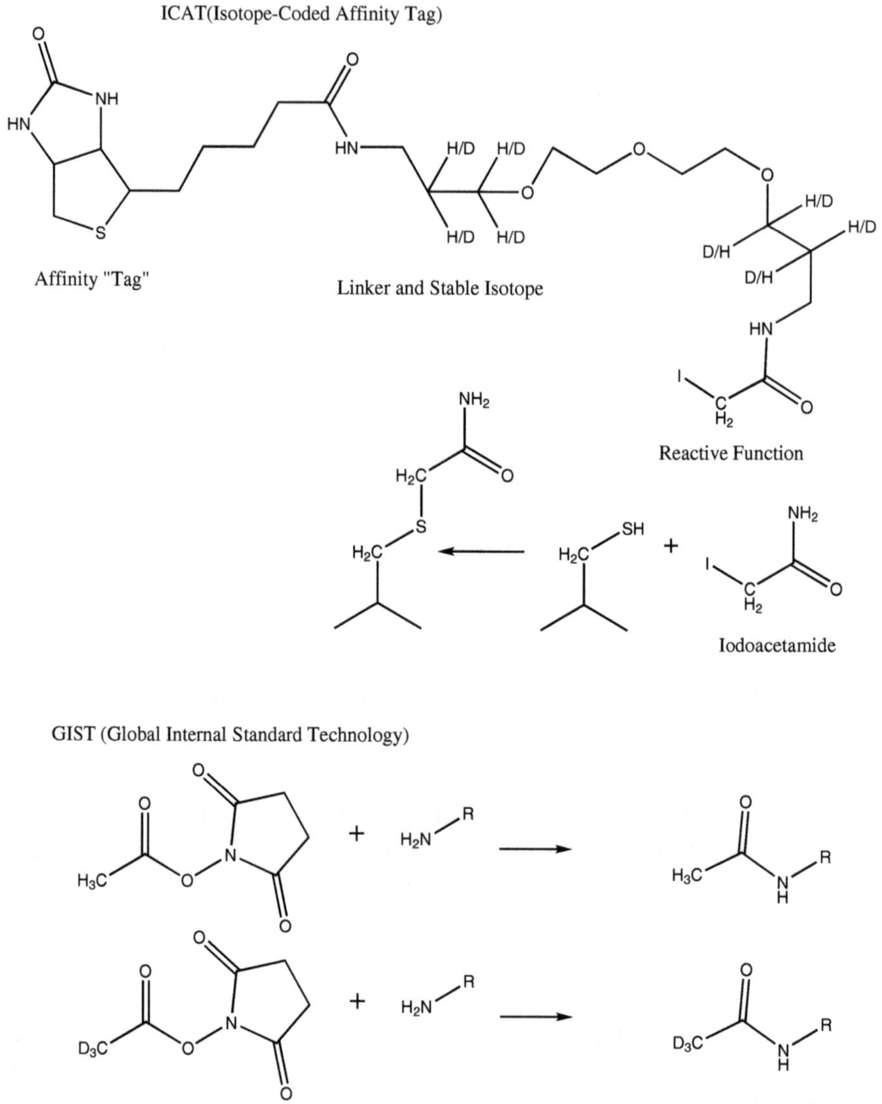

FIGURE 3.24 Labeling of proteins with stable isotope chemical reagents for proteomic research.

and identification. This reaction is shown in Figure 29. This group[190] has also used the Staudinger reaction advanced by Bertozzi and coworkers[186,187] to characterize farnesyl-modified proteins. An azide-farnesyl intermediate is coupled to the recipient protein. Subsequent condensation with a phosphine derivative allows the placement of a probe such as biotin. The chemistry for this process is outlined in Figure 3.30.

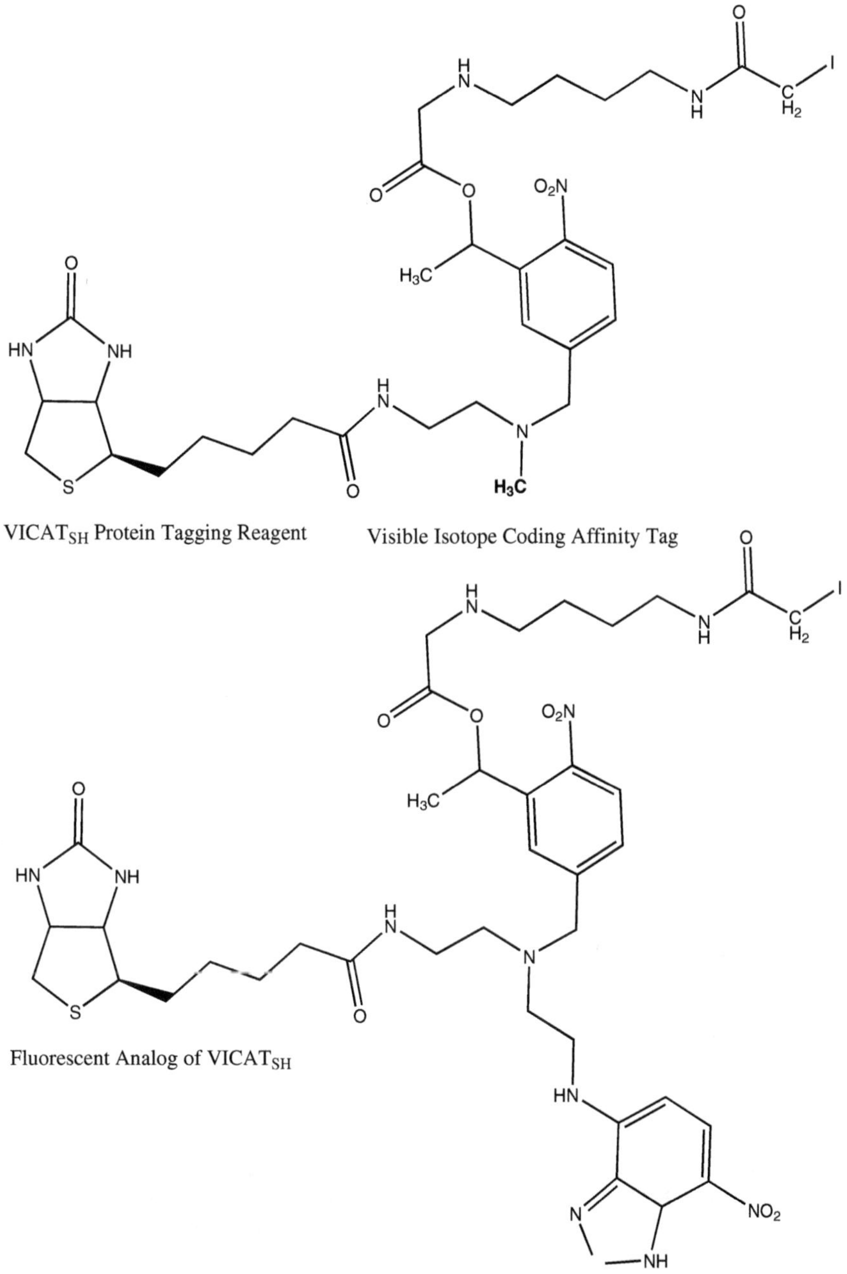

VICAT$_{SH}$ Protein Tagging Reagent   Visible Isotope Coding Affinity Tag

Fluorescent Analog of VICAT$_{SH}$

FIGURE 3.25 Visible isotope coding affinity tag.

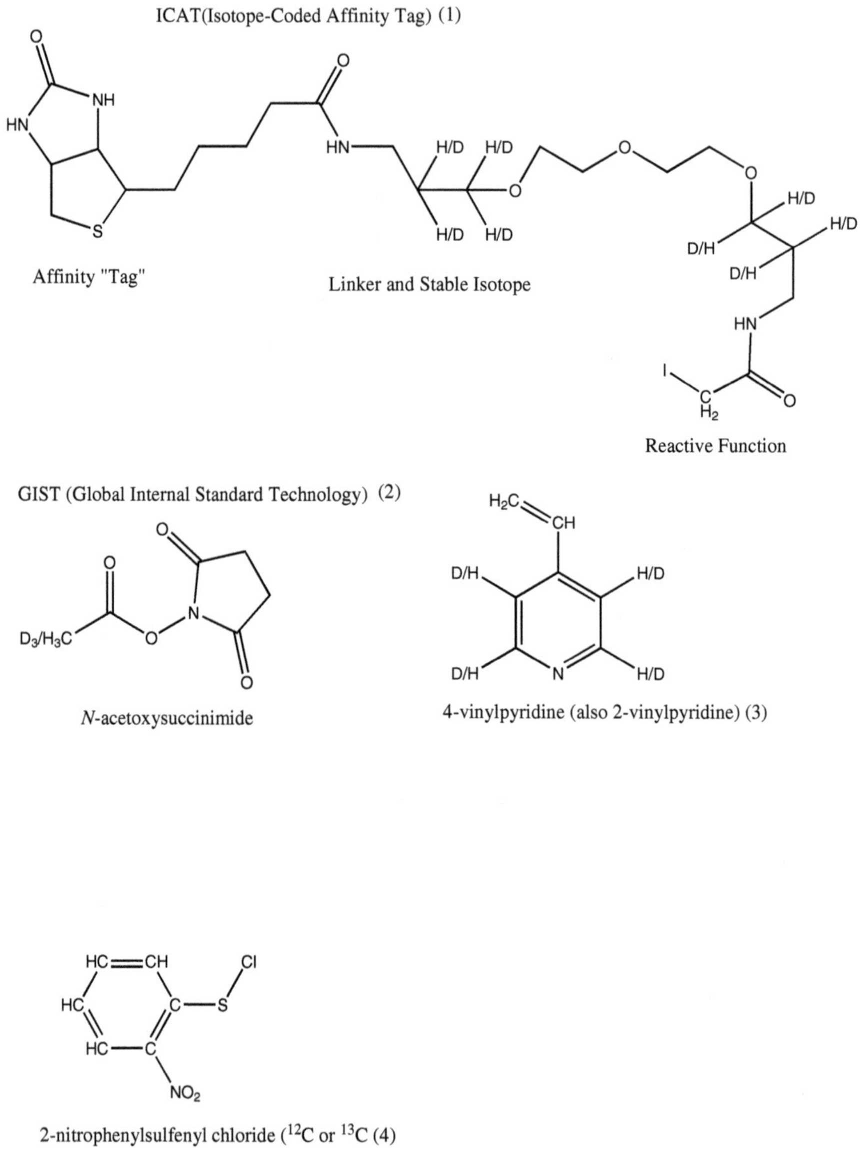

ICAT(Isotope-Coded Affinity Tag) (1)

Affinity "Tag"   Linker and Stable Isotope

Reactive Function

GIST (Global Internal Standard Technology) (2)

N-acetoxysuccinimide

4-vinylpyridine (also 2-vinylpyridine) (3)

2-nitrophenylsulfenyl chloride (^{12}C or ^{13}C (4)

FIGURE 3.26 A selection of stable isotope chemical reagents. The use of these reagents for the determination of differential protein expression is based on the inclusion of "light" or "heavy" isotopes. (1) Gygi, S.P., Rist, B., Gerber, S.A., Turecek, F., Gels, M.H. and Aebersold, R., Quantitative analysis of complex protein mixtures using isotope-coded affinity tags, *Nat. Biotech.*, 17, 994–999, 1999. (2) Chakraborty, A. and Regnier, F.E., Global internal standard technology for comparative proteomics, *J. Chrom. A*, 949, 173–184, 2002. (3) Sebastino, R., Cirreria, A., Lapadula, M. and Righetti. P.G., A new deuterated alkylating agents for quantitative proteomics, *Rapid Comm. Mass Spec.*, 17, 2380–2386, 2003. (4) Kuyama, H., Watanabe, M., Todo, C., Ando, E., Tanaka, K. and Nishimura, O., An approach to quantitative proteome analysis by labeling tryptophan residues, *Rapid Comm. Mass Spec.*, 17, 1642–1650, 2003.

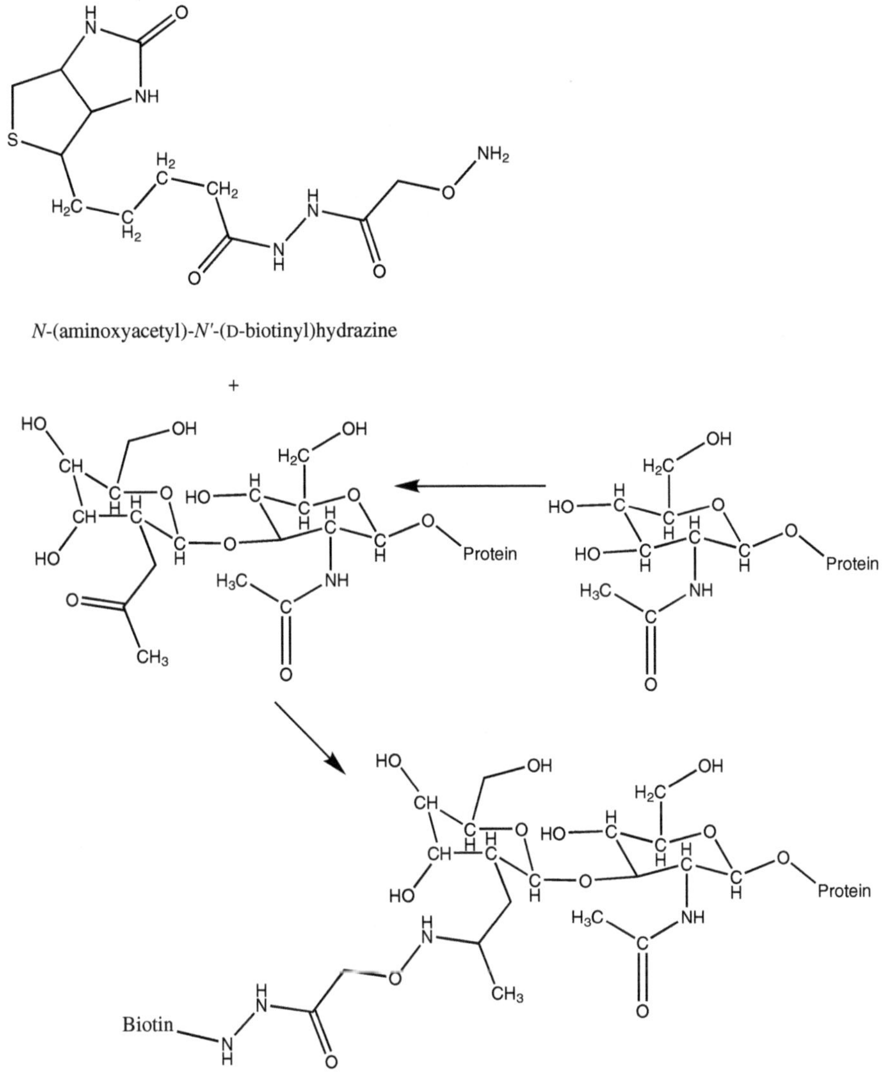

N-(aminoxyacetyl)-*N'*-(D-biotinyl)hydrazine

+

FIGURE 3.27 The combination of a chemical and enzymatic approach to the detection of *O*-acetylglucosamine in proteins.

LABELING OF PROTEIN PRIOR TO FRACTIONATION — FLUORESCENT DYES

Fluorescent dyes have proved extremely useful in proteomics in pre-labeling samples prior to proteomic analysis. The best known example is the use of the Cy™ fluorescent dyes in two-dimensional difference gel electrophoresis (2D-DIGE).[4,191–193] The modification of proteins with fluorescent dyes has a long history in protein chemistry[194] and continues to be used extensively in the site-specific modification

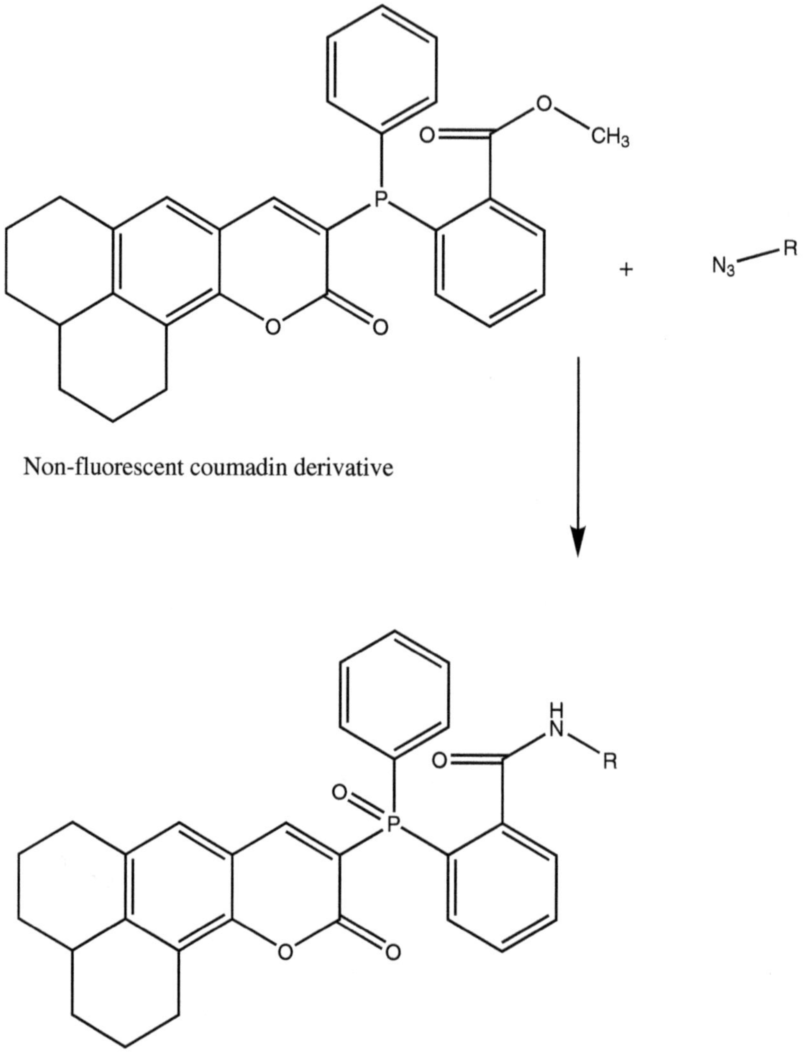

Non-fluorescent coumadin derivative

Fluorescent coumadin derivative

FIGURE 3.28 An example of a fluorogenic dye activated by the staudinger ligation. See Lemieux, G.A., De Graffenried, C.L. and Bertozzi, C.R., A fluorogenic dye activated by the Staudinger ligation, *J. Am. Chem. Soc.*, 125, 4708–4709, 2003.

of proteins.[195] The Cy™ dyes are cyanine dyes (Figure 3.31) which have a long history as photographic sensitizers[196–198] and have more recent uses as probes for biological systems.[199]

Cyanine dyes are one of a series of dyes consisting of two heterocyclic groups (usually a quinoline nucleus) connected by a chain of conjugated double bonds containing an odd number of carbon atoms (Figure 3.31).[198] Their spectral properties are derived from the resonance between the two nitrogen atoms. Distinct advantages

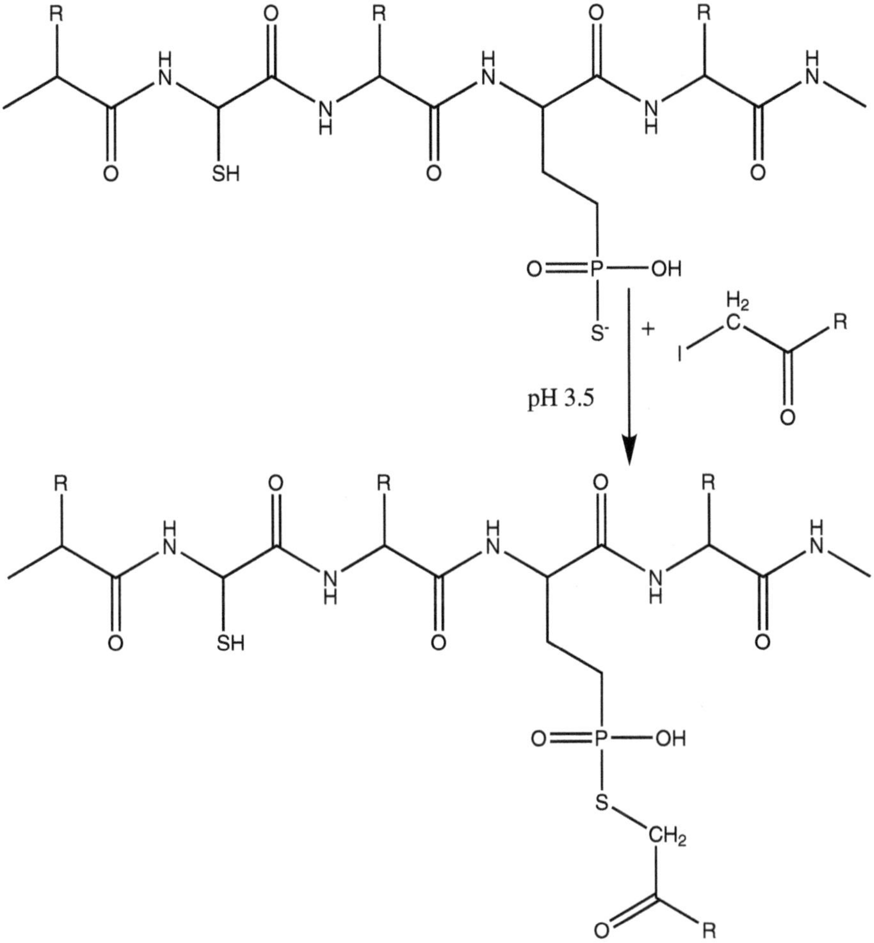

FIGURE 3.29 Modification of a thiophosphorylated serine residues in proteins. Selectively of alkylation is obtained by reaction at low pH.

exist in working at longer wavelengths (> 600 nm) as problems with sample degradation are reduced and sensitivity of measurement is improved.[197]

The signature paper in 2D-DIGE was published in 1997.[4] The general concept is similar to that described above for ICAT and related approaches using stable isotope labeling. Two-dimensional difference gel electrophoresis detects the qualitative difference in the composition of two protein samples with a single two-dimentional electrophoretogram; the qualitative changes are striking while a quantitative measure is more difficult. Two different fluorescent dyes with very similar chemical structures but differing wavelengths for absorption (excitation). Thus, one sample can be labeled with one dye and the other sample with another dye, then combined and subjected to electrophoretic analysis. Excitation at one wavelength will show the proteins labeled in the first sample while excitation at another wavelength will demonstrate the proteins present in the other sample. The reader is directed to a

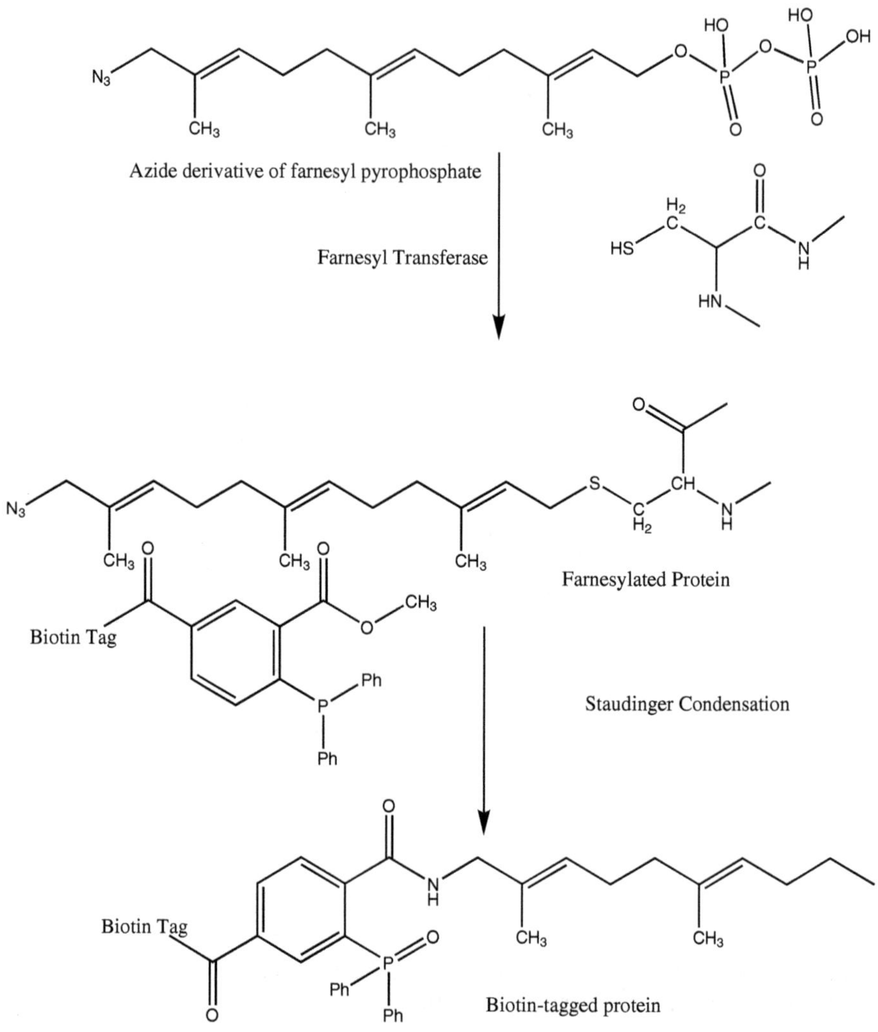

FIGURE 3.30 A process for the detection and isolation of farnesylated proteins. See Kuo, Y. Kim, S.C., Jiang, C. et al., A tagging-via-substrate technology for detection and proteomics of farnesylated proteins, *Proc. Natl. Acad. Sci. USA*, 101, 12479–12484, 2004.

recent review[197] for a more detailed discussion of the experimental techniques. The two dyes used for detection are Cy3 and Cy5. The derivatives shown in Figure 3.32 and Figure 3.33 are the *N*-hydroxysuccinimide derivatives. A third dye, C2 (Figure 3.34), can be used to normalize a series of electrophoretograms.[195] The various cyanine dyes can be used for modification of proteins in capillary electrophoresis[195] and for the modification of nucleic acids.[200,201] Cyanine dyes are also used for the colorimetric determination of nucleic acids by non-covalent binding (Aldrich DISC2(5), Figure 3.35).

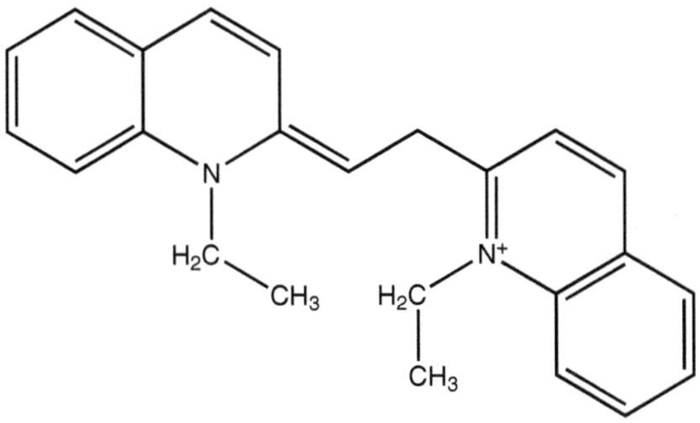

pinacyanal A_{max} = 604 nm

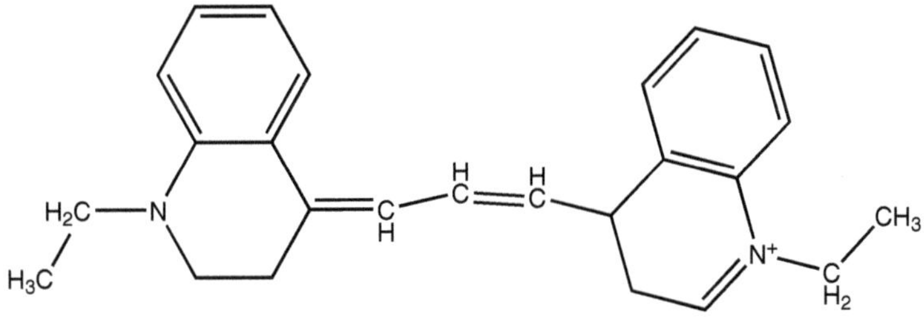

kryptocyanine, A_{max} = 704 nm

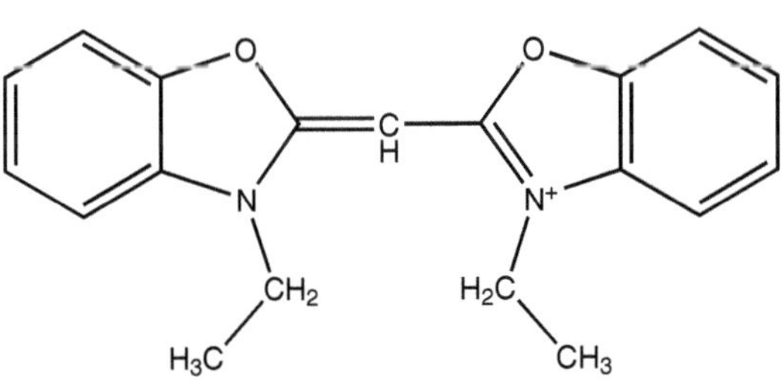

oxacyanine

FIGURE 3.31 Some representative members of the cyanine dye class. Cyanine dyes are one of a series of dyes consisting of two heterocyclic groups (usually quinoline nuclei) connected by a chain of conjugated double bonds containing an odd number of carbon atoms (see *Hawley's Condensed Chemical Dictionary*, 12th Edition, R.J. Lewis, Jr.. Ed., Van Nostrand Reinhold Company, New York, NY, 1993). These dyes are used as sensitizers for photographic emulsions. See also Hamer, F.M., The Cyanine Dyes and Related Compounds, in *Chemistry of Heterocyclic Compounds*, Vol. 18, Interscience (John Wiley) New York, NY, 1964.

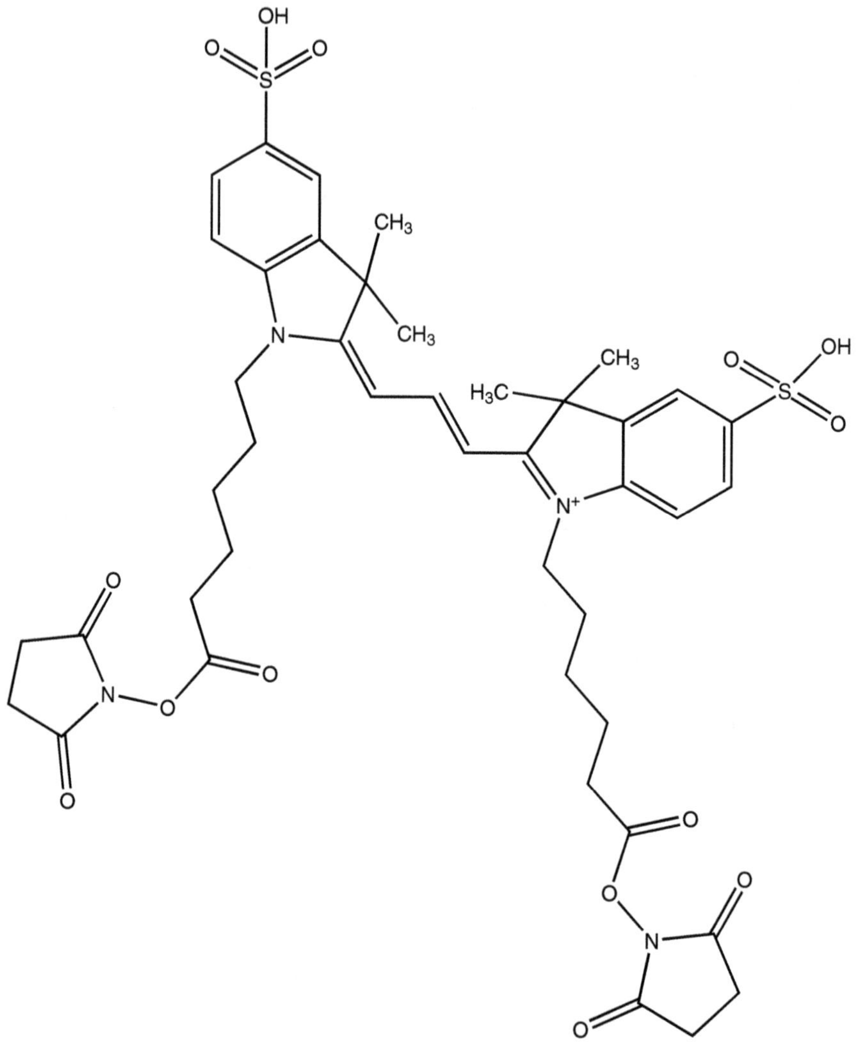

FIGURE 3.32 Cy3, a carboxymethylcyanine dye. The *N*-hydroxysuccinimide form is shown.

The cyanine dye derivatives currently used in proteomics can be traced back to the early work by Waggoner and colleagues at Carnegie Mellon.[202–204] The first compounds were iodoacetamide derivatives designed to react with sulfydryl groups (cysteine residues) in proteins.[202] The nomenclature currently used for these compounds was also advanced in this publication: "XX#ZZ" where XX is the polymethine dye type (Figure 3.36), # is the number of carbon atoms in the conjugated chain and ZZ is the reactive group. Waggoner and coworkers also developed isothiocyanate derivatives[203] that reacted with lysine residues. Later work[204,206] resulted in the synthesis of *N*-hydroxysuccinimide derivatives. More recently, other thiol-reactive Cy3 and Cy5 derivatives have been obtained.[207] Gruber and coworkers[208] have described the preparation of a thiol-reactive Cy5 derivative

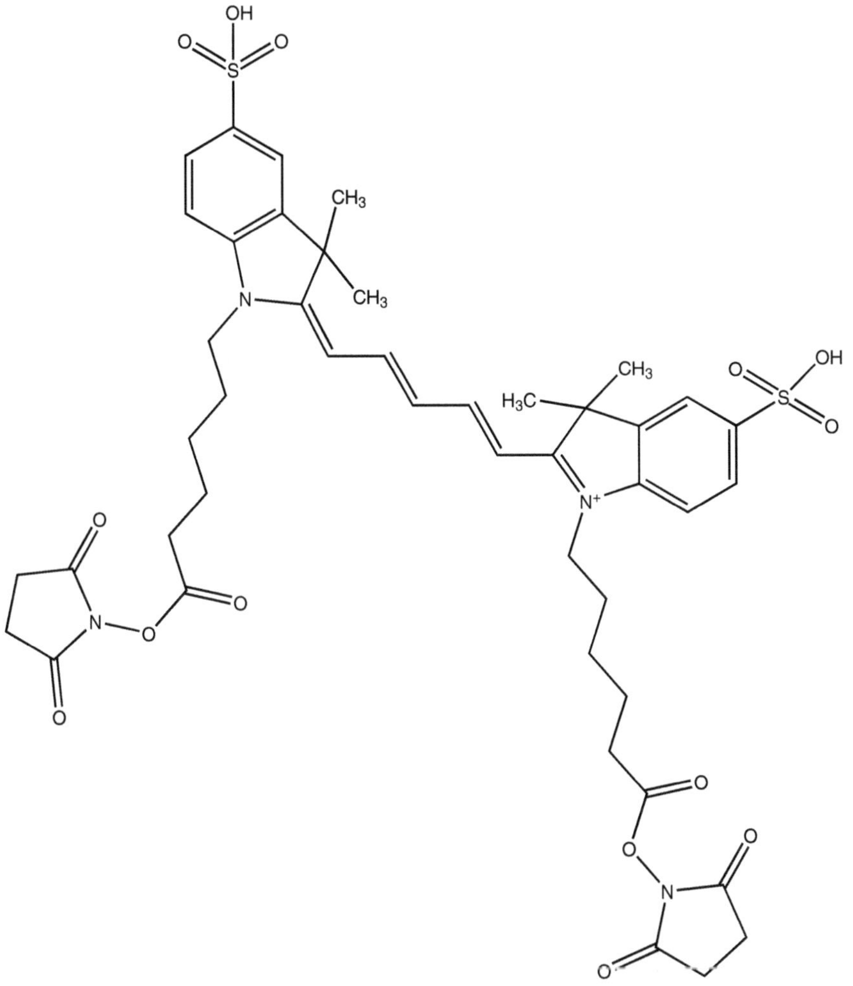

FIGURE 3.33 Cy5, a carboxymethylcyanine dye. The *N*-hydroxysuccinimide form is shown.

(maleimide or pyridyldithio) from the commercially available amine-reactive (succinimidyl) derivative.

2D-DIGE has proved to be a very useful approach for the definition of biomarkers. Tonge and coworkers[209] have used this approach for the study of changes in mouse liver tissue after administration of paracetamol (*N*-acetyl-*p*-aminophenol). Protein determination was performed before transfer of the homogenates to lysis buffer (4 *M* urea, 2 *M* thiourea, 2% CHAPS, 2% sulfobetaine 3-10 and 0.5% Triton X-100). The pH was adjusted to 8.5 with pH 9.5 Tris-HCl and the Cy dye derivatives added to give a ration of 200 pmol dye to 50 µg protein. After mixing, the reaction was allowed to proceed for 10 minutes at 0 to 4°C in the dark. The reaction was terminated and the sample taken to isoelectric focusing, intended to restrict modification to approximately 1 mole per mole protein. This approach is referred to as minimal labeling, as opposed to saturation labeling at cysteine residues described

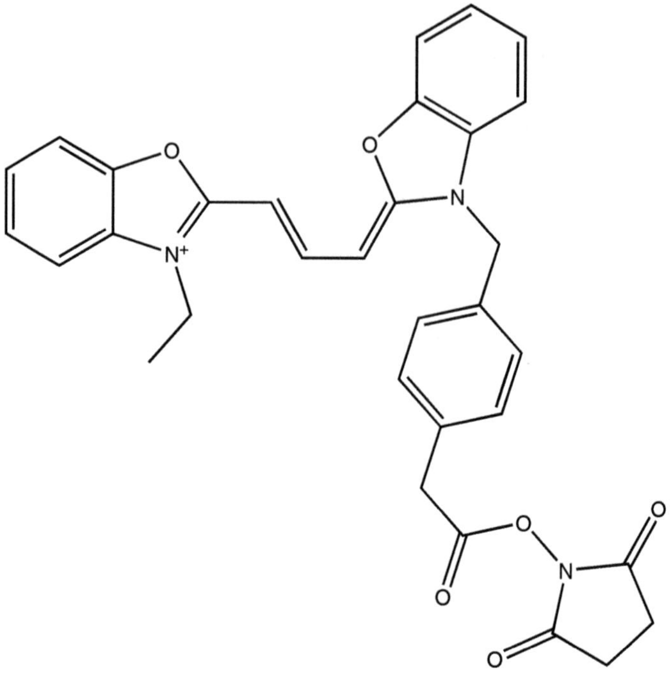

Cy2

FIGURE 3.34 Cy2, a carboxymethylcyanine dye. The *N*-hydroxysuccinimide form is shown.

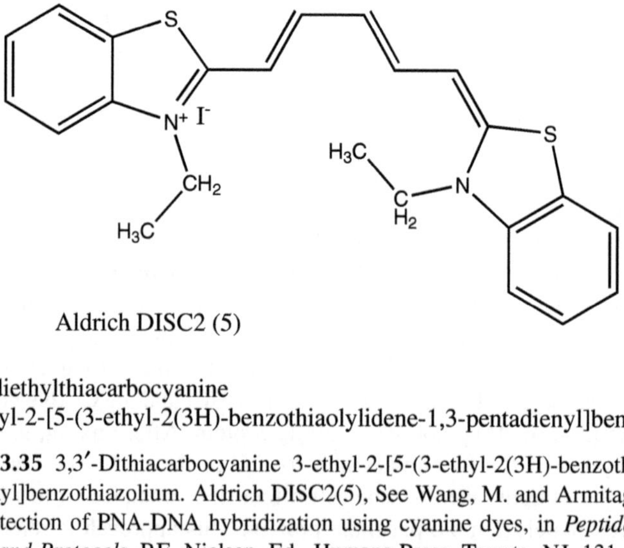

Aldrich DISC2 (5)

3,3'-diethylthiacarbocyanine
3-ethyl-2-[5-(3-ethyl-2(3H)-benzothiaolylidene-1,3-pentadienyl]benzot hiazoliu

FIGURE 3.35 3,3'-Dithiacarbocyanine 3-ethyl-2-[5-(3-ethyl-2(3H)-benzothiaolylidene-1,3-pentadienyl]benzothiazolium. Aldrich DISC2(5), See Wang, M. and Armitage, B.A., Colorimetric detection of PNA-DNA hybridization using cyanine dyes, in *Peptide Nucleic Acids: Methods and Protocols*, P.E. Nielsen, Ed., Humana Press, Towata, NJ, 131–142, 2002.

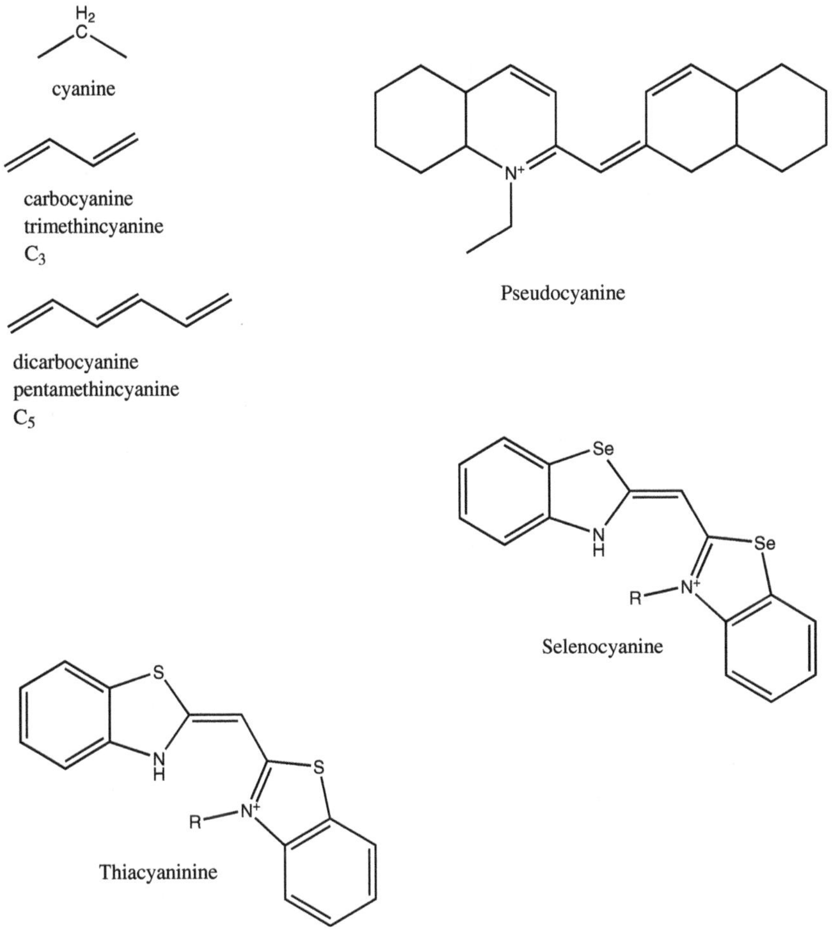

FIGURE 3.36 Some background chemistry on cyanine dyes.

below. This paper also contains relevant information on the properties of the various dyes and are found in Table 3.1.

The AstraZeneca group extended this work with a substantial analysis of acetaminophen in a murine system.[210] The focus in this chapter is the chemistry of the modification and little discussion of use is intended. The reader is, however, directed to a rather nice study by Bahn and coworkers[211] who used the N-hydroxysuccinimide derivatives with 2D-DIGE in a study of human brain tissue. Their use of the Cy2 dye for normalization of electrophoretograms is quite useful. Friedman and coworkers[212] used the triple dye (Cy2, Cy3 and Cy5) with N-hydroxysuccinimide chemistry for the proteomic analysis of human tumor tissue. Homogenates of individual tissue samples were suspended in a Tris-based buffer at pH 7.5 and the supernatant fraction precipitated (after determination of protein concentration with bicinchoninic acid) with chloroform/methanol.[213] Precipitates corresponding to 100 μg protein were solubilized in a urea/thiourea/CHAPS solvent and modified

TABLE 3.1

Properties of Cyanine Dyes Used in Proteomic Research

Property	Dye		
	Cy2	Cy3	Cy5
Excitation (nm)	491	553	645
Emission (nm)	509	569	664

with 200 pmol of Cy3 (normal tissue) or Cy5 (tumor tissue). Each subject served as its own control. A mixture of 50 μg each of the 12 samples (six normal and six colon cancer, 600 μg protein) was modified with 1200 pmol Cy2. The Cy3- and Cy5-labeled samples were combined with an equal amount of the Cy2-labeled material and subjected to two-dimensional electrophoresis. The use of the Cy2 internal standard permitted the identification of some potentially useful biomarkers. Further validation of the triple-dye approach is provided from the work of Lilley and coworkers.[214] Individuals wishing to use the triple-dye approach are strongly suggested to consult this very useful article and the Friedman article with respect to experimental design.

While cyanine dye derivatives reacting at cysteine residues have been available for some time, studies using this technology have only recently appeared. The rationale for the use of saturation labeling at cysteine residues with maleimide derivatives is described by Tonge and coworkers.[215] The modification reaction is performed at pH 7.5 with 20 nmol dye/50 μg protein for 30 minutes at 37°C after reduction of the protein with tri(carboxymethyl)phosphine (TCP). These authors correctly emphasize the necessity of evaluating reaction conditions with each protein sample. While extensive labeling is achieved, complete labeling of all cysteine residues is extremely difficult with these large reagents. Even with smaller derivatives, extended reaction periods are required.[216,217] Maleimide derivatives are a good choice, as haloacetyl derivatives such as those originally developed by Waggoner and colleagues react with thiourea,[218,219] a common component of "lysis" buffers used before the isoelectric focusing step. (Thiourea is included to promote solubilization of membrane proteins.[220]) Hirohashi and colleagues[221] have used saturation labeling to identify biomarkers in tumor tissue following laser microdissection. Finnie and coworkers have shown that reaction with the Cy5 maleimide derivative is more sensitive than monobromobimane in the detection of free thiols in seed proteins.[222]

Other approaches exist to the use of fluorescent probes in proteomic analysis. Monobromobimane (Figure 3.37) has been used to label proteins prior to electrophoresis,[223–226] a useful fluorescent probe which is non-fluorescent until reaction with a nucleophile such as cysteine. This reagent is reasonably specific for cysteine[227]; modification is usually performed at pH 7.0 to 8.0 as the thiolate is the reactive species. While not necessarily relevant under the denaturing solvent conditions usually used for modification prior to the isoelectric focusing step, monobromobimane is an affinity reagent in reaction with some proteins such as glutathione S-transferase.[228] Alexa dyes[229] have spectral properties similar to the Cy

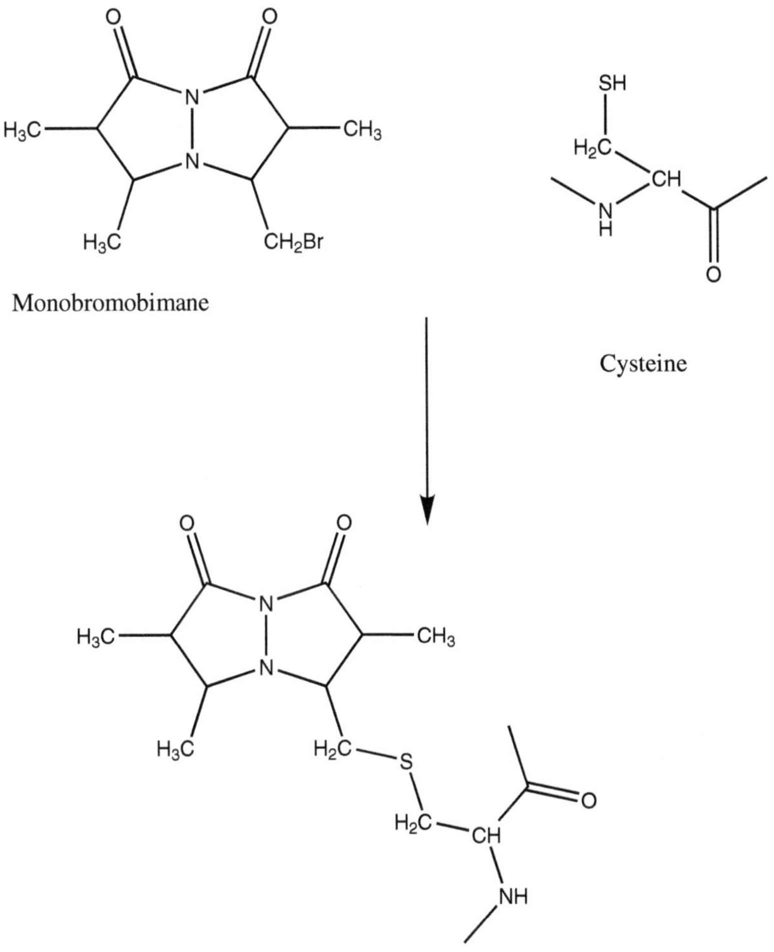

FIGURE 3.37 Monobromobimane and its reaction with cysteine.

dyes and have seen some use in proteomic analysis.[191,230–232] 3-(2-Furoyl)quinoline-2-carboxaldehyde has been used to label proteins for detection in multi-dimensional capillary electrophoresis.[223,224]

Tyagarajan and coworkers[218] have characterized the reaction of several commercially available dyes used for pre-labeling proteins prior to two-dimensional electrophoresis. These dyes would be expected to preferentially react with sulfydryl groups and include BODIPY® TMR Cadavarine IA and BODIPY® Fl Cl IA; maleimide derivatives include Thio Glo 1 and Rhodamine Red C$_2$ maleimide. The structures for these compounds are shown in Figure 3.38. The protein samples were reduced with Tris(carboxyethyl) phosphine (TCEP) in 8.0 M urea, 50 mM Tris, pH 7.5 or pH 8.0 for reaction with the iodoacetamide derivatives. Reaction with the maleimide derivatives was performed at pH 7.2 and the reactions were quenched

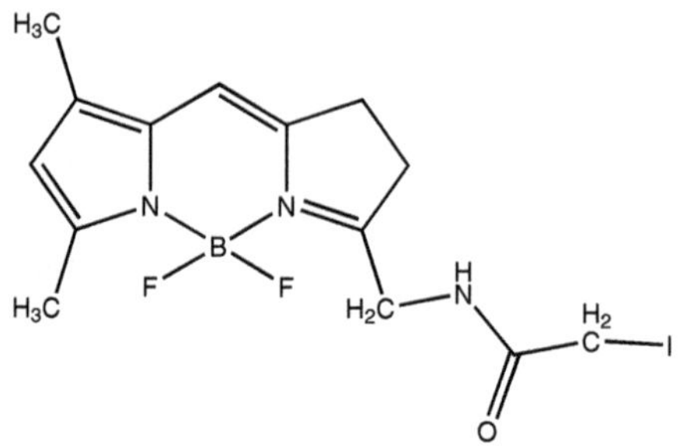

I N-(4.4-difluro-5.7-dimethyl-4-bora-3a,4a-diaza-s-indacene-3-yl)methyl)

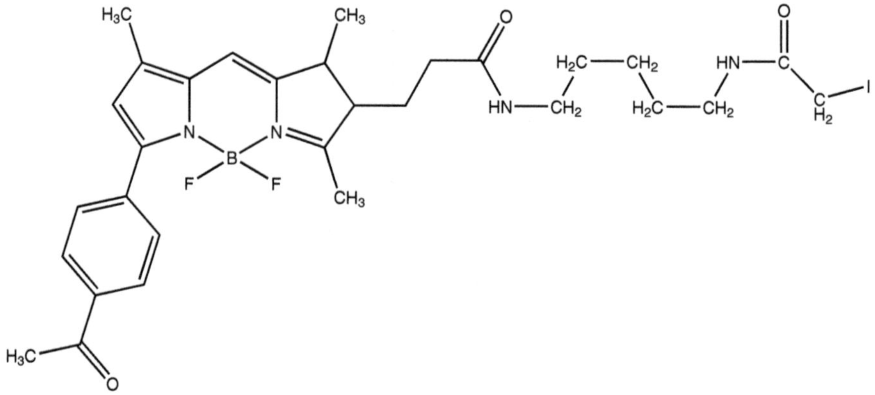

II N-(5-((4,4-difluoro-1,3-dimethyl-5-(4-methoxyphenyl)-4-bora-3a,4a-di aza-s-indecene-
2-propionyl)amino)pentyl)idoacetamide

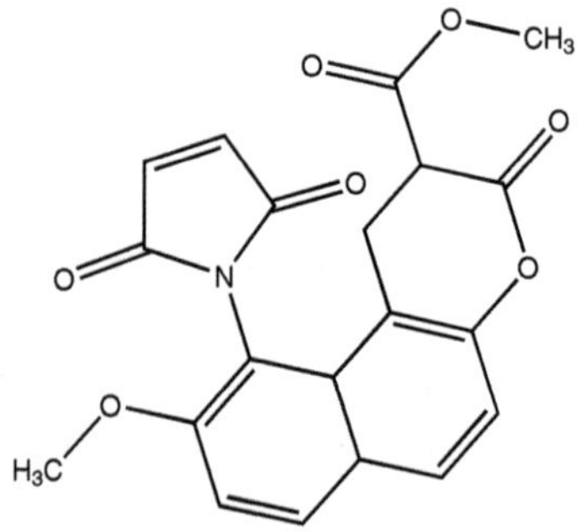

Thio Glo I; 10-(2,5-dihydro-2,5-dioxo-(1H-pyrrol-1-yl-9-methoxy-3-oxom ethyester 3H napthol [2,1-b]
pyran-s-carboxylic acid

FIGURE 3.38 BODIPY and ThioGlo®.

with 2-mercaptoethanol. The following observations are of critical importance in the use of dyes reacting with sulfydryl groups in proteins:

- The iodoacetamide derivatives retain specificity for modification of cysteine at high dye/protein ratios; the maleimide derivatives appear to be somewhat less specific.
- TCEP at 1 mM does not appear to influence the reaction of either iodoacetamide or maleimide derivatives with proteins; reduction of modification is observed at 2.5 mM TCEP.
- As observed by others,[235] the iodoacetamide derivates react with thiourea reduced modification of proteins. Reaction of maleimide derivatives with thiourea is not observed. Thiourea is included on occasion to promote solubilization of membrane proteins.[236]

Imai and coworkers[237] have proposed the use of 4-(dimethylaminoethyl-aminosulfonyl)-7-chloro-2,1,3-benzoxadiazole (DAABD-Cl) and 7-chloro-2,1,3-benzoxadiazole-4-sulfonylaminoethyltrimethylammonium chloride (TAABD-Cl) as fluorogenic probes (Figure 3.39) for use in proteomics. Their major use would be in detection in multi-dimensional HPLC analyses. These reagents are not necessarily specific for cysteine[238] but will be useful with a careful choice of reaction conditions

ENZYMATIC MODIFICATION OF TARGETS FOR PROTEOMIC ANALYSIS

Another related approach uses intrinsic tissue transglutaminases activity to modify specific peptide sequences.[239] A discussion of the usefulness of site-specific enzyme modification is included as it is considered complementary to the chemical approaches described above. Transglutaminases catalyze the formation of an isopeptide bond between a glutamine residue and a lysine residue (Figure 3.40). Blood Coagulation Factor XIII/XIIIa (fibrin-stabilizing factor) is an example of a transglutaminases; factor XIIIa "stabilizes" fibrin clots by forming cross-links between specific regions in fibrin monomers.[240,241] Tissue transglutaminases serve a variety of functions including adhesion and maturation of matrix proteins.[242,243] Transglutaminases have a high specificity for glutamine residues but are considerably more promiscuous with respect to the amine donor. For example, transglutaminases can catalyze the incorporation of dansylcadavarine into proteins (Figure 3.41).[244] Esposito and colleagues[245] have used two approaches to identify transglutaminase substrates by proteomic technology. 5-(Biotinamido) pentylamine (biotinylated cadavarine) (Figure 3.42) was used to identify glutamine sites while a biotinylated peptide (biotin-TVQQEL-OH) was used to identify lysine residues. A mixture of seminal vesicle protein IV, vasoactive intestinal peptide and β-casein was used as the test mixture. After transglutaminase treatment, the modified proteins were separated by reversed-phase HPLC. After hydrolysis with trypsin, the peptides were separated by HPLC and submitted to MS/MS analysis. In a subsequent study,[246] this group used the same technical approach to identify transglutaminase substrates in human intestinal epithelial cells

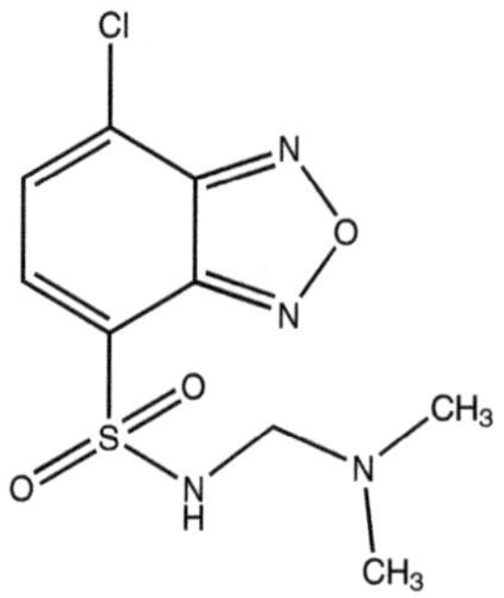

DAABD-Cl;[4-(dimethylaminoethylaminosulfonyl)-7-chloro-2,1,3-benzoxadiazole]

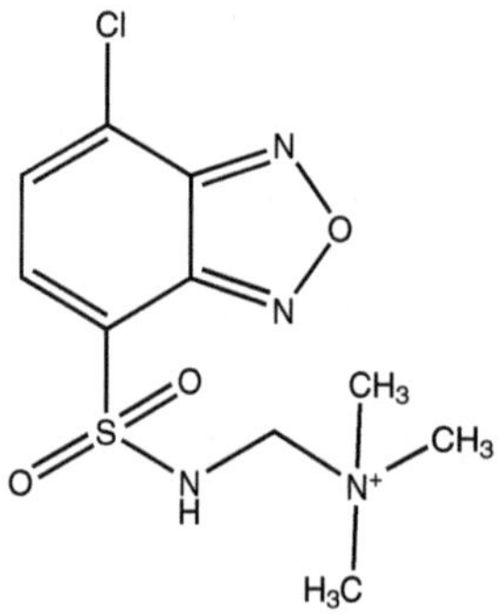

TAABD-Cl; [7-chloro-2,1,3-benzoxadiazole-4-sulfonylaminoethyltrimethylammonium chlorid e

FIGURE 3.39 4-(Dimethylaminoethylaminosulfonyl)-7-chloro-2,1,3-benzoxadiazole (DAABD-CL) and 4-(trimethylammoniumethylaminosulfonyl)-7-chloro-2,1,3-benzoxadiaz-ole (TAABD-Cl). Fluorogenic reagents for labeling cysteine peptides prior HPLC/MS analysis. See Matsuda, M. et al., Fluorogenic derivatization reagents suitable for isolaton and identification of cysteine-containing proteins utilizing high-performance liquid chromatograpy-tandem mass spectrometry, *Anal. Chem.*, 76, 728–735, 2004.

as part of a study on celiac disease. Transglutaminase is thought to be involved in the etiology of celiac disease.[247] In the above study identifying transglutaminase substrates in human intestinal epithelial cells,[247] more than 25 proteins were identified. The reader is directed to several recent publications[248,249] for information on technical approaches to the study of transglutaminase substrates.

PROTEIN MICROARRAYS

Nucleic acid-based microarrays (DNA microarrays) have been useful in measuring gene expression via transcription. While DNA microarray technology is distinct from genome sequence analysis, technical applications depend on the knowledge of linear gene structure.[250] DNA microarrays measure gene function by measuring RNA

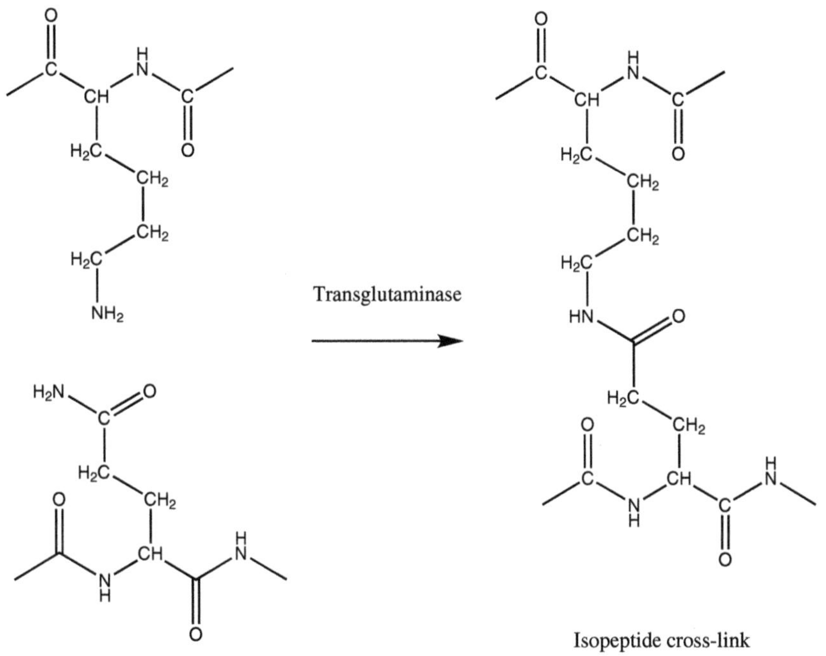

Lysine

Transglutaminase

Isopeptide cross-link

Glutamine

FIGURE 3.40 The formation of an isopeptide bond between a lysine and glutamine in a protein.

formation.[251] Two general experimental approaches exist to transcriptional analysis (transcriptomics). One involves the direct use of DNA microarrays to measure the population of messenger RNA at a point in time via hybridization to a DNA microarray.[252] The other, serial analysis of gene expression (SAGE),[253] is based on the principle that a relative nucleotide sequence contains sufficient information to identify a gene and that concatenation of these short sequence tags will allow the efficient analysis of a population of transcripts. Two excellent recent reviews in this area are worth serious consideration.[254,255] While there may or may not be a direct relationship between transcription and translation,[256,257] DNA microarray analysis is definitely useful for the analysis of susceptibility and prognosis. Ample rationale is also found for the comparison of DNA microarray data and proteomics.[258,259]

DNA microarrays are "easy" in the sense that the rules for hybridization between polynucleotides is well-understood, permitting the relatively facile manufacture of large microarrays with high specificity.[260,261] Unfortunately, the same is not true for proteins where interactions are not as predictable and specific probes, such as antibodies or antibody fragments, are not as resilient as oligonucleotides.[262]

Notwithstanding the formidable technical issues such as probe selectivity, probe stability and the comparative paucity of probes, substantial interest exists in the development of microarrays as an analytical tool for proteomic research. In principle,

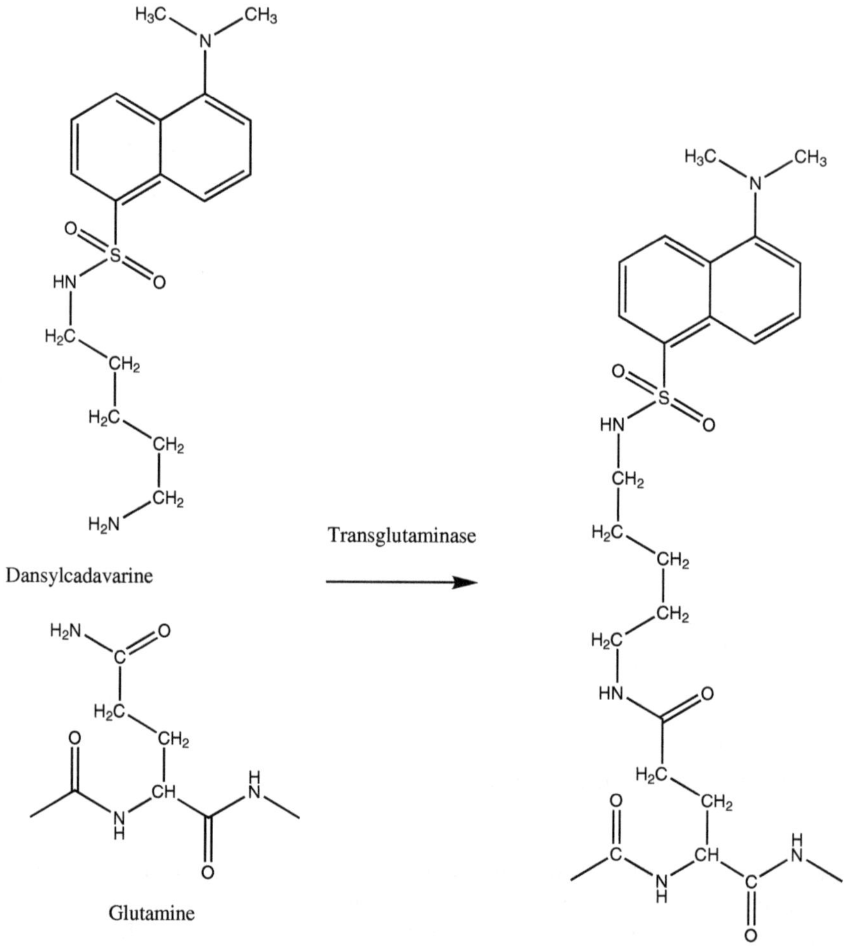

FIGURE 3.41 The incorporation of dansylcadavarine into a protein via the action of transglutaminase. This forms the basis for an assay for transglutaminase.

this development would allow the high-throughput screening of proteins in complex samples. In simplest terms, a probe is bound to a surface and the probe will specifically bind to an analyte and the analyte:probe complex will be detected, permitting the identification of the analyte.[262,263] In the process of developing these assays for potential commercial use, the probes are most likely analyte specific reagents (ASR).[264–266] Regardless of regulatory issues, the success of microarray technology will depend on the number of analyte-specific probes. A variety of approaches exist including antibodies and antibody fragments,[267,268] aptamers,[269] somewhat simple synthetic organic compounds[270,271] or more complex structures derived from combinatorial chemistry[272,273] or phage display.[274,275] The latter technologies are of particular value in the preparation of high-throughput systems. For the purpose of the present discussion, chemical proteomics does not include protein/antibody microarray technology. The basic chemistry used for the attachment of proteins/antibodies

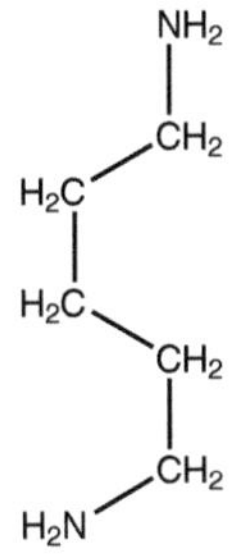

Cadaverine (1,5-pentanediamine

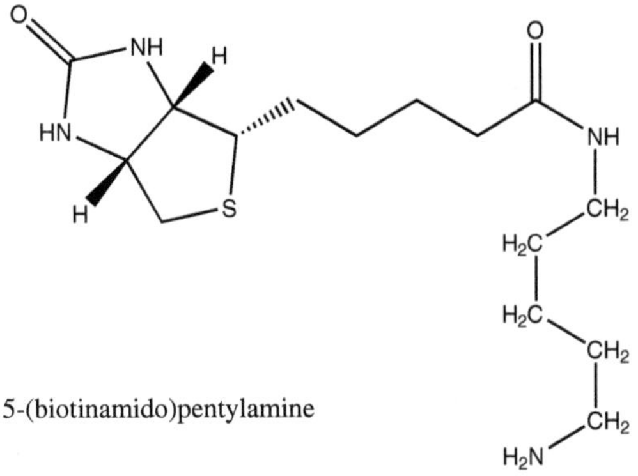

5-(biotinamido)pentylamine

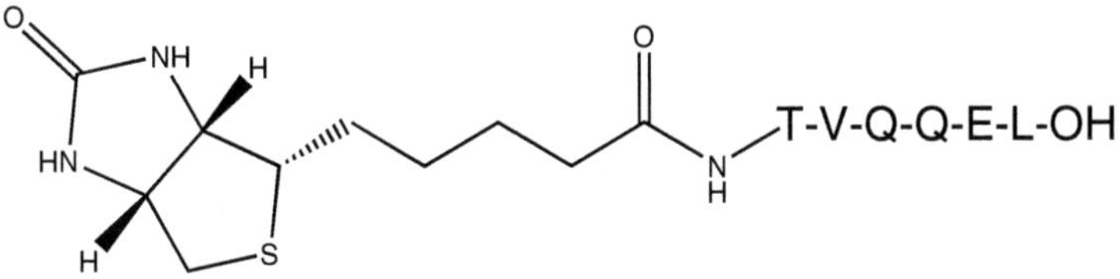

T-V-Q-Q-E-L-OH

FIGURE 3.42 Some donors and acceptors for transglutaminase which are useful in proteomic analysis. See Ruoppolo, M., Orru, S., D'Amato, A. et al., Analysis of transglutaminase protein substrates by functional proteomics, *Protein Sci.*, 14, 1290–1297, 2003.

to microarray matrices is discussed in Chapter 2. A related area is enzyme micro-kinetics, which could be used for high-throughput screening in a microarray format.[276] Pohl and coworkers[277] have developed microarrays with carbohydrate libraries for the study of glycosidase activity. Products of the reactions are analyzed by mass spectrometry and this provides a considerable advantage over substrates based on the release of *p*-nitrophenol. However, newly developed synthetic substrates have been developed[278] that show considerable promise for a broad spectrum of enzymes. These include fluorogenic and chromogenic substrates that generate a signal by β-elimination. With the anticipated density of microarrays, more sophis-ticated methods of analysis will be required.[279–283] Investigators (and investors) need to clearly understand that protein microarrays will be far more difficult than nucleic acid-based microarrays. Investigators are also strongly advised to carefully consider statistical design in the design of their analytical matrices.[284]

Much of the basic technology underlying the development of protein microarray assays is, in fact, based on earlier work with solid-phase assay systems. Most notable are the traditional enzyme-linked immunoassays. While these systems are mostly larger than the current "macrosystems,"[285] an understanding of the development of

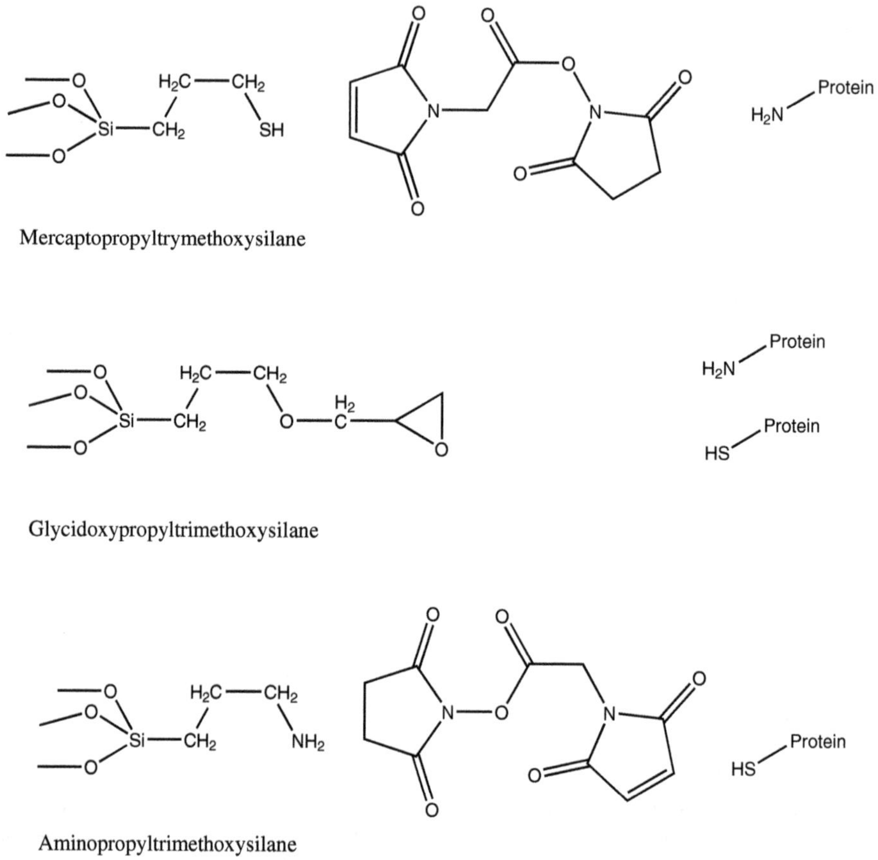

Mercaptopropyltrymethoxysilane

Glycidoxypropyltrimethoxysilane

Aminopropyltrimethoxysilane

FIGURE 3.43 Trimethoxysilane chemistry useful in building protein array matrices.

this technology is useful.[286–293] The reader is also referred to several recent articles on solid-phase assay systems.[294–297]

The above discussion was intended to provide a brief cursory overview of protein microarray technology. The rest of this section will concern some of the chemistry involved with coupling probes to matrices. The chemistry is neither new nor novel and the reader is directed to several recent reviews.[19,298] A consideration of the wealth of technical data in bioconjugate chemistry[299] will be useful. The following discussion will mention several studies that are useful for consideration.

Cahill and coworkers have carefully examined the various matrix materials available for the production of antibody microarrays.[300] Both noncovalent and covalent binding modes are examined. An aldehyde-containing matrix that would react with protein amino groups was useful. In addition to the aldehyde matrix, a polystyrene matrix, a polylysine matrix, a modified polyacrylamide matrix and a 2-aminopropyltriethoxysilane matrix were examined. In total, 11 different array surfaces were examined with five different antibodies. Kusnezow and

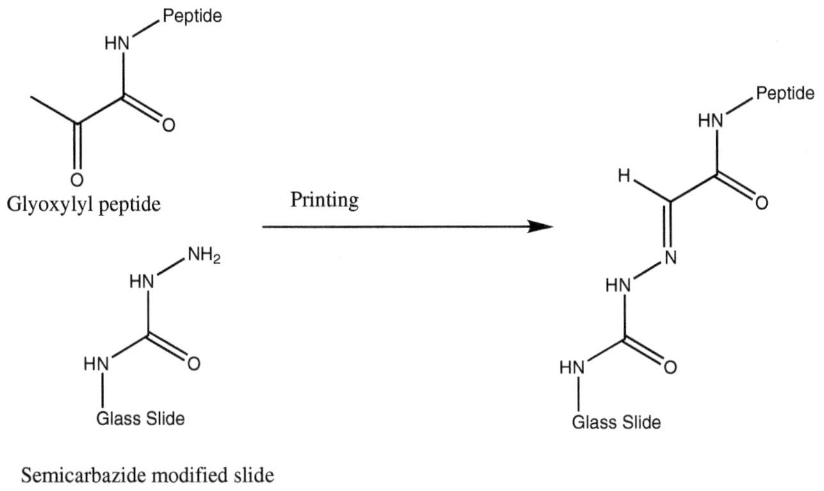

FIGURE 3.44 Semicarbizide coupling to glyoxylyl peptides in the construction of protein microarray matrices.

coworkers[301] have evaluated various production parameters in the preparation of antibody microarrays, including modification of the glass slide surface, linker quality, pH and antibody concentration. Chemistries included amine coupling (aminopropyl-trimethoxysilane with N-hydroxysuccinimide and N-ethylmaleimide for coupling to protein), epoxide (glycidoxypropyltrimethoxysilane) and sulfydryl (mercaptopropyl-trimethoxysilane with N-ethylmaleimide and N-hydroxysuccinimide for coupling to protein amino groups (see Figure 3.43). Nock and coworkers[302] modified IgG and Fab fragments with biotin and subsequently bound these modified proteins to a streptavidin matrix to optimize orientation of the binding domain. A clever approach that does not utilize chemical modification places a hexahistidine "tail" on an engineered protein A that is then used for the immobilization of IgG on a histidine matrix,[303] permitting optimal presentation of the binding site. This approach would have value for use with other recombinant probes.

Melnyk and coworkers[304] have used coupling of glyoxylyl peptides to semicarbazide-modified glass slides to prepare microarrays (Figure 3.44), permitting site-specific ligation of the peptides. These microarrays can be printed using commercially available instrumentation (Affymetrix, Zenopore). DeNardo and coworkers[305] produced recombinant single chain Fv fragments with a terminal free thiol for coupling to a matrix. An approach using intein-mediated coupling of a C-terminal cysteine residue for subsequent attachment of an antibody fragment to a biochip has been reported by Sidor and coworkers.[306] Gold is frequently used as a matrix for biosensors. The process of protein coupling to a matrix via "cross-linkage" process as developed by Butt and coworkers[307] is shown in Figure 3.45.

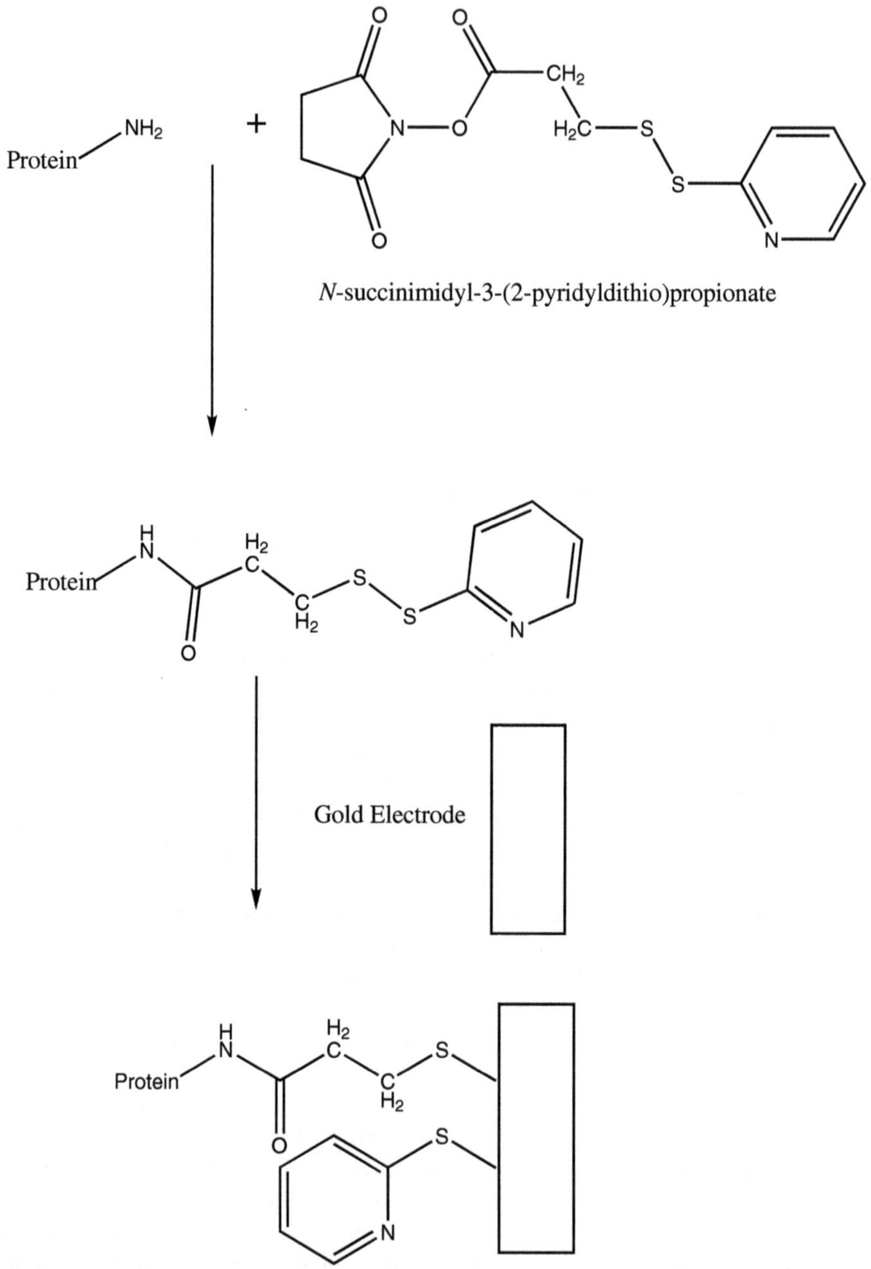

FIGURE 3.45 Gold electrode technology for sensor design: Protein/peptide coupling via amino groups after chemical modification. This technology could also be used for free sulfydryl groups and disulfide bonds.

REFERENCES

1. Smolka, M.B., Zhou, H., Purkayastha, S. and Aebersold, R., Optimization of the isotope-coded affinity tag-labeling procedure for quantitative proteome analysis, *Anal. Biochem.*, 297, 25, 2001.
2. Watt, S.A., Patschkowski, T., Kalinowski, J. and Niehaus, K., Qualitative and quantitative proteomics by two-dimensional gel electrophoresis peptide mass fingerprinting and a chemical coded affinity tag (CCAT), *J. Biotech.*, 106, 287, 2003.
3. Parker, K.C., Patterson, D., Williamson, B. et al., Depth of proteome issues: A yeast isotope-coded affinity tag reagent study, *Mol. Cell. Proteomics*, 3, 625, 2004.
4. Ünlü, M., Morgan, M.E. and Minden, J.S., Difference gel electrophoresis: A single gel method for detecting changes in protein extracts, *Electrophoresis*, 18, 2071, 1997.
5. Tyagarajan, K., Pretzer, E. and Wiktorowicz, J.E., Thiol-reactive dyes for fluorescence labeling of proteomic samples, *Electrophoresis*, 24, 2346, 2003.
6. Jeffrey, D.A. and Bogyo, M., Chemical proteomics and its application to drug discovery, *Curr. Opin. Biotech.*, 14, 87, 2003.
7. Kocks, C., Maehr, R., Overlkeeft, H.S. et al., Functional proteomics of the active cysteine protease content in Drosophila S2 cells, *Mol. Cell. Proteomics*, 2, 1188, 2003.
8. Hemelaar, J., Galardy, P.J., Borodovsky, A. et al., Chemistry-based functional proteomics: Mechanism-based activity-profiling tools for ubiquitin and ubiquitin-like specific proteases, *J. Proteome Res.*, 3, 268, 2004.
9. Spears, A.E. and Cravatt, B.F., Chemical strategies for activity-based proteomics, *ChemBioChem*, 5, 41, 2004.
10. Figeys, D., Novel approaches to map protein interactions, *Curr. Opin. Biotech.*, 14, 119, 2003.
11. Cuatrecasas, P., Wilchek, M. and Anfinsen, C.B., Selective enzyme purification by affinity chromatography, *Proc. Nat. Acad. Sci. USA*, 61, 636, 1968.
12. Porath, J. and Kristiansen, T., Biospecific affinity chromatography and related methods, in *The Proteins*, 3rd Edition, H. Neurath and R.L. Hill, Eds., Academic Press, New York, NY, Vol. 1, Ch. 2, 95, 1975.
13. Lolli, G., Thaler, F., Valsanina, B. et al., Inhibitor affinity chromatography: Profiling the specific reactivity of the proteome with immobilized molecules, *Proteomics*, 3, 1287, 2003.
14. Lee, W.-C. and Lee, K.H., Applications of affinity chromatography in proteomics, *Anal. Biochem.*, 324, 1, 2004.
15. Oda, Y., Owa, T., Sato, T. et al., Quantitative chemical proteomics for identifying candidate drug targets, *Anal. Chem.*, 75, 2159, 2003.
16. Lefkovits, I., Functional and structural proteomics: A critical appraisal, *J. Chrom. B*, 787, 1, 2003.
17. Yanagadi, M., Functional proteomics; current achievements, *J. Chrom.*, 771, 89, 2002.
18. Righetti, P.G., Campostrini, N., Pascali, J., Herndon, M. and Astner, H., Quantitative proteomics: A review of the different methodologies, *Eur. J. Mass Spec.*, 10, 335, 2004.
19. Masselson, C., Paša-Toli, L., Toli, N. et al., Targeted comparative proteomics by liquid chromatography-tandem Fourier ion cyclotron resonance mass spectrometry, *Anal. Chem.*, 77, 400, 2005.
20. Kubota, K., Kosaka, T. and Ichikawa,, K., Combination of two-dimensional electrophoresis and shotgun peptide sequences in comparative proteomics, *J. Chrom. B*, 815, 2, 2005.

21. Whiteleggee, J.P., Mass spectrometry for high throughput quantitative proteomics in plant research: Lessons from thylakoid membranes, *Plant Phys. Biochem.*, 42, 919, 2004.
22. Adam, G.C., Sorensen, E.J. and Cravatt, B.F., Proteomic profiling of mechanistically distinct enzyme classes using a common chemotype, *Nat. Biotech.*, 20, 805, 2002.
23. Burbaum, J. and Topal, G.M., Proteomics in drug discovery, *Curr. Opin. Chem. Biol.*, 6, 427, 2002.
24. Jessani, N., Liu, Y., Humphrey, M. and Cravatt, B.F., Enzyme activity profiles of the secreted and membrane proteome that depict cancer cell invasiveness, *Proc. Natl. Acad. Sci. USA*, 99, 10335, 2002.
25. Kozarich, J.W., Activity-based proteomics: Enzyme chemistry redux, *Curr. Opin. Chem. Biol.*, 7, 78, 2003.
26. Speers, A.E. and Cravatt, B.F., Chemical strategies for activity-based proteomics, *ChemBioChem*, 5, 41, 2004.
27. Kumar, S., Zhou, B., Liang, F. et al., Activity-based probes for protein tyrosine phosphatases, *Proc. Natl. Acad. Sci. USA*, 101, 7943, 2004.
28. van Swieten, P.F., Maehr, R., van den Nieuwendijk, A.M. et al., Development of an isotope-coded activity-based probe for the quantitative profiling of cysteine proteases, *Bioorg. Med. Chem. Letters*, 14, 3131, 2004.
29. Kridel, S.J., Alexrod, F., Rozenkrantz, N. and Smith, J.W., Orlistat is a novel inhibitor of fatty acid synthase with antitumor activity, *Cancer Res.*, 64, 2070, 2004.
30. Patricelli, M.P., Activity-based probes for functional proteomics, *Brief Funct. Genomic Proteomic*, 1, 151, 2002.
31. Wang, S., Sim, T.B., Kim, Y.S. and Chang, Y.T., Tools for target identification and validation, *Curr. Opin. Chem. Biol.*, 8, 371, 2004.
32. Comess, K.M. and Schurdak, M.E., Affinity-based screening techniques for enhancing lead discovery, *Curr. Opin. Drug. Disc. Dev.*, 7, 411, 2004.
33. Plapp, B.V., Application of affinity labeling for studying structure and function of enzymes, *Meth. Enzymology*, 87, 469, 1982.
34. Colman, R.F., Affinity labeling of purine nucleotide sites in proteins, *Ann. Rev. Biochem.*, 52, 67, 1983.
35. Sweet, F. and Murdock, G.L., Affinity labeling of hormone-specific proteins, *Endocrin. Rev.*, 8, 154, 1987.
36. Cooperman, B.S., Affinity labeling of ribosomes, *Meth. Enzymol.*, 164, 341, 1988.
37. Ji, T.H., Nishimura, R. and Ji, I., Affinity labeling of binding proteins for the study of endocytic pathways, *Meth. Cell. Biol.*, 32, 277, 1989.
38. Hohenegger, M., Freissmuth, M. and Nanoff, C., Covalent modification of G proteins by affinity labeling, *Meth. Mol. Biol.*, 83, 179, 1997.
39. Marcotte, P. and Walsh, C., Vinylglycinie and proparglyglycine — Complementary suicide substrates for L-amino acid oxidase and D-amino acid oxidase, *Biochemistry*, 15, 3070, 1976.
40. Hanzlik, R.P., Kishore, V. and Tullman, R., Cyclopropylamides as suicide substrates for cytochromes P-450, *J. Med. Chem.*, 22, 759, 1979.
41. Alston, T.A., Suicide substratesfor mitochondrial enzymes, *Pharm. Ther.*, 12, 1, 1981.
42. Walsh, C.T., Suicide substrates, mechanism-based enzyme inactivations: Recent developments, *Ann. Rev. Biochem.*, 53, 493, 1984.
43. Decodts, G. and Wakselman, M., Suicide inhibitors of proteases. Lack of activity of halomethyl derivatives of some aromatic lactams, *Eur. J. Med. Chem.*, 18, 107, 1983.
44. Chowdhry, V. and Westheimer, F.H., Photoaffinity labeling of biological systems, *Ann. Rev. Biochem.*, 48, 293, 1979.

45. Dorman, G. and Prestwich, G.D., Using photolabile ligands in drug discovery and development, *Trends Biotech.*, 18, 64, 2000.
46. Evans, S.J. and Moore, F.L., Nonradioactive photoaffinity labeling of steroid receptors using western blot detection systems, *Meth. Mol. Biol.*, 76, 261, 2001.
47. Hatanaka, Y. and Sadkane, Y., Photoaffinity labeling in drug discovery and developments: Chemical gateway for entering proteomic frontier, *Curr. Topics Med. Chem.*, 2, 271, 2002.
48. Gartner, C.A., Photoaffinity ligands in the study of cytochrome p450 active site structure, *Curr. Med. Chem.*, 10, 671, 2003.
49. Knorre, D.G. and Godovikova, T.S., Photoaffinity labeling as an approach to study supramolecular structure, *FEBS Letters*, 433, 9, 1998.
50. Karwatsky, J.M. and Georges, E., Drug binding domains of MRP1 (ABCC1) as revealed by photoaffinity labeling, *Curr. Med. Chem. Anti-Cancer Agents*, 4, 19, 2004.
51. Braunschweig, T., Chung, J.Y. and Hewitt, S.M., Perspectives in tissue microarrays, *Comb. Chem. High Throughput Screen*, 7, 575, 2004.
52. Asen, A., Progress in focus: Recent advances in histochemisry and cell biology, *Histochem. Cell. Biol.*, 118, 507, 2002.
53. Brown, R.E. and Boyle, J.L., Mesenchymal chondrosarcoma: Molecular characterization by a proteomic approach with morphogenic and therapeutic implications, *Ann. Clin. Lab. Sci.*, 33, 131, 2003.
54. Mobasheri, A., Airley, R., Foster, C.S. et al., Post-genomic applications of tissue microarrays: Basic research, prognostic oncology, clinical genomics and drug discovery, *Histol. Histopath.*, 19, 325, 2004.
55. Gygli, S.P., Rist, B., Gerber, S.A. et al., Quantitative analysis of complex protein mixtures using isotope-coded affinity tags, *Nature Biotech.*, 17, 994, 1999.
56. Mann, M., Quantitative proteomics? *Nat. Biotech.*, 17, 954, 1999.
57. Jansen, E.P., Nutting, M.D.F., Jong, R. and Balls, A.K., Inhibition of the proteinase and esterase activities of trypsin and chymotrypsin by diisopropyl fluorophosphate: Crystallization of the inhibited chymotrypsin, *J. Biol. Chem.*, 179. 189, 1949.
58. Oosterbaan, R.A., Kunst, P. and Cohen, J.A., The nature of the reaction between diisuprapyl fluorophosphate and chymstrypsin, *Biochim. Biophys. Acta*, 16, 299, 1955.
59. Jansen, E.P., Nutting, M.-D.F. and Balls, A.K., Mode of inhibition of chymotrypsin by diisopropyl fluorophosphate, I. Introduction of phosphorous, *J. Biol. Chem.*, 179, 201, 1949.
60. Bender, M.L. and Clement, G.E., Kinetic evidence for an acyl-enzyme intermediate in the alpha-chymotrypsin-catalyzed hydrolysis of *N*-acetyl-L-tryptophan ethyl ester, *Biochem. Biophys. Res. Comm.*, 12, 339, 1969.
61. Greenzaid, P. and Jencks, W.P., Pig liver esterase. Reactions with alcohols, structure-reactivity correlations, and the acyl-enzyme intermediate. *Biochemistry*, 10, 1210, 1971.
62. Miller, C.G. and Bender, M.L., The direct observation of an acyl-enzyme intermediate in the alpha-chymotrypsin-catalyzed hydrolysis of a specific substrate at neutral pH, *J. Am. Chem. Soc.*, 90, 6850, 1968.
63. Kirby, E.P., Niewiarowski, S., Stocker, K. et al., Thrombocytin, a serine protease from *Bothrops atrox* venom. 1. Purification and characterization of the enzyme, *Biochemistry*, 18, 3564, 1979.
64. Cook, R.R. and Powers, J.C., Benzyl-*p*-guanidinothiobenzoate hydrochloride, a new active-site titrant for trypsin and trypsin-like enzymes, *Biochem. J.*, 215, 287, 1983.
65. Radisky, E.S. and Koshland, D.E., Jr., A clogged gutter mechanism for serine proteases, *Proc. Natl. Acad. Sci. USA*, 99, 10316, 2002.

66. Mazur, A. and Bodensky, O., The mechanism of *in vitro* and *in vivo* inhibition of cholinesterase activity by diisopropyl fluorophosphate, *J. Biol. Chem.*, 163, 261, 1945.

67. McAndrew, R.S. and Chapman, K.D., Enzymology of cottonseed microsomal *N*-acylphosphatidylethanolamine synthase: Kinetic properties and mechanism-based inactivation, *Biochim. Biophys. Acta*, 1390, 21, 1998.

68. Liu, Y., Patricelli, M.P. and Cravatt, B.F., Activity-based protein profiling: The serine hydrolases, *Proc. Natl. Acad. Sci. USA*, 96, 14694, 1999.

69. Kidd, D., Liu, Y. and Cravatt, B.F., Profiling serine hydrolase activity in complex proteomes, *Biochemistry*, 40, 4005, 2001.

70. Kridel, S.J., Axelrod, F., Rozenkrantz, N. and Smith, J.W., Orlistat is a novel inhibitor of fatty acid synthase with antitumor activity, *Cancer Res.*, 64, 2070, 2004.

71. Perez-Gil, J., Martin, J.J., Acebal, C. and Arche, R., Essential residues in lysolecithin:lysolecithin acyltransferase from rabbit lung: Assessment by chemical modification, *Arch. Biochem. Biophys.*, 277, 180, 1990.

72. Keshavarz-Shokri, A., Suntornwat, O. and Kitos, P.A., Identification of serine esterases in tissue homogenates, *Anal. Biochem.*, 267, 406, 1999.

73. Green, A.L. and Nichols, J.D., The reactivation of phosphorylated chymotrypsin, *Biochem. J.*, 72, 70, 1959.

74. Cohen, W. and Erlanger, B.F., Studies on the reactivation of diethylphosphoryl-chymotrypsin, *J. Am. Chem. Soc.* 82, 3928, 1960.

75. Cohen, W., Lache, M. and Erlanger, B.F., The reactivation of diethylphosphoryltrypsin, *Biochemistry*, 1, 686, 1962.

76. Caplow, M. and Jencks, W.P., The effect of substituents on the deacylation of benzoyl-chymotrypsin, *Biochemistry*, 1, 883, 1962.

77. Moutier, L.A., Reactivation of organophosphate-inhibited trypsin by hydroxylamine, *Biochim. Biophys. Acta*, 77, 301, 1963.

78. Gray, A.P., Design and structure-activity relationships of antidotes to organophosphorous anticholinesterase agents, *Drug Metab. Rev.*, 15, 557, 1984.

79. Clothier, B. and Johnson, M.K., Rapid aging of neurotoxic esterase after inhibition by di-isopropyl phosphorofluoridate, *Biochem. J.*, 177, 549, 1979.

80. Johnson, M.K., Organophosphorous esters causing delayed neurotoxic effects: Mechanism of action and structure activity studies, *Arch. Tox.*, 34, 259, 1975.

81. Kropp, T.J., Glynn, P. and Richardson, R.J., The Mipafox-inhibited catalytic domain of human neuropathy target esterase ages by reversible proton loss, *Biochemistry*, 43, 3716, 2004.

82. Li, Y, Dinsdale, D. and Glynn, P., Protein domains, catalytic activity, and subcellular distribution of neuropathy target esterase in mammalian cells, *J. Biol. Chem.*, 278, 8820, 2003.

83. Zaccheo, O., Dinsdale, D., Meacock, P.A. and Glynn, P., Neutropathy target esterase and its yeast homologue degrade phosphatidylcholine to glycerophosphocholine in living cells, *J. Biol. Chem.*, 279, 24024, 2004.

84. Estevez, J. Garcia-Perez, A.G., Barril, J., Pellin, M. and Vilanova, E., The inhibition of the high sensitive peripheral nerve soluble esterases by Mipafox. A new mathematical processing for the kinetics of inhibition of esterases by organophosphorous compounds, *Tox. Letters*, 151, 171, 2004.

85. Williams, D.G. and Johnson, M.K., Gel-electrophoretic identification of hen brain neurotoxic esterase, labelled with tritiated di-isopropyl phosphorofluoridate, *Biochem. J.*, 199, 323, 1981.

86. Williams, D.G., Intramolecular group transfer is a characteristic of neurotoxic esterase and is independent of the tissue source of the enzyme, *Biochem. J.*, 209, 817, 1983.

87. Planque, S., Bangale, Y., Song, X.-T. et al., Ontogeny of proteolytic immunity. IgM serine protease, *J. Biol. Chem.*, 279, 14024, 2004.

88. Planque, S., Taguchi, H., Burr, G. et al., Broadly distributed chemical reactivity of natural antibodies expressed in coordination with specific antigen binding activities, *J. Biol. Chem.*, 278, 20436–20443, 2003.

89. Paul, S., Tramontano, A., Gololobov, G. et al., Phosphonate ester probes for proteolytic antibodies, *J. Biol. Chem.*, 276, 28314, 2001.

90. Shaw, E. and Glover, G., Further observations on substrate-derived chloromethyl ketones that inactivate trypsin, *Arch. Biochem. Biophys.*, 139, 298, 1970.

91. Grabarek, J., Dragan, M., Lee, M.W. et al., Activation of chymotrypsin-like serine protease(s) during apoptosis detected by affinity-labeling of the enzymatic center with fluoresceinated inhibitor, *Intl. J. Oncology*, 20, 225, 2002.

92. Schoellmann, G. and Shaw, E., Direct evidence for the presence of histidine in the active center of chymotrypsin, *Biochemistry*, 2, 252, 1963.

93. Williams, E.B., Krishnaswamy, S. and Mann, K.G., Zymogen/enzyme discrimination using peptide chloromethyl ketones, *J. Biol. Chem.*, 263 7536, 1989.

94. Williams, E.B. and Mann, K.G., Peptide chloromethyl ketones as labelling reagents, *Meth. Enzymology*, 222, 503, 1993.

95. Tracy, R.P., Jenny, R., Williams, E.B. and Mann, K.G., Active site-specific assays for enzymes of coagulation and fibrinolytic pathways, *Meth. Enzymology*, 222, 514, 1993.

96. Lundblad, R.L., Bergstrom, J., De Vreker, R. et al., Measurement of active coagulation factors in Autoplex®-T with colorimetric active site-specific assay technology, *Thromb. Haemost.*, 80, 811, 1998.

97. Mack, A., Furmann, C. and Hacker, G., Detection of capase-activation in intact lymphoid cells using standard capase substrates and inhibitors, *J. Immunol. Meth.*, 241, 19, 2000.

98. Jayaraman, S., Intracellular determination of activated capases (IDAC) by flow cytometry using a pancapase inhibitor labeled with FITC. *Cytometry*, 56A, 104, 2003.

99. Cohen, G.M., Capases: The executioners of apoptosis, *Biochem. J.*, 326, 1, 1997.

100. Wolthers, B.C., Kinetics of inhibition of papain by TLCK and TPCK in the presence of BAEE as substrate, *FEBS Letters*, 2, 143, 1969.

101. Jonák, J., Rychlík, I., Smrt, J. and Holý, A., The binding site for the 3-terminus of aminoacyl-tRNA in the molecule of elongation factor Tu for *Escherichia coli*, *FEBS Letters*, 98, 329, 1979.

102. Jonák, J., Smrt, J., Holý, A. and Rychlík, I., Interaction of *Escherichia coli* EF-Tu, GTP, EF-Tu-GDP with analogues of the 3-terminus of aminoacyl-tRNA, *Eur. J. Biochem.*, 105, 315, 1980.

103. Arroyo, R. and Alderete, J.F., *Trichaomonas vaginalis* surface proteinase activity is necessary for parasite adherence to epithelial cells, *Infect. Immunit.*, 57, 2991, 1989.

104. Wikstrom, P., Kirschhe, H., Stone, S. and Shaw, E., The properties of peptidyl diazoethanes and chloroethanes as protease inhibitors, *Arch. Biochem. Biophys.*, 270, 286, 1989.

105. Stoppler, H., Stoppler, M.C., Adduci, A., Kov, D. and Schlegel, R., The serine protease inhibitors TLCK and TPCK react with the RB-binding core of HPV-18 E7 protein and abolish its RB-binding capability, *Virology*, 217, 5432, 1996.

106. Ott, G., Janák, J., Abrahams, I.P. and Sprinzl, M., The influence of different modification of elongation factor Tu from *Escherichia coli* on ternary complex function investigated by fluorescence spectroscopy, *Nucl. Acid Res.*, 18, 437, 1990.

107. Shaw, E., Angliker, H., Rauber, P. et al., Peptidyl fluoromethyl ketones as thiol protease inhibitors, *Biomed. Biochem. Acta*, 45, 1397, 1986.

108. Falk, M., Ussat, S., Reiling, N. et al., Capase inhibition blocks human T cell proliferation by suppressing appropriate regulation of IL-2, CD25, and cell-cycle associated proteins, *J. Immunol.*, 173, 5077, 2004

109. Bergen, H.R, III, Klug, M.G., Bolander, M.E. and Muddiman, D.C., Informed use of proteolytic inhibitors in biomarker discovery, *Rapid Comm. Mass Spec.*, 18, 1001, 2004.

110. Hunter, T.C., Andon, N.L., Koller, A. et al., The functional proteomics toolbox: Methods and applications, *J. Chrom.*, 782, 165, 2002.

111. Patterson, S.D. and Aebersold, R.H., Proteomics: The first decade and beyond, *Nat. Genetics*, Suppl. 33, 311, 2003.

112. He, Q.-Y. and Chiu, J.-F., Proteomics in biomarker discovery and drug development, *J. Cell. Biochem.*, 89, 868, 2003.

113. Kumar, S., Zhou, B., Liang, F. et al., Activity based probes for protein tyrosine phophatases, *Proc. Natl. Acad. Sci. USA*, 101, 7942, 2004.

114. Taylor, W.P., Zhang, Z.-Y. and Widlanski, T.S., Quiescent affinity inactivators of protein tyrosine phosphatases, *Bioorg. Med. Chem.*, 4, 1515, 1996.

115. Hemelaar, J., Galardy, P.J., Borodovsky, A. et al., Chemistry-based functional proteomics: Mechanism-based activity-profiling tools for ubiquitin and ubiquitin-like specific proteases, *J. Proteome Res.*, 3, 268, 2004.

116. Greenbaum, D., Medzihradszky, K.F., Burlingame, A. and Bogyo, M., Epoxide electrophiles as activity-dependent cysteine protease profiling and discovery tools, *Chem. & Bio.*, 7, 569, 2000.

117. Albeck, A. and Kliper, S., Inactivation of cysteine proteases by peptidyl epoxides: Characterization of the alkylation sites on the enzyme and the inactivator, *Biochem. J.*, 346, 71, 2000.

118. James, K.E., Asgian, J.L., Li, Z.Z. et al., Design, synthesis, and evaluation of azapeptide epoxides as selective and potent inhibitors of capases-1, -3, -6, and -8, *J. Med. Chem.*, 47, 1553, 2004.

119. Dupont, D., Rolet-Repecaud, O. and Senocq, D., A new approach to monitoring proteolysis phenomena using antibodies specifically directed against the enzyme cleavage site on its substrate, *Anal. Biochem.*, 317, 240, 2003.

120. Bredemeyer, A.J., Lewis, R.M., Malone, J.P., Davis, A.E., Gross, J., Townsend, R.R. and Ley, T.J., A proteomic approach for the discovery of proteases substrates, *Proc. Natl. Acad. Sci. USA*, 101, 11785, 2004.

121. Adam, G.C., Cravatt, B.F. and Sorensen, E.J., Profiling the specific reactivity of the proteome with non-directed activity-based probes, *Chem. & Biol.*, 8, 81, 2001.

122. Chang, L.S.-H., An, E.H.-H., Oh, H.-K. and Li, B.F.-L., Selective depletion of human DNA-methyltransferase DNMT1 proteins by sulfonate-derived methylating agents, *Cancer Res.*, 62, 1592, 2002.

123. Cloutier, J.-F., Castonguay, A., O'Connor, T.R. and Drouin, R., Alkylating agent and chromatin structure determine sequence context-dependent formation of alkylpurines, *J. Mol. Biol.*, 306, 169, 2001.

124. Roberts, D.W., Goodwin, B.F.J. and Basketter, D., Methyl groups as antigenic determinants in skin sensitization, *Contact Dermatitis*, 18, 219, 1988.

125. Roberts, D.W. and Basketter, D., A quantitative structure activity/dose response relationship for contact allergic potential of alkyl group transfer agents, *Contact Dermatitis*, 23, 331, 1990.

126. Roberts, D.W. and Basketter, D.A., Further evaluation of the quantitative structure-activity relationship for skin-sensitizing alkyl transfer agents, *Contact Dermatitis*, 37, 107, 1997.

127. Roberts, D.W. and Basketter, D.A., Quantitative structure-activity relationships in sulfonate esters in the local lymph node assay, *Contact Dermatitis*, 42, 154, 2000.

128. Roberts, D.W., Ward, R.S. and Hughes, P.J., Relative reactivities in aminolysis reactions of alkyl alkanesulfonates, *J. Chem. Res. (S)*, 20, 70, 1995.

129. Larsen, D. and Shaw, E., Active-site directed alkylation of chymotrypsin by reagents utilizing various departing groups, *J. Med. Chem.*, 19, 1284, 1976.

130. Nakagawa, Y. and Bender, M.L., Methylation of histidine 57 in α-chymotrypsin by methyl-*p*-nitrobenzenesulfonate: A new approach to enzyme modification, *Biochemistry*, 9, 259, 1970.

131. Porter, M.A. and Hartman, F.C., Catalytic nonessentiality of an active-site cysteinyl residue of phosphoribulokinase, *J. Biol. Chem.*, 263, 14846, 1988.

132. Marcus, J.P. and Dekker, E.E., Identification of a second active-site residue in *Escherichia coli* L-threonine dehydrogenase — Methylation of histidine-90 with methyl-*p*-nitrobenzenesulfonate, *Arch. Biochem. Biophys.*, 316, 413, 1995.

133. Berna, P.P., Mrabet, N.T., van Beeumen, J. et al., Residue accessibility, hydrogen bonding, and molecular recognition: Metal-chelate probing of active site histidines in chymotrypsins, *Biochemistry*, 3, 6896, 1997.

134. Grybel, J. and Wilusz, T., The use of immobilized methylchymotrypsin for the purification of human and sheep alpha-1-proteinase inhibitor (α_1-PI), *Cell. Mol. Bio. Letters*, 8, 363, 2003.

135. Heinrikson, R.L., The selective *S*-methylation of sulfhydryl groups in proteins and peptides with methyl-*p*-nitrobenzene sulfonate, *J. Biol. Chem.*, 246, 4090, 1971.

136. Plapp, B.V., Mechanisms of carboxymethylation of bovine pancreatic nucleases by haloacetates and tosylglycolate, *J. Biol. Chem.*, 248, 4896, 1973.

137. Oshima, H. and Takahashi, K., The structure and function of ribonuclease T_1. XX. Specific inactivation of ribonuclease T_1 by reaction with tosylglycolate, *J. Biochem.*, 80, 1259, 1976.

138. Rochat, C., Rochat, H. and Edman, P., Some *S*-alkyl derivatives of cysteine suitable for sequence determination by the phenylisothiocyanate technique, *Anal. Biochem.*, 37, 259, 1970.

139. Eyem, J., Sjödahl, J. and Sjöquist, J., *S*-methylation of cysteine residues in peptides and proteins with dimethyl sulfate, *Anal. Biochem.*, 74, 359, 1976.

140. Adam, G.C., Sorensen, E.J. and Cravatt, B.F., Proteomic profiling of mechanistically distinct enzyme classes using a common chemotype, *Nat. Biotech.*, 20, 805, 2002.

141. Adam, G.C., Sorensen, E.J. and Cravatt, B.F., Trifunctional chemical probes for the consolidated detection and identification of enzyme activities from complex proteomes, *Mol. Cell. Proteomics*, 1, 828, 2002.

142. Adam, G.C., Burbaum, J., Kozarich, J.W. et al., Mapping enzyme active sites in complex proteomes, *J. Am. Chem. Soc.*, 126, 1363, 2004.

143. Speers, A.E., Adam, G.C. and Cravatt, B.F., Activity- based protein profiling *in vivo* using a copper(I)-catalyzed azide-alkyne [3 + 2] cycloaddition, *J. Am. Chem. Soc.* 125, 4686, 2003.

144. Speers, A.E. and Cravatt, B.F., Profiling enzyme activities *in vivo* using click chemistry methods, *Chem. & Biol.*, 11, 535, 2004.

145. Breinbauer, R. and Köhn, M., Azide-alkyne coupling: A powerful reaction for bioconjugate chemistry, *ChemBioChem*, 4, 1147, 2003.

146. Wang, Q., Chan, T.R., Hilgraf, R., Fokin, V.V., Sharpless, K.B. and Finn, M.G., Bioconjugation by copper(I)-catalyzed azide-alkyne [3 + 2] cycloaddition, *J. Am. Chem. Soc.*, 125, 3192, 2003.

147. Levine, R.L., Williams, J.A., Stadtman, E.R. and Shacter, E., Carbonyl assays for determination of oxidatively modified proteins, *Meth. Enzymology*, 233, 346, 1994.
148. Ghezzi, P. and Bonetto, V., Redox proteomics: Identification of oxidatively modified proteins, *Proteomics*, 3, 1145, 2003.
149. Nikov, G., Bhat, V., Wishnok, J.S. and Tannenbaum, S.R., Analysis of nitrated proteins by nitrotyrosine-specific affinity probes and mass spectrometry, *Anal. Biochem.*, 320, 214, 2003.
150. Goshe, M.B., Conrads, T.P., Panisko, E.A. et al., Phosphoprotein isotope coded affinity tag approach for isolating and quantitating phosphopeptides in proteome-wide analyses, *Anal. Chem.*, 73, 2578, 2001.
151. Zhou, H., Watts, J.D. and Aebersold, R., A systematic approach to the analysis of protein phosphorylation, *Nat. Biotech.*, 19, 375, 2001.
152. Oda, Y., Nagasu, T. and Chait, B.T., Enrichment analysis of phosphorylated proteins as a tool for probing the phosphoproteome, *Nat. Biotech.*, 19, 379, 2001.
153. Conrads, T.P., Issaq, H.J. and Veenstra, T.D., New tools for quantitative phosphoproteome analysis, *Biochem. Biophys. Res. Comm.*, 290, 885, 2002.
154. Li, W., Backlund, P.S., Boykins, R.A. et al., Susceptibility of the hydroxyl groups in serine and threonine to β-elimination/Michael addition under commonly used conditions moderately high-temperature conditions, *Anal. Biochem.*, 323, 94, 2003.
155. Thaler, F., Valsasina, B., Baldi, R. et al., A new approach to phosphoserine and phosphothreonine analysis in peptides and proteins: chemical modification, enrichment via solid-phase reversible binding, and analysis by mass spectrometry, *Anal. BioAnal. Chem.*, 376, 366, 2003.
156. Goshe, M.B., Blonder, J. and Smith, R.D., Affinity labeling of highly hydrophobic integral membrane proteins for proteome-wide analysis, *J. Proteome Res.*, 2, 153, 2003.
157. Zhang, H., Yan, W. and Aebersold, R., Chemical probes and tandem mass spectrometry: A strategy for the quantitative analysis of proteomes and subproteomes, *Curr. Opin. Chem. Biol.*, 8, 66, 2004.
158. Whitehead, J.K. and Dean, H.G., The isotope derivative method in biochemical analysis, *Meth. Biochem. Anal.*, 16, 1, 1968.
159. Szpunar, J., Advances in analytical methodology for bioinorganic speciation analysis: Metallomics, metalloproteomics and heteroatom-tagged proteomics and metabolomics, *Analyst*, 130, 442, 2005.
160. Gygi, S.P., Rist, B., Gerber, S.A. et al., Quantitative analysis of complex protein mixtures using isotope-coded affinity tags, *Nat. Biotech.*, 17, 994, 1999.
161. Griffin, T.J. and Aebersold, R., Advances in proteome analysis by mass spectrometry, *J. Biol. Chem.*, 276, 45479, 2001.
162. Smolka, M.B., Zhou, H., Purkayastha, S. and Aebersold, R., Optimization of the isotope-coded affinity tag-labeling procedure for quantitative proteome analysis, *Anal. Biochem.*, 297, 25, 2001.
163. Smolka, M., Zhou, H. and Aebersold, R., Quantitative protein profiling using two-dimensional gel electrophoresis, isotope-coded affinity tag labeling, and mass spectrometry, *Mol. Cell. Proteomics*, 1, 19, 2002.
164. Tao, W.A. and Aebersold, R., Advances in quantitative proteomics via stable isotope tagging and mass spectrometry, *Curr. Opin. Biotech.*, 14, 110, 2003.
165. Zhang, Z., Edwards, P.J., Roeske, R.W. and Guo, L., Synthesis and self-alkylation of isotope-coded affinity tag reagents, *Bioconjugate Chem.*, 16, 458, 2005.
166. Goshe, M.B., Conrads, T.P., Panisko, E.A. et al., Phosphoprotein isotope-coded affinity tag approach for isolating and quantitating phosphopeptides in proteome-wide analyses, *Anal. Chem.*, 73, 2578, 2001.

167. Kuyama, H., Watanabe, M., Toda, C. et al., An approach to quantitative proteome analysis by labeling tryptophan residues, *Rapid Comm. Mass Spec.*, 17, 1642, 2003.

168. Goshe, M.B. and Smith, R.D., Stable isotope-coded proteomic mass spectrometry, *Curr. Opin. Biotech.*, 14, 101, 2003.

169. Qiu, Y., Sousa, E.A., Hewick, R.M. and Wang, J.H., Acid-labile isotope-coded extractants: A class of reagents for quantitative mass spectrometric analysis of complex protein mixtures, *Anal. Chem.*, 74, 4969, 2002.

170. Lu, Y., Bottari, P., Turecek, F. et al., Absolute quantitation of specific proteins in complex mixtures using visible isotope-coded affinity tags, *Anal. Chem.*, 76, 4104, 2004.

171. Ong, S.-E., Blagoev, B., Kratchmarova, I. et al., Stable isotope labeling by amino acids in cell culture, SILAC, as a simple and accurate approach to expression proteomics, *Mol. Cell. Proteomics*, 1, 376, 2002.

172. Loyet, K.M., Ouyang, W., Eaton, D.L. and Stults, J.L., Proteomic profiling of surface proteins on Th1 and Th2 cells, *J. Proteome Res.*, 4, 400, 2005.

173. Zhang, R., Sioma, C.S., Wang, S. and Regnier, F.E., Fractionation of isotopically labeled peptides in quantitative proteomics, *Anal. Chem.*, 73, 5142, 2001.

174. Chakraborty, A. and Regnier, F.E., Global internal standard technology for comparative proteomics, *J. Chrom. A.*, 949, 173, 2002.

175. Hansen, K.C., Schmitt-Ulms, G., Chalkey, R.J., et al, Mass spectrometric analysis of protein mixtures at low levels using cleavable ^{13}C-isotope-coded affinity tag and multidimensional chromatography, *Mol. Cell. Proteomics*, 2, 229, 2003.

176. Gevaert, K., Van Damme, J., Goethals, M. et al., Chromatographic isolation of methionine-containing peptides for gel-free proteome analysis, *Mol. Cell. Proteomics*, 1, 896, 2002.

177. Savige, W.E. and Fontana, A., Interconversion of methionine and methionine sulfoxide, *Meth. Enzymology*, 47, 453, 1977.

178. Tang, J. and Hartley, B.S., A diagonal electrophoretic method for the purification of methionine peptides, *Biochem. J.*, 102, 593, 1967.

179. Gundlach, G., Moore, S. and Stein, W.H., The reaction of iodoacetate with methionine, *J. Biol. Chem.*, 234, 1761, 1959.

180. Weinberger, S.R., Viner, R.I. and Ho, P., Tagless extraction-retentate chromatography: A new global protein digestion strategy for monitoring differential protein expression, *Electrophoresis*, 23, 3182, 2002.

181. Shen, M., Guo, L., Wallace, A. et al., Isolation and isotope labeling of cysteine- and methionine-containing tryptic peptides. Application to the study of cell surface proteolysis, *Mol. Cell. Proteomics*, 2, 315, 2003.

182. Grunert, T., Pock, K., Buchacher, A. and Allmaier, G., Selective solid-phase isolation of methionine-containing peptides and subsequent matrix-assisted laser desorption mass spectrometric detection of methionine- and methionine-sulfoxide-containing peptides, *Rapid Comm. Mass Spec.*, 17, 1815, 2003.

183. Khidekel, N., Ficarro, S.B., Peters, E.C. and Hsieh-Wilson, L.C., Exploring the *O*-GlaNac-modified proteins from the brain, *Proc. Natl. Acad. Sci. USA*, 101, 13132, 2004.

184. Datta, D., Wang, P., Carrico, I.S., Mayo, S.L. and Tirrell, D.A., A designed phenylalanyl-tRNA synthetase variant allows efficient *in vivo* incorporation of aryl ketone functionality into proteins, *J. Am. Chem. Soc.*, 124, 5652, 2002.

185. Khidekel, N., Arndt, S., Lamarre-Vincent, N. et al., A chemoenzymatic approach toward the rapid and sensitive detection of *O*-GlcNAc posttranslational modifications, *J. Am. Chem. Soc.*, 125, 16162, 2003.

186. Saxon, E. and Bertozzi, C.R., Cell surface engineering by a modified Staudinger reaction, *Science*, 287, 2007, 2000.

187. Vacadlo, D.J., Hang, H.C., Kim, E.-J. et al., A chemical approach for identifying *O*-GlcNAc-modified proteins in cells, *Proc. Natl. Acad. Sci. USA*, 100, 9116, 2003.

188. Lemieux, G.A., de Graffenried, C.L. and Bertozzi, C.R., A fluorogenic dye activated by the Staudinger ligation, *J. Am. Chem. Soc.*, 125, 4708, 2003.

189. Kwon, S.W., Kim, S.C., Jaunbergs, J. et al., Selective enrichment of thiophosphorylated polypeptides as a tool for the analysis of protein phosphorylation, *Mol. Cell. Proteomics*, 2, 242, 2003.

190. Kho, Y., Kim, S.C., Jiang, C. et al., A tagging-via-substrate technology for detection and proteomics of farnesylated proteins, *Proc. Natl. Acad. Sci. USA*, 101, 12479, 2004.

191. Von Eggling, F., Gawriljuk, A., Fiedler, W. et al., Fluorescent dual colour 2D-protein gel electrophoresis for rapid detection of differences in protein pattern with standard image analysis software, *Int. J. Mol. Med.*, 8, 373, 2001.

192. Van den Bergh, G. and Arckens, L., Fluorescent two-dimensional difference gel electrophoresis unveils the potential of gel-based proteomics, *Curr. Opin. Biotech.*, 15, 38, 2004.

193. Alban, A., David, S.O., Bjorkesten, L, Andersson, C., Sloge, E., Lewis, A. and Currie, I., A novel experimental design for comparative two-dimensional gel analysis. Two-dimensional difference gel electrophoresis incorporating a pooled internal standard, *Proteomics*, 3, 36, 2003.

194. Jameson, D.M., Croney, J.C. and Moens, P.D.J., Fluorescence: Basic concepts, practical aspects, and some anecdotes, *Meth. Enzymology*, 360, 1, 2003.

195. Sowell, J., Salon J., Shekowski, L. and Patonay, G., Covalent and noncovalent labeling schemes for near-infrared dyes in capillary electrophoresis protein applications, in *Capillary Electrophoresis of Proteins and Peptides*, M.A. Strege, and A.L. Luge, Eds., Humana Press, Towata, NJ, Ch. 4, 39, 2004.

196. Venkataramen, K., *The Chemistry of Synthetic Dyes*, Vol. 1, Academic Press, New York, NY, 1952.

197. Gómez-Hens, A. and Aguilar-Caballos, M.P., Long-wavelength fluorophores: New trends in their analytical use, *Trends Anal. Chem.*, 23, 127, 2004.

198. Hamer, F.M., The cyanine dyes and related compounds, *Chemistry of Heterocyclic Compounds*, Vol. 19, Interscience (John Wiley), New York, NY, 1964.

199. *Hawley's Condensed Chemical Dictionary*, 12th Edition, R.J. Lewis, Sr., Ed., Van Nostrand Reinhold Company, New York, NY, 1993.

200. Wang, M. and Armitage, B.A., Colorimetric detection of PNA-DNA hybridization using cyanine dyes, in *Peptide Nucleic Acids: Methods and Protocols*, P.E. Nielsen, Ed., Humana Press, Towata, NJ, Ch. 9, 131, 2002.

201. Kane, M.D., Jatkoe, T.A., Stumpf, C.R., Lu, J., Thomas, J.D. and Madre, S.J., Assessment of the sensitivity and specificity oligonucleotide (50 mer) microarrays, *Nucleic Acid Res.*, 28, 4552, 2000.

202. Ernst, L.A., Gupta, R.K., Mujubdar, R.B. and Waggoner, A.S., Cyanine dye labeling reagents for sulfydryl groups, *Cytometry*, 10, 3, 1989.

203. Mujumdar, R.B., Ernst, L.A., Mujumdar, S.R. and Waggoner, A.S., Cyanine dye labeling reagents containing isothiocyanate groups, *Cytometry*, 10, 11, 1989.

204. Southwick, P.L., Ernst, L.A., Tauriello, E.W., Parker, S.R., Mujumdar, R.B., Mujumdar, S.R., Clever, H.A. and Waggoner, A.S., Cyanine dye labeling reagents — carboxymethylindocyanine succinimidyl esters, *Cytometry*, 11, 418, 1990.

205. Majumdar, S.R., Majumdar, R.B., Grant, C.M. and Waggoner, A.S., Cyanine-labeling reagents: Sulfobenidocyanine succinimidyl esters, *Bioconjugate Chem.*, 7, 356, 1996.

206. Coopes, M., Ebner, A., Briggs, M., et al., Cy3B: improving the performance of cyanine dyes, *J. Flucresc.*, 14, 145–150, 2004.
207. Toutchkine, A., Nalbant, P. and Hahn, K.M., Facile synthesis of thiol-reactive Cy3 and Cy5 derivatives with enhanced water solubility, *Bioconjugate Chem.*, 13, 387–391, 2002.
208. Gruber, H.J., Kada, G., Pragl, B. et al., Preparation of thiol-reactive Cy5 derivatives from commercial Cy5 succinimidyl ester, *Bioconjugate Chem.*, 11, 161, 2000.
209. Tonge, R., Shaw, J., Middelton, B. et al., Validation and development of fluorescence two-dimensional differential gel electrophoresis proteomics technology, *Proteomics*, 1, 377, 2001.
210. Ruepp, S.V., Tonge, R.P., Shaw, J. et al., Genomics and proteomics analysis of acetaminophen toxicity in mouse liver, *Tox. Sci.*, 65, 135, 2003.
211. Swatton, J.E., Prabakaren, S., Karp, N.A. et al., Protein profiling of human postmortem brain using 2-dimensional fluorescence difference gel electrophoresis (2-D DIGE), *Mol. Psych.*, 9, 1181, 2003.
212. Friedman, D. B., Hill, S., Keller, J.W. et al., Proteomic analysis of human colon cancer by two-dimensional difference gel electrophoresis and mass spectrometry, *Proteomics*, 4, 793, 2004.
213. Wessel, D. and Flügge, U.I., A method for the quantitative recovery of protein in dilute solution in the presence of detergents and lipids, *Anal. Chem.*, 138, 141, 1984.
214. Karp, N.A., Kreill, D.P. and Lilley, K.S., Determining a significant change in protein expression with DeCyder™ during a pair-wise comparison using two-dimensional difference gel electrophoresis, *Proteomics*, 4, 1421, 2004.
215. Shaw, J., Rowlinson, R., Nickson, J., Stone, T., Sweet, A., Williams, K. and Tonge, R., Evaluation of saturation labelling two-dimensional difference gel electrophoresis fluorescent dyes, *Proteomics*, 3, 1181, 2003.
216. Britto, P.J., Knipling, L. and Wolff, J., The local electrostatic environment determines cysteine reactivity of tubulin, *J. Biol. Chem.*, 277, 29018, 2002.
217. Kim, Y.J., Pannell, L.K. and Sackett, D.L., Mass spectrometric measurement of differential reactivity of cysteine to localize protein-ligand binding sites. Application to tubulin-binding drugs, *Anal. Biochem.*, 332, 376, 2004.
218. Tyagarajan, K., Pretzer, E. and Wiktorowicz, J.E., Thiol-reactive dyes for fluorescence labeling of proteomic samples, *Electrophoresis*, 24, 2348, 2003.
219. Galvani, M., Rovath, L., Hamden, M., Herbert, B. and Righetti, P.G., Protein alkylation in the presence/absence of thiourea in proteome analysis: A matrix-assisted laser desorption-ionization time-of-flight mass spectrometry investigation, *Electrophoresis*, 22, 2066, 2002.
220. Shaw, M.M. and Riederer, B.M., Sample preparation for two-dimensional gel electrophoresis, *Proteomics*, 3, 1408, 2003.
221. Kondo, T., Seiko, M., Mori, Y. et al., Application of a sensitive fluorescent dyes in linkage of microdissection and two-dimensional gel electrophoresis as a cancer proteomics study tool, *Proteomics*, 3, 1758, 2003.
222. Maeda, K., Finnie, C. and Svensson, B., Cy5 maleimide labelling for sensitive detection of free thiols in native protein extracts: Identification of seed proteins targeted by barley thioredoxin isoforms, *Biochem. J.*, 378, 497, 2004.
223. Urwin, V.F. and Jackson, P., Two-dimensional polyacrylamide gel electrophoresis of proteins labeled with the fluorophore monobromobimane prior to first-dimensional isoelectric focusing: Imaging of the fluorescent protein spot patterns using a cooled change-coupled devices, *Anal. Biochem.*, 209, 57, 1993.

224. Herick, K., Jackson, P., Weisch, G. and Burkowski, A., Detection of fluorescence dye-labeled proteins in 2-D gels using an Arthur 1442 multiwavelength fluoroimager, *BioTechniques*, 31, 146, 2001.

225. Yano, H., Kurado, S. and Buchanan, B.B., Disulfide proteome in the analysis of protein function and structure, *Proteomics*, 2, 1090, 2002.

226. Yano, H., Fluorescent labeling of disulfide proteins on 2D gels for screening allergens: A preliminary study, *Anal. Chem.*, 75, 4682, 2003.

227. Wang, J.H., Cai, N., Balmer, Y. et al., Thioredoxin targets of developing wheat seeds identified by complementary proteomic approaches, *Phytochemistry*, 65, 1629, 2004.

228. Ralat, L.A. and Colman, R.F., Monobromobimane occupies a distinct xenobiotic substrate site in glutathione *S*-transferase, *Protein Sci.*, 12, 2575, 2003.

229. Buschmann, V., Weston, K.D. and Sauer, M., Spectroscopic study and evaluation of red-absorbing fluorescent dyes, *Bioconjugate Chem.*, 14, 195, 2003.

230. Martin, K., Hart, C., Schulenberg, B. et al., Simultaneous red/green dual fluorescence detection on electroblots using BODIPY TR-X succinimidyl ester and ELF 39 phosphate, *Proteomics*, 2, 499, 2002.

231. Hitt, A.L, Laing, S.D. and Olson, S., Development of a fluorescent F-actin blot overlay assay for detection of F-actin binding proteins, *Anal. Biochem.*, 310, 67, 2002.

232. Huang, S., Wang, H., Carroll, C.A. et al., Analysis of proteins stained with Alexa dyes, *Electrophoresis*, 25, 799, 2004.

233. Michels, D.A., Hu, S., Schoenherr, R.M. et al., Fully automated two-dimensional capillary electrophoresis for high sensitivity protein analysis, *Mol. Cell. Proteomics*, 1, 69, 2002.

234. Michels, D.A., Hu, S. and Dambrowitz, K.A., Capillary sieving electrophoresis micellar electrokinetic chromatography fully automated two-dimensional capillary electrophoresis analysis of *Deinococcus radiodurans* protein homogenate, *Electrophoresis*, 25, 3090, 2004.

235. Galvani, M., Rovatti, L., Hamden, M., et al., Protein alkylation in the presence/absence of thiourea in proteome analysis: A matrix-assisted laster desorption/ionization time-of-flight mass spectrometry investigation, *Electrophoresis*, 22, 2066, 2002.

236. Rabilloud, T., Adessi, C., Giraudel, A. and Lunardi, J., Improvement of the solubilization of proteins in two-dimensional electrophoresis with immobilized pH gradients, *Electrophoresis*, 18, 307, 1997.

237. Masuda, M., Toriumi, C., Santa, T. and Imai, K., Fluorogenic derivatization reagents suitable for isolation and identification of cysteine-containing proteins utilizing high performance liquid chromatography-tandem mass spectrometry, *Anal. Chem.*, 76, 728, 2004.

238. Lundblad, R.L., *Chemical Reagents for Protein Modification*, 3rd Edition, CRC Press, Boca Raton, FL, 2004.

239. Collier, J.H. and Messersmith, P.B., Enzymatic modification of self-assembled peptide structures with tissue transglutaminases, *Bioconjugate Chem.*, 14, 748, 2003.

240. Mosesson, M.W., Siebenlist, K.R., Hernandez, I., Wall, J.S. and Hainfeld, J.F., Fibrinogen assembly and crosslinking on a fibrin fragment E template, *Thromb. Haemostas.*, 87, 651, 2002.

241. Colman, R.W., Clowes, A.W., George, J.N., Hirsh, J. and Marder, V.J., Overview of Hemostasis, in *Hemostasis and Thrombosis: Basic Principles and Clinical Practice*, R.W. Colman, J. Hirsh, V.J. Marder, A.W. Clowes and J.N. George, Eds., Lippincott Williams & Wilkins, Philadelphia, PA, Ch. 1, 3, 2001.

242. Lorand, L. and Conrad, S.M., Transglutaminases, *Mol. Cell. Biochem.*, 58, 9, 1984.

243. Griffin, M., Casadio, R. and Bergamini, C.M., Transglutamimases: Nature's biological glues, *Biochem. J.*, 368, 377, 2002.
244. Lorand, L., Siefring, G.E., Tong, Y.S. et al., Dansylcadavarine specific staining for transamidating enzymes, *Anal. Biochem.*, 93, 453, 1979.
245. Ruopppolo, M., Orru, S., D'Amato, A. et al., Analysis of transglutaminase protein substrates by functional proteomics, *Protein Sci.*, 12, 1290, 2003.
246. Orru, S., Caputo, I., D'Amato, A. et al., Proteomics identification of acyl-acceptor and acyl-donor substrates for transglutaminase in a human epithelial cell line. Implications for celiac disease, *J. Biol. Chem.*, 278, 31766, 2003.
247. Caputo, I., D'Amato, A., Troncone, R. et al., Transglutaminase 2 in celiac disease: Minireview article, *Amino Acids*, 26, 381, 2004.
248. Nieuwenhuizen, W.F., Dekker, H.L., Gronevald, T., de Koster, C.G. and de Jong, G.A., Transglutaminase-mediated modification of glutamine and lysine residues in native bovine beta-lactoglobulin, *Biotech. Bioeng.*, 85, 248, 2004.
249. Meusel, M., Synthesis of hapten-protein conjugates using microbial transglutaminase, *Meth. Mol. Biol.*, 283, 109, 2004.
250. Venter, J.C., Myers, E.W., Li, P.W. et al., The sequence of the human genome, *Science*, 291, 1304, 2001.
251. Lee, P.S. and Lee, K.H., Genomic analysis, *Curr. Opin. Biotech.*, 11, 171, 2000.
252. Chee, M., Yang, R., Hubbell, E. et al., Accessing genetic information with high-density DNA arrays, *Science*, 274, 610, 1996.
253. Velculescu, V.E., Zhang, L., Vogelstein, B. and Kinzler, K.W., Serial analysis of gene expression, *Science*, 270, 484, 1995.
254. Polyak, K. and Riggins, G.J., Gene discovery using the serial analysis of gene expression technique: Implications for cancer research, *J. Clin. Oncology*, 19, 2948, 2001.
255. Dennis, J.L. and Oien, K.A., Hunting the primary: Novel strategies for defining the origin of tumours, *J. Pathol.*, 205, 236, 2005.
256. Gygi, S.P., Rochon, Y., Franza, B.R. and Aebersold, R., Correlation between protein and mRNA abundance in yeast, *Mol. Cell. Biol.*, 19, 1720, 1999.
257. Werner, T., Proteomics and regulomics: The yin and yang of functional genomics, *Mass Spec. Rev.*, 23, 25, 2004.
258. Hosack, D.A., Dennis, G., Jr., Sherman, B.T. et al., Identifying biological themes within lists of genes with EASE, *Genome Biol.*, 4, R70, 2003.
259. Holloway, A.J., van Laar, R.K., Tothill, R.W. and Bowtell, D.D., Options available — from start to finish — for obtaining data from DNA microarrays II. *Nat. Genetics*, Suppl. 32, 481, 2002.
260. Jordan, B., Historical background and anticipated developments, *Ann. N.Y. Acad. Sci.*, 975, 24, 2002.
261. Hu, Z., Troester, M. and Perou, C.M., High reproducibility using sodium hydroxide-stripped long oligonucleotide DNA microarrays, *BioTechniques*, 38, 121, 2005.
262. Petach, H. and Gold, L., Dimensionality is the issue: Use of photoaptamers in protein microarrays, *Curr. Opin. Biotech.*, 13, 309, 2002.
263. Krska, R. and Janotta, M., Technology and applications of protein microarrays, *Anal. BioAnal. Chem.*, 379, 339, 2004.
264. FDA Guidance for Industry In the Manufacture and Clinical Evaluation of *In Vitro* Tests to Detect Nucleic Acid Sequences of Human Immunodeficiency Viruses Types 1 and 2, http://www.fda.gov/cber/gdlns/hivnas.pdf.
265. FDA Guidance for Industry Analyte Specific Reagents; Small Entity Compliance Guidance; Guidance for Industry, http://www.fda.gov/crdh/oivd/guidance/1205.html.

266. Gibbs, J.N., The past, present, and future of ASR's, IVD Technology, November-December, 2003, http://www.devicelink.com/ivdt.

267. Huang, R.-P., Detection of multiple proteins in an antibody-based protein microarray system, *J. Immunol. Meth.*, 255, 1, 2001.

268. Peluso, P., Wilson, D.S., Do, D. et al., Optimizing antibody immobilization strategies for the construction of protein microarrays, *Anal. Biochem.*, 312, 113, 2003.

269. McCauley, T.G., Hamaguchi, N. and Stanton, M., Aptamer-based biosensor arrays for detection and quantification of biological macromolecules, *Anal. Biochem.*, 319, 244, 2003.

270. Xu, Q. and Lan, K.S., Protein and chemical microarrays: Powerful tools for proteomics, *J. Biomed. Technol.*, 5, 257, 2003.

271. Hughes, K.A., Use of a small molecule-based affinity system for the preparation of protein microarrays, *Meth. Mol. Biol.*, 264, 111, 2004.

272. Lan, K.S. and Renil, M., From combinatorial chemistry to chemical microarray, *Curr. Opin. Chem. Biol.*, 6, 353, 2002.

273. Gosalia, D.N. and Diamond, S.L., Printing chemical libraries on microarrays for fluid phase nanoliter reactions, *Proc. Natl. Acad. Sci. USA*, 100, 8721, 2003.

274. Smothers, J.F. and Henikoff, S., Predicting *in vivo* protein peptide interactions with random phage display, *Comb. Chem. High Throughput Screen.*, 4, 585, 2001.

275. Lopez, M.F. and Pluskal, M.G., Protein micro- and macroarrays: Digitizing the proteome, *J. Chrom. B*, 787, 19, 2004.

276. Gleason, N.J. and Carbeck, J.D., Measurement of enzyme kinetics using microscale steady-state kinetic analysis, *Langmuir*, 20, 6374, 2004.

277. Yu, Y., Ko, K.S., Zea, C.J. and Pohl, N.L., Discovery of the chemical function of glycosidases: Design, synthesis, and evaluation of mass-differentiated carbohydrate libraries, *Organic Letters*, 6, 2031, 2004.

278. Goddard, J.-P. and Reymond, J.-L., Recent advances in enzyme assays, *Trends Biotech.*, 22, 363, 2004.

279. Debusschere, B.J., Najm, H.N., Matta, A., Knio, O.M., Ghanem, R.G. and Le Maître, O.P., Protein labeling reactions in electrochemical microchannel flow: Numerical simulation and uncertainty propagation, *Physics and Fluids*, 15, 2238, 2003.

280. Graham. D.L., Ferreira, H.A., Freitas, P.P. and Cabral, J.M., High sensitivity detection of molecular recognition using magnetically labelled biomolecules and magnetoresistive sensors, *Biosens. Bioelec.*, 18, 483, 2003.

281. Sukhanova, A., Devy, J., Venteo, L., Kaplan, H., Aremyev, M., Oleinikov, V., Klinov, D., Pluot, M., Cohen, J.H. and Nabiev, I., Biocompatible fluorescent nanocrystals for immuolabeling of membrane proteins and cells, *Anal. Biochem.*, 324, 60, 2004.

282. Kapanidis, A.N., Lee, N.K., Laurence, T.A., Doose, S., Margeat, E. and Weiss, S., Fluorescence-aided molecule sorting: Analysis of structure and interactions by alternating-laser excitation of single molecules, *Proc. Natl. Acad. Sci. USA*, 101, 8936, 2004.

283. Biebricher, A., Paul, A., Tinnefeld, P., Golzhauser, A. and Sauer, M., Controlled three-dimensional immobilization of biomolecules on chemically patterned surfaces, *J. Biotech.*, 112, 97, 2004.

284. Mehta, T., Tanik, M. and Allison, D.B., Towards sound epistemological foundations of statistical methods for high-dimensional biology, *Nat. Genetics*, 36, 943, 2004.

285. Reimer, U., Reineke, U. and Schneider-Mergener, J., Peptide arrays; from macro to micro, *Curr. Opin. Biotech.*, 13, 315, 2002.

286. Schuurs, A.H. and Van Weemen, B.K., Enzyme immunoassay, *Clin. Chim. Acta*, 81, 1, 1977.

287. Engvall, E., Quantitative enzyme immunoassay (ELISA) in microbiology, *Med. Biol.*, 55, 193, 1977.
288. O'Beirne, A.J. and Cooper, H.R., Heterogeneous enzyme immunoassay, *J. Histochem. Cytochem.*, 27, 1148, 1979.
289. Voller, A., Bartlett, A. and Bidwell, D.E., Enzyme immunoassays with special reference to ELISA techniques, *J. Clin. Pathol.*, 31, 507, 1978.
290. Hicks, J.M., Fluorescence immunoassay, *Human Path.*, 15, 112, 1984.
291. Oellerich, M., Enzyme-immunoassay: A review, *J. Clin. Chem. Clin. Biochem.*, 22, 895, 1984.
292. Bulinski, J.C., Peptide antibodies: New tools for cell biology, *Int. Rev. Cytol.*, 103, 281, 1986.
293. Hashida, S., Hashinkaka, K. and Ishikawa, E., Ultrasensitive enzyme immunoassay, *Biotech. Ann. Rev.*, 1, 403, 1995.
294. Meldal, M., Properties of solid supports, *Meth. Enzymology*, 289, 83, 1997.
295. Mould, A.P., Solid phase assay for studying ECM protein–protein interactions, *Meth. Mol. Biol.*, 139, 295, 2000.
296. Kusnezow, W. and Hoheisel, J.D., Solid supports for microarray immunoassays, *J. Mol. Recog.*, 16, 165, 2003.
297. Nielson, U.B. and Geierstanger, B.H., Multiplexed sandwich assays in microarray format, *J. Immunol. Meth.*, 290, 107, 2004.
298. Hermansen, G.T., *Bioconjugate Techniques*, Academic Press, San Diego, CA, 1996.
299. Vreeland, W.N. and Barron, A.E., Functional materials for microscale genomic and proteomic analysis, *Curr. Opin. Biotech.*, 13, 87, 2002.
300. Angenendt, P., Glöcker, J., Murphy, D., Lehrach, H. and Cahill, D.J., Toward optimized antibody microarrays: A comparison of current microarray support materials, *Anal. Biochem.*, 309, 253, 2002.
301. Kusnezow, W., Jacob, A., Walijew, A., Diehl, F. and Hoheisel, J.D., Antibody microarrays: An evaluation of production parameters, *Proteomics*, 3, 254, 2003.
302. Peluso, P., Wilson, D.S., Do, D. et al., Optimizing antibody immobilization strategies for the contruction of protein microarrays, *Anal. Biochem.*, 312, 113, 2003.
303. Johnson, C.P., Jensen, I.E., Prakasam, A. et al., Engineered protein A for the orientational control of immobilized proteins, *Bioconjugate Chem.*, 14, 974, 2003.
304. Duburcq, X., Olivier, C., Malingue, F., Gras-Masse, H. and Melnyk, O., Peptide-protein microarrays for the simultaneous detection of pathogen infections, *Bioconjugate Chem.*, 15, 307, 2004.
305. Albrecht, H., Burke, P.A., Natarajan, A., Xiong, C.-Y., Kalicinsky, M., DeNardo, G.L. and DeNardo, S.J., Production of soluble SCFv with C-terminal free thiol for site-specific conjugation or stable dimeric ScFvs on demand, *Bioconjugate Chem.*, 15, 16, 2004.
306. Sidor, J.P., Mariano, M., Sideris, S. and Nock, S., Establishment of intein-mediated protein ligation under denaturing conditions: C-terminal labeling of a single-chain antibody for biochip screening, *Bioconjugate Chem.*, 13, 707, 2002.
307. Butt, J.N., Thorton, J., Richardson, D.J. and Dobbis, P.S., Voltammetry of a flavoprotein c_3: The lowest heme mediates fumarate reduction rates, *Biophysical J.*, 78, 1001, 2000.

4 Sample Preparation for Proteomic Studies

CONTENTS

INTRODUCTION

The preparation of a sample for analysis by proteomic technologies is a critical part of the overall analytical process. Issues in sample preparation can add as much variance as biological variation.[1,2] When a identical sample is used by three different laboratories for two-dimensional gel electrophoresis, a high degree of reproducibility exists between the various laboratories, further suggesting that the sample preparation step is the primary component of analytical variability.[3,4] This circumstance is true for most current methods of proteomic analysis, including multidimensional chromatography. The information presented herein is also valuable for the preparation of samples for microarray analysis. Issues in sample preparation are not unique to proteomics but are also concerns in clinical chemistry and analytical chemistry. Knowing where a sample came from, what happened to it along the way and having the documentation to describe these two issues is critical. Sample preparation in proteomics is a commercially important area as several studies[5-7] suggest that the "World Proteomics Sample Preparation Market" had a revenue value of $57.8 million (US) in 2000 and is expected to grow to $367.7 million (US) by 2009. This growth likely reflects the anticipation of increased activity as well as the increasing use of "prepackaged" reagents in black box approaches. Assay validation issues are discussed in Chapter 8 but all processes with sample preparation including sample prefractionation are subject to such validation. Also, although the separation is somewhat artificial, issues associated with sample prefractionation are discussed in Chapter 5 with particular emphasis on biological fluids. Although proteomics is intended to be a global assessment, current technology does not permit separation of more than 1000 or so separate protein species from the 30,000+ total number; a new paradigm for analysis is required. Most of the discussion in this chapter focuses on the acquisition of a solution sample from a cellular/tissue source.

CONTRIBUTION OF SAMPLE PREPARATION TO EXPERIMENTAL VARIANCE

Variance in the data can arise from multiple sources[1] including biological variation, sample preparation, the analytical process including separation and quantitation and data analysis. Individual (single subject) variation is suggested to be minor compared to intraindividual variation.[8,9] The reader is directed to an expanded consideration of subject and population variation (Index of Individuality) in Chapter 8.

Several studies have attempted to "isolate" the variance in sample preparation from "downstream" variance in the analytical instrumentation and data analysis. Anderle and coworkers[10] have presented an analysis of proteomic expression profiling by HPLC-MS. The model sample was human serum. Sample preparation involved several steps: removal of albumin and IgG, reduction/alkylation and tryptic digestion in a process known as shotgun proteomics.[11] Choe and Lee[12] observed a high CV in their studies on the reproducibility of two-dimensional gel electrophoresis and reported a method for assessing variability.

BIOLOGICAL FLUIDS

Some issues with plasma and serum samples are addressed in the chapter on prefractionation. That material is concerned with the fractionation of whole blood into cellular elements and blood plasma and the subsequent fraction of plasma to remove high-abundance proteins such as albumin and immunoglobulin G (IgG). This separation is somewhat arbitrary but our sense is that this is appropriate. We will briefly discuss several techniques developed for blood plasma and other fluids but are generally applicable to biological fluids such as cerebrospinal fluid, lymph and saliva. General issues with sample solvent/lysis solution and reduction/alkylation are discussed below and apply equally to plasma/serum samples and tissue derived samples.

We will briefly discuss some specific applications to serum/plasma with the understanding that such applications are applicable to other biological fluids and tissue extracts. One approach uses a precipitation step prior to taking the protein sample into sample buffer/lysis buffer.[13] This study also noted that ammonium sulfate fractionation could remove albumin; however, it is not clear that other proteins are also not removed in this process. Of the four precipitation methods evaluated, trichloroacetic acid and acetone were the most efficient; ultrafiltration was also useful in sample concentrations. In all cases, the protein precipitate was taken up in sample buffer (7.0 M urea, 2.0 M thiourea, 20 mM Tris, pH 7.5, 2% CHAPS, 1,4-dithioerythritol). The samples were not alkylated prior to the two-dimensional gel electrophoresis. In studies on human serum,[14] samples depleted of albumin and IgG were dialyzed into 10 mM NH_4HCO_3, 5% acetonitrile, pH 7.5 prior to digestion with trypsin. In a study with another biological fluid,[15] whole saliva (as differentiated from separate glandular secretions such as parotid saliva or submandibular saliva) was dialyzed into 12 mM ammonium bicarbonate, concentrated by vacuum centrifugation and dissolved into 8.0 M urea, 0.4 mM ammonium bicarbonate, 80 mM methylamine prior to reduction with dithiothreitol followed with iodoacetamide. Serum samples can also be taken into sample buffer by dilution; Ahmed and coworkers[16] took 25 µg protein (depleted serum, protein concentration determined by a technique not described) into rehydration buffer (7.0 M urea, 2 M thiourea, 100 mM dithiothreitol, 4% CHAPS, 0.5% carrier amphopholines in 40 mM Tris with bromphenol blue) to a final volume of 200 µL and then taken to two-dimensional gel electrophoresis. Samples for SELDI (ProteinChip®) systems may or may not be taken into a chaotropic solvent prior to analysis; if the samples are taken into a chaotropic solvent such as urea, urea/thiourea or guanidine, dilution of the sample is required prior to analysis. Wadsworth and coworkers[17] "pretreated" serum with 8.0 M urea, 1% CHAPS and then diluted 1.0 M urea, 0.125% CHAPS in phosphate-buffer saline prior to analysis. Petricoin and coworkers[18] took the serum sample directly to the hydrophobic protein chip. Protein microarray technology is an event in-process but the physiological solvent systems used for ELISA systems should prove to be the satisfactory approach.[19] Samples for multidimensional protein identification are generally taken into a urea-containing solvent, subjected to reduction and carboxymethyl followed by proteolysis for shotgun proteomics.[20]

The above examples and the studies cited below suggest that there are a variety of approaches in the preparation of fluid samples for proteomic analysis (Table 4.1). The critical issue is to clearly understand and record the various steps in the preparation of samples. The increasing availability of "black box" approaches to proteomic analysis is useful to the investigator in providing (hopefully) standardized reagents for use, but this practice should not release the investigator from understanding the processes within the black box. The sample may or may not be alkylated after the reduction step.[47] However, there are issues common to biological fluids, tissue cells and bacterial/yeast cells. For tissue samples, the sample may be intact "whole" tissue or a specific fraction of cells, such as tumor cells obtained by laser capture microdissection, or a subcellular fraction.[48–50]

CELL AND TISSUE EXTRACTION

We now want to explore some specific examples using plant and animal tissues with the understanding that the same considerations apply to studies on bacterial cells and yeast cells. Furthermore, the majority of these studies focus on the study of membrane-associated proteins and, hence, most of the effort is focused on various approaches to solubilization of proteins in preparation for subsequent electrophoretic or chromatographic separation. The focus on membrane proteins reflects, in part, the desire to identify new receptors for pharmaceutical targeting. Issues with sample variation is briefly discussed above and extensively elsewhere (Chapter 8) but, considering the importance of variance, it is again discussed. Variance in intrinsic

TABLE 4.1

Technical Approaches for the Preparation of Samples for Proteomic Analysis

Native

- Function retain
- Antigenic epitopes retained
- Functional domains retained
- Limited (native) proteolysis yields some structural information
- Native electrophoresis — evaluates ligand binding
- Structural analysis and identification problematic using mass spectrometry although possible via SELDI technology
- Microarray binding

Reduced in the Absence of Chaotropic Agents and Detergents

- Secondary and tertiary structure largely retained
- Function may be retained
- Antigenic epitopes may be retained
- Altered binding to microarray platforms
- Disulfide bond exchange studies possible which will yield structural information
- Alkylation will yield structural information
- More extensive digestion than that observed with native proteolysis
- Alkylation will be required for further processing to prevent aggregation/oxidation
- Possible value for shotgun proteomics — would, in principle, yield longer peptides

Reduced with Chaotropic Agents and Detergents

- Loss of function
- Likely loss of antigenic determinants unless such determinants are strictly linear
- Markedly altered binding to microarray platforms
- Likely loss of functional domains
- Necessary for two-dimensional gel electrophoresis or shotgun proteomics

sample quality can be reduced by decreasing the number of steps in the process before analysis. Variance can also be decreased by minimizing the number of transfer steps. Ideally, the sample preparation process should occur in a closed system using a single reaction vessel. With most studies on biomarker identification, since the target is not known prior to the completion of the analysis, calculating a yield is therefore not possible as one could do with the purification of an enzyme from a tissue source. However, once the biomarker is identified, calculating the yield of biomarker resulting from the sample preparation process is possible. This exercise is not expressly value-added but would be a useful piece of documentation to include in a regulatory filing.

The majority of the work described below is intended for the preparation of materials for analysis by two-dimensional gel electrophoresis. Two-dimensional gel electrophoresis is comprised of two steps: an isoelectric focusing (IEF) step and an SDS-electrophoretic separation (SDS-PAGE). This process is an example of an orthogonal separation; the IEF step separates proteins on the basis of charge while the SDS-PAGE separates proteins on the basis of molecular size. The IEF step is performed in the presence of an uncharged chaotropic agent (most often urea) to promote protein solubility, an uncharged or zwitterionic detergent such as CHAPS (also for protein solubility), a reducing agent and ampholines.[21-24] CHAPS (3[(3-cholamidopropyl) dimethylammonio]-1-propanesulfonate, Figure 4.1) is derived from a natural detergent, cholic acid, found in bile.[25,26] Few rigorous studies are found that compare detergents. Fountoulakis and Takács[27] have evaluated sodium dodecyl sulfate, lithium dodecyl sulfate and guanidine hydrochloride in the solubilization of proteins from *Haemophilus influenzae* as compared to the standard urea/CHAPS system. Sodium dodecyl sulfate, lithium dodecyl sulfate or guanidine hydrochloride all provide a quantitative and

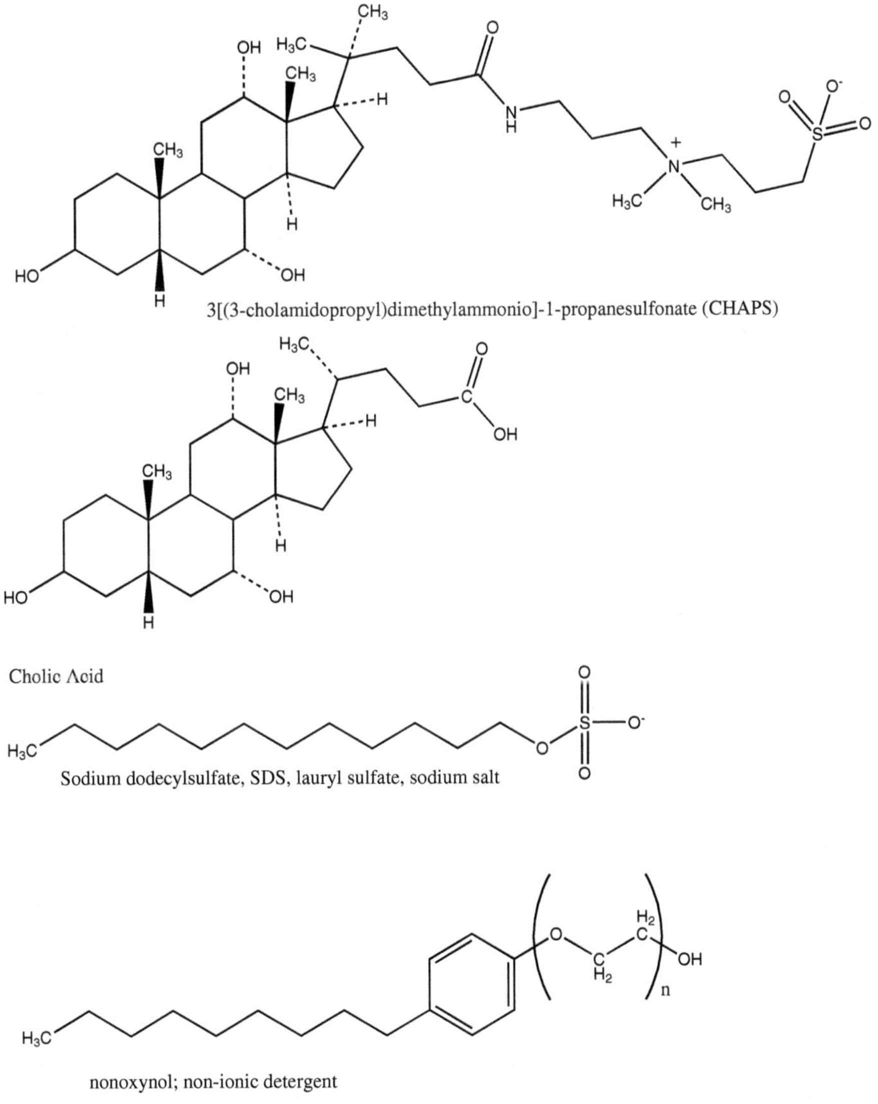

3[(3-cholamidopropyl)dimethylammonio]-1-propanesulfonate (CHAPS)

Cholic Acid

Sodium dodecylsulfate, SDS, lauryl sulfate, sodium salt

nonoxynol; non-ionic detergent

FIGURE 4.1 Some detergents used in proteomic analysis.

qualitative increase in protein; however, issues exist with taking sample of these solvents into the IEF step. Van Eyk and coworkers[28] have evaluated sulfobetaine 3-10 (*N*-decyl-*N,N′*-dimethyl-3-ammonio-1-propane sulfonate (SB3-10) and amidosulfobetaine-14 (ASB-14). With ventricular tissue, the inclusion of ASB-14 in the initial homogenization buffer is useful when followed by IEF in the presence of CHAPS. ASB-14 and SB3-10 were introduced by Chevalet and coworkers[29] in 1997.

Most of the original work on the use of detergents in electrophoretic analysis used sodium dodecyl sulfate (SDS) and consideration of some of the earlier work in detail is useful for the comparison of SDS and the current zwitterionic detergents such as CHAPS; the reader is directed to some additional review articles.[30–41] Historically, sodium dodecyl sulfate is the detergent of choice for separations based on size using poly-acrylamide gel electrophoresis (PAGE);[42] hence, the widely used acronym, SDS-PAGE. SDS was the detergent of choice in O'Farrell's seminal study[22] describing two-dimensional gel electrophoresis. Although SDS has been replaced with either non-ionic or zwitterionic detergents, some intriguing early observations exist that seem to have been lost. Wilson and coworkers[43] adapted O'Farrell's method for the study of eukaryotic cells. They observed that the inclusion of SDS in the solubilization solvent enhanced the reproducibility of gel electrophoretograms but also appeared to cause some artifactual spots as SDS might interfere with the IEF step; the possibility of increased cyanate formation from the urea in the solubilization solvent could not be excluded. Ames and Nikaido[44] also used SDS to solubilize bacterial membrane proteins. The membranes were solubilized in 50 mM Tris-HCl, 2% sodium dodecyl sulate, 20 mM $MgCl_2$ and diluted (1:2) sample buffer (9.5 M urea, 2% ampholines, 5% 2-mercaptoethanol, 8% NP-40). On occasion, additional urea was added to the solubilized material prior to dilution into the sample buffer. These investigators stated that a ratio of NP-40 to SDS of 8:1 is critical to avoid streaking due to charge interactions during IEF. Subsequently, Rubin and Milikowski[45] applied two-dimensional electrophoresis to the analysis of erythrocyte membranes and resolved more that 200 spots. These investigators also noted the importance of SDS in the preparation of the solvent. The reader is referred to Rubin and Leonardi[46] for a review of some of this work.

The IEF solvent (lysis buffer, sample buffer) developed by O'Farrell[22] consisted of 9.5 M urea, 2% NP-40 (Nonidet®-P40, a non-ionic detergent developed by Shell Chemicals; no longer commercially available but there are NP-40 substitutes), 2% ampholines and 5% 2-mercaptoethanol. The basic quality of this solvent has not changed but specific components have changed. In a recent review, Görg and coworkers[51] report a composition for the lysis (sample) buffer of 9.0 M urea, 1% dithiothreitol, 2 to 4% CHAPS and ampholines. The introduction of thiourea (Figure 4.2) to improve protein solubilization was begun by Rabilloud and colleagues in 1997.[52] These investigators explored a variety of reagents including thiourea, formamide, butylurea and methylurea as well as some novel detergents (e.g., myristylamido-bis-(propyldimethyl)-ammoniopropane sulfonate) with the goal of improving protein solubilization. Improved solubility was observed with thiourea; thiourea had to be used in combination with urea because of solubility issues. These investigators found that 2.0 M thiourea with 5 M or 7 M urea with CHAPS seemed to be the most useful combination. These investigators suggest that it is useful to

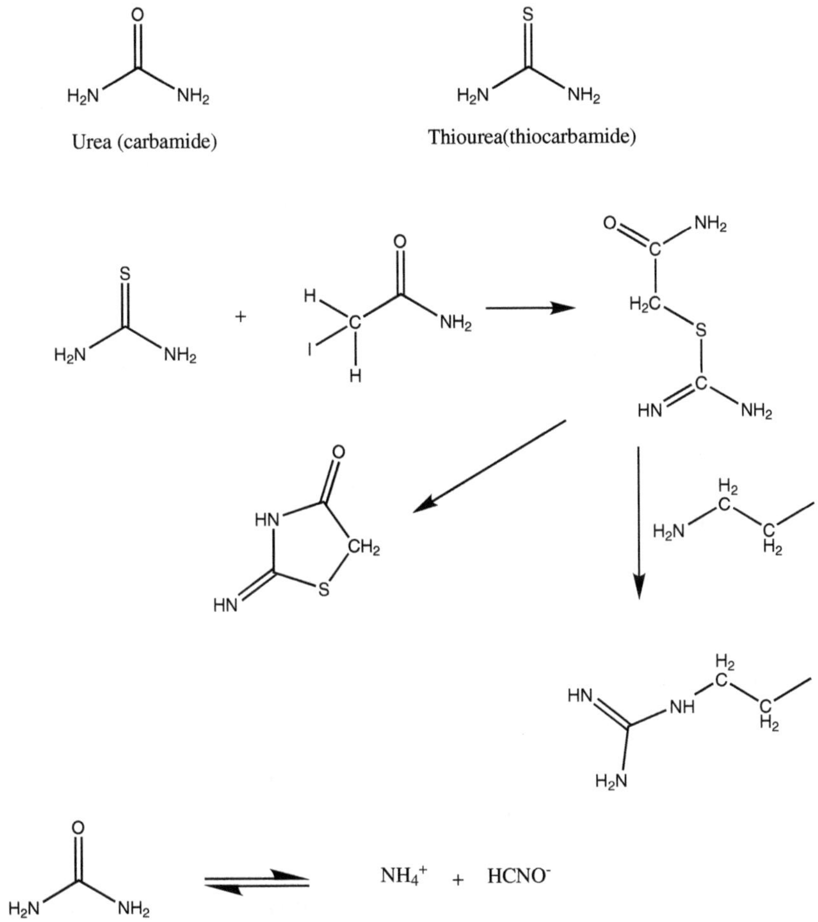

FIGURE 4.2 Thiourea and urea. The reaction of thiourea with iodacetamide.

evaluate several combinations of detergents and chaotropic agents to optimize solubilization for a specific sample. Lanna and coworkers[53] have reported that thiourea improves the proteomic analysis of adipose tissue. Murine adipose tissue was extracted with 7.0 M urea, 2.0 M thiourea with 4% CHAPS or 8.0 M urea with 4% CHAPS. Analysis of two-dimensional gel electrophoretograms with PDQuest V (BioRad Laboratories). Using a 14.0% gel with a load of 25 μg protein, 444 ± 31 spots were found with the thiourea-containing solvent while 299 ± 110 spots with the urea-only solvent ($p = 0.014$); similar results were observed with other gel porosities and protein loads. Herbert has prepared an excellent review of the protein solubilization issue.[54]

Thiourea does present a problem for the processing of samples for proteomic analysis in that it interferes with the alkylation of reduced proteins with haloalkyl compounds.[55] Several possible reaction mechanisms are possible for the reaction of thiourea and iodoacetamide (Figure 4.2). In the absence of a protein nucleophile, the formation of the thiazolinidone monimine is favored; with a protein nucleophile such as lysine, the formation of guanidinated derivative is a possibility; this latter possibility has not been demonstrated. In any event, alkylation of the sulfydryl groups with iodoacetamide would not proceed in the presence of thiourea. Alkylation of the sulfydryl groups can be achieved with acrylamide[56] or maleimide derivatives[57] in the presence of thiourea. Why there is such a difference between the haloalkyl compounds and the alkene derivatives is not immediately clear. Both types of reagents react via a S_N2 mechanism; with acrylamide and maleimides, this is a Michael-type addition. A mechanistic difference exists in that the haloalkyl compounds react with the sulfydryl groups via a nucleophilic displacement reaction while the maleimide react via a nucleophilic addition to the double bond. Consideration of a study comparing the reaction of haloalkyl compounds and maleimide compounds with tubulin[57] would provide additional support for the use of maleimide derivatives in the preparation of samples for two-dimensional gel electrophoresis. Thiourea does provide an added advantage in that it does serve as an inhibitor of proteolytic enzymes via the guanidine-like tautomeric form. This property and the ability of guanidine to serve as an inhibitor of proteases is discussed below.

2-Mercaptoethanol, the reagent for reducing disulfide bonds as described by O'Farrell,[22] has been largely replaced by other agents such as dithiothreitol[58] or the phosphines (Figure 4.3).[59] 2-Mercaptoethanol forms a charged derivative that interfered with the ampholines and is a less potent reducing agent than dithiothreitol, Tris(2-carboxyethyl) phosphine or tributylphosphine. Dithiothreitol is a more effective reducing agent for proteins than 2-mercapoethanol but does present a problem in that it will react with alkylation agents. Tri-*n*-butyl phosphine has been demonstrated to be useful for the solubilization of membrane proteins[59,60] and has been used for the determination of aminothiols in plasma.[61–63] Phosphines are subject to oxidation[64,65] with the rate of oxidation increasing with increasing pH. In a somewhat unrelated study[66] on the analysis of aminothiols (cysteine, homocysteine, glutathione) in plasma, assays utilizing tris(2-carboxyethyl)phosphine are suggested to be more robust and reproducible than those with tri-*n*-butyl phosphine. These authors correctly note that the different polarities of the two phosphine derivatives may influence their reactivity with disulfide bonds. The vapor phase reduction and alkylation of proteins with tri-*n*-butylphosphine and 4-vinylpyridine has been reported.[67] Other earlier studies exist on the use of tri-*n*-butylphosphine that might be useful.[68–70] Cobb and Novotny[69] used a tenfold molar excess of tri-*n*-butylphosphine over disulfide bonds and a 40-fold molar excess of 2-methylaziridine over sulfydryl groups. Tri-*n*-butylphosphine has also been used as a catalyst for the synthesis of *N*-substituted aziridines and epoxides[71,72] and in the synthesis of thizolines from the annulation of of thioamides and 2-alkynoates.[73]

Bal and coworkers[74] reported on the chelating properties of TCEP and note that TCEP is less potent as a chelating agent for metal ions such as Ni^{2+}, Cu^{2+} and Zn^{2+} than other reducing agents. Cupric ions can catalyze the oxidation of TCEP to the oxide. TCEP does have some buffering capacity at neutral pH. (These investigators note that TCEP is an acronym for other organic compounds.) Tris(2-carboxyethyl)phosphine is more effective than dithiothreitol in reducing disulfide bonds and is suggested to be more effective than tri-*n*-butylphosphine.[75] TCEP is more "socially

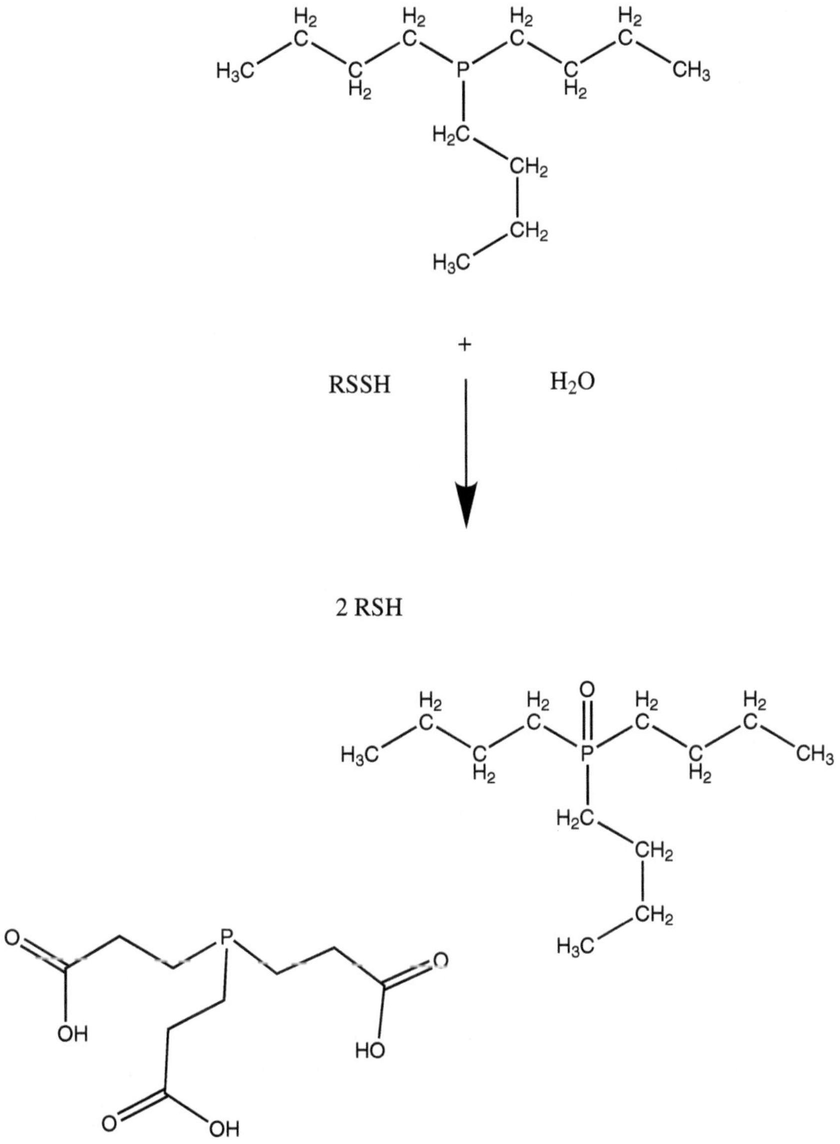

Tris(2-carboxyethyl)phosphine

FIGURE 4.3 Phosphines and their reaction with disulfide bonds.

acceptable" because of the stench properties of tri-*n*-butylphosphine,[76] and TCEP has been used extensively for the reduction of disulfide bonds in proteins.[77–84] Work from Watson's laboratory[83] provides an alternative strategy for protein cleavage by cleavage at cysteine residues after cyanylation providing a structure determination strategy similar to the "top-down" approach with cyanogen bromide.[84] TCEP is also used for the determination of aminothiols in plasma.[66,85,86]

Tris(2-carboxyethyl)phosphine can inhibit the alkylation reaction of reduced proteins by either haloalkyl compounds or maleimide derivatives,[55,56] although one group of investigators did not observe this problem.[87] On the other hand, Cobb and Novotny[69] showed that tri-*n*-butylphosphine and 2-methylaziridine could be used together *in situ* for reduction and alkylation of disulfide bonds. The resulting derivative, unlike the product derived by reaction with aziridine, is not cleaved by trypsin. Granted that the systems are different, this observation is somewhat contrary to the aforementioned observations of Wiktorowicz and coworkers[56] who observed that tris(2-carboxyethyl)phosphine did interfere with the reaction of haloalkyl or maleimide derivatives with reduced protein. The differences in the nucleophilicity of two phosphine derivatives could possibly account for the differences in reactivity and the tri-*n*-butylphosphine derivative would not interfere with the modification of cysteine residues with alkylating agents.

Phosphines and other reducing agents such as dithiothreitol, as well as thiourea, do have the potential to negatively influence the subsequent alkylation reaction. This problem can be surmounted by assuring a sufficient excess of alkylating agent to reducing agent and free protein sulfydryl groups. This approach is useful and economical when common alkylating agents such as iodoacetamide or *N*-ethylmaleimide are used. However, as in the case of fluorescent alkylating reagents such as Cy dyes, this approach could get somewhat expensive. One possible solution to this problem would be the removal of the reduced protein prior to the alkylation step. Crestfield, Moore and Stein[88] used ethanol precipitation of reduced protein prior to carboxymethylation.[89]

The above section provides some information about the properties of the various components included in the solvents for two-dimensional gel electrophoresis. An optimistic goal is to solubilize the proteins in a tissue sample for subsequent analysis; this, of course, is not an issue with serum or plasma samples. An immense challenge is provided for the analytic systems in the dynamic range of analytes and reproducibility of the systems. Few studies exist on the optimization of sample preparation for proteomic analysis, one reason being that the time and expense required for the evaluation, requiring two-dimensional electrophoresis or orthogonal chromatography, is considerable. Sample complexity can be reduced by prefractionation, covered in great detail in Chapter 5. Some examples of tissue extraction, preparation of acetone powers, subcellular fractionation, protein fractionation, affinity chromatography and ion-exchange chromatography are discussed or referenced in this chapter. In addition to the references in Chapter 5, the reader is specifically directed to Volume 104 of *Methods in Enzymology*, for it would appear to be the last comprehensive review of the more classical techniques for protein purification with particular reference to the purification of membrane proteins.

The sheer number of proteins in a proteome preclude resolution of all components with existing systems. As such, most systems will require some type of prefractionation prior to analysis. This procedure is discussed in Chapter 5 for plasma/serum and for laser microdissection capture technology as well as little discussion of ProteinChip® chip technology (surface-enhance laser desorption/ionization [SELDI]); this application is considered to be of an affinity technology also discussed in Chapter 6. A further complication in the review of this area is that each study appears to be unique with respect to sample preparation technology, making the comparison of studies extremely difficult. Complication is provided by the paucity of experimental detail in many papers. If proteomics is to continue to move forward as a discipline, a more rigorous approach will be required.

We are going to focus on one specific approach — sequential extraction, as introduced by Molloy, Herbert and coworkers.[90] This method is, in fact, an example of sample prefractionation but is considered separately to emphasize the various consideration in sample preparation from tissues. In principle, this approach could be used with samples obtained with laser capture prefractionation but the small initial sample size for such samples might present complications. This study has been extensively cited since its publication in 1999. We have selected the seminal study as well as several of the cited studies for consideration in detail. This study[90] used *Escherichia coli* as a test material and, as such, is an example of the study of an intact cell by proteomic technology. A combination of vortex mixing and sonication in 40 mM Tris, pH 9.5 at 10°C yielded 40% of the total protein in the supernatant fraction. Extraction of the pellet from this step with 8.0 M urea, 4% CHAPS, 100 mM DTT, 0.5% ampholines, 150 U endonuclease in 40 mM Tris, pH 9.5 followed by centrifugation yielded 49% of the total protein in the supernatant fraction. Extraction of the pellet fraction from this step with 5 M urea, 2 M thiourea, 2% CHAPS, 2% sulfobetane (3-10), 2 mM tributylphosphine, 0.5% ampholines, 150 U endonuclease in 40 mM Tris, pH 9.5 followed by centrifugation yielded an additional 5% of total protein in the supernatant fraction. The final pellet was solubilized in 1% SDS, 0.375 M Tris-HCl, pH 8.9, 50 mM DTT and 25% (v/v) glycerol with boiling for 5 minutes. Qualitative analysis demonstrated more overlap between the first Tris base extraction and the second extraction with urea; less overlap existed between the urea and urea/thiourea extractions. The terminal extraction of the pellet fraction remaining after the thio/urea extraction with SDS showed less overlap with the previous fractions. A preponderance of acidic proteins was observed in this fraction and clear enrichment of some highly expressed proteins that were not detected in earlier extractions was noted.

Several studies have been selected from the numerous citations to the Molloy et al. study[90] for consideration in detail. Most of the studies in this area have focused on the development of solvent preparation formats designed to improve the recovery of membrane proteins. Focusing on a specific protein fraction would appear to be somewhat contrary to the goal of proteomics but proteomics appears to be as much about technology as it is concept. The general approach has been somewhat described above in that there is first an extraction with a non-denaturing buffer with less than physiological ionic strength to promote cell lysis by removal of the "membrane fraction" by centrifugation. The isolated membrane fractions thus obtained might be a chloroplast membrane fraction,[91] a plant plasma membrane,[92,93] a mitochondrial preparation[94] or intact cells[95] and represent an example of subcellular fractionation to obtain a homogenous population. Membranes are attractive targets for study because of the relevance to pharmaceutical development that targets receptors.

Two groups have used a variation on sequential extraction where chloroform/methanol has been used to extract *Escherichia coli*[95] and chloroplast membrane proteins.[91] Some difference exists in the technical aspects of these two studies;

however, the basic thrust is the initial extraction with chloroform/methanol followed by washing this extract with water to remove the soluble proteins, leaving some unique proteins in the chloroform/methanol fraction. The chloroform/methanol fraction is taken to dryness *in vacuo* and then taken into a sample buffer for electrophoretic analysis. In the study on *Escherichia coli*,[95] lyophilized cells (40 mg) were incubated with 1.5 mL chloroform/methanol (1:1), sonicated and placed at 2 to 4°C for 30 minutes; 600 μl ultrapure water was added with mixing and then placed at 25°C and centrifuged. The aqueous phase is removed and the organic phase concentrated to dryness and taken up in one of two sample solvents. The sample solvent was either 5 M urea, 2 M thiourea, 2% (w/v) CHAPS, 2% (w/v) sulfobetaine 3-10 (*N*-decyl-*N,N*-dimethyl-3-ammonio-1-propane sulfonate), 2 mM TBP, 0.5% ampholines in 40 mM Tris base or 7 M urea, 2 M thiourea, 1% amidosulfobetaine 14, 2 mM TBP, 0.5% ampholines in 40 mM Tris. The pH of either solution was not defined but is likely to 9.5 (not adjusted). The samples were then analyzed by two-dimensional gel electrophoresis. More protein (qualitative evaluation of the electrophoretograms) was obtained with the aminobetaine-14 solvent than with the standard CHAPS/sulfobetaine solvent, however, the resolution of spots is less; more lipid was noticed with the amidosulfobetaine 14 detergent. These data should be compared with those obtained with the above observations of Van Eyk and coworkers[28] on the use of amidobetaine-14 in the proteomic analysis of ventricular tissue. These investigators noted that the amidobetaine-14 was useful when CHAPS was included in the IEF buffer; thus, there is a difference between the solubilization solvent and the rehydration buffer.

The study of chloroplast membranes[91] evaluated a series of chloroform/methanol mixtures with varying ratios for the extraction of of hydrophobic proteins. Isolated chloroplast membrane vesicles (5 to 10 mg/mL) in 50 mM MOPS, 0.5 mM ε-aminocaproic acid, 1 mM benzamidine, pH 7.8 (100 μl) was diluted into 900 μl cold chloroform/methanol (1:8, 2:7. 3:6, 4:5, 5:4, 6:3, 7:2 and 8:1). After 15 minutes at 2 to 4°C, the mixtures were separated by centrifugation. Protein insoluble appeared as a white pellet with the 1:8, 2:7, 3:6, 4:5 and 5:4 samples (v/v, chloroform/methanol); above this ratio (6:3, 7:2 and 8:1, chloroform/methanol), the insoluble protein appeared as a white layer at the interface between the organic phase and the aqueous phase. The organic phase was removed, taken to dryness and taken into 0.1 M Na$_2$CO$_3$, 0.1 M DTT for subsequent analysis by SDS-PAGE. One-dimensional SDS-PAGE provided sufficient resolution to provide samples for mass spectrometry analysis. Chloroform/methanol extraction provides enrichment of hydrophobic proteins and the exclusion of hydrophilic proteins from both chloroplast envelope and thyalkoid membranes. The yield of protein into the chloroform/methanol was 5 to 10%. Chloroform/methanol ratios of 4:5 to 6:3 provided the highest recovery of hydrophobic proteins. Note that the studies on *Escherichia coli*[95] discussed above used chloroform/methanol (1:1). In a subsequent study,[92] Ephritikhine and colleagues used chloroform/methanol (5:4) to extract hydrophobic proteins from purified plasma membrane preparations from *Arabidopsis thaliana*, a small plant in the mustard family which is used as a model organism for genomic studies.[93] This extract was compared to those obtained with 1% Triton-X100 in 50 mM MOPS, pH 7.8 containing 1.0 mM DTT or 0.1 M NaOH in 50 mM MOPS, pH 7.8 containing

1.0 mM DTT. The different solvent systems extracted different classes of proteins (chloroform/methanol: membrane transporters, metabolism; NaOH/MOPS: signaling and cellular components), although there is significant overlap between the separate fractions.

Plant tissues provide a considerable challenge for the preparation of samples for proteomic analysis. Saravanan and Rose[94] state that most studies on plant tissues have used a procedure where proteins are precipitated from a solution of trichloroacetic acid in acetone.[96–102] Jacobs and coworkers[100] reported the proteomic analysis of *Catharanthus roseus*, a medicinal plant used in traditional medicine and as a source for viblastine and vincristin. A quantity of 1 g of frozen cells (or seedlings) was ground in a mortar and pestle previously cooled with liquid nitrogen. Cold (–20°C) 10% trichloroacetic acid in acetone containing 0.07% 2-mercaptoethanol (10 mL) and the suspension allowed to stand at –20°C for 2 hours. A precipitate was obtained by low speed centrifugation and washed (three times) with acetone containing 0.07% 2-mercaptoethanol (–20°C) and dried by lyophilization. The precipitate was extracted first with 9.5 M urea, 2% CHAPS, 0.5% ampholines, 65 mM DTT. After washing the residual precipitate with the same solvent, it was then extracted with 7.0 M urea, 2.0 M thiourea, 4% CHAPS, 0.5% ampholines, 65 mM DTT. Thus, two distinct samples were obtained and the efficacy compared with a single extraction with the urea/thiourea combination. Analysis of these samples with two-dimensional gel electrophoresis demonstrated 918 spots with the urea solvent and 1162 spots with the urea/thiourea combination with 192 spots duplicated in the two extracts. The single urea/thiourea extract yielded 1238 spots. These data suggest that sequential extraction is more effective than a single extraction step.

Saravanan and Rose[94] compared the trichloroacetic acid/acetone extraction/precipitation with phenol extraction. The test tissue was tomato (*Lycopersicon esculentum* cv. Alisa Craig). Frozen plant tissue was ground in liquid nitrogen and the resultant powdered extracted with five volumes of 10% trichloroacetic acid in acetone containing 1% polyvinylpyrrolidone and 2% 2-mercaptoethanol. An alternative trichloroacetic acid/acetone process first extracted the ground plant tissue power with 0.5 M Tris, 0.1 M KCl, 500 mM EDTA, 1 mM PMSF, pH 7.5 with 1% polyvinylpyrrolidone and 2% 2-mercaptoethanol. Four volumes of the trichloroacetic acid/acetone as above were added and the resulting precipitate obtained by centrifugation. The phenol extraction procedure used the preparation of an extract of the ground plant power with 0.5 M Tris, 0.1 M KCl, 500 mM EDTA, 1mM PMSF, 0.7 M sucrose, pH 7.5, the extraction of this solution with an equal volume of phenol saturated with 0.5 M Tris-HCl, pH 7.5, the precipitation of protein from the phenol extract with five volumes of saturated ammonium acetate. The various precipitates were washed with cold methanol followed by cold acetone and then taken up in 7.0 M urea, 2.0 M thiourea, 4% CHAPS, 10 mM DTT, 1.0% ampholines and analyzed by two-dimensional gel electrophoresis. Regardless of tissue (e.g., roots, stems, flowers, leaves), the phenol extract yielded more protein; qualitative differences were also found between the phenol extract and the TCA precipitates. The investigators suggested that both methods be used to obtain a maximum number of spots. As an added bonus, ribulose biphosphate decarboxylase/oxygenase (Rubisco) is reduced in the phenol extract permitting greater resolution of low-abundance proteins. Rubisco is an extremely abundant plant protein presenting similar problems to those observed with albumin in the electrophoretic analysis of serum.

Analysis of integral cell wall proteins is always complicated by the issue of differentiating between non-specifically adsorbed proteins and "free" proteins in the cell from proteins embedded in membranes. Pont-Lezica and coworkers[103] used a sequential extraction process with various combinations of salts and a chelating agent (EDTA) to obtain samples of cell wall proteins from *Aribidopis thaliana*. Cells were obtained from culture and subjected to plasmolysis (25% glycerol, 50% glycerol; final wash with 50% glycerol). The cells were then sequentially exhausted with salt solutions (NaCl, LiCl, Na_2BO_3 or $CaCl_2$) and a chelating agent (EDTA) in several combinations. The protein extracts were prefractionated on anion-exchange or cation-exchange matrices prior to electrophoretic analysis. Both $CaCl_2$ and EDTA caused membrane damage with the concomitant release of intracellular proteins. Evaluation of the various combinations suggested that an initial extraction with 0.15 M NaCl followed by 1.0 M NaCl provided the most effective approach.

Lo and coworkers[104] reported what can only be described as a Herculean effort in the optimization of sample preparation for the proteomic analysis of *Prorocentrum triestinum* I. These investigators compared the sequential extraction process (90) with independent single extractions with four solvents (40 mM Tris, 1 mM DTT, 1 mM ascorbic acid, 5 mM $MgCl_2$ containing polyvinylpyrrolidone, pH 8.7; 8.0 M urea, 2 mM tributylphosphine, 4% CHAPS, 40 mM Tris, 0.2% ampholines; 5.0 M urea, 2.0 M thiourea, 2 mM tributylphosphine, 2% CHAPS, 2%sulfobetaine (3-10), 40 mM Tris, 0.2% ampholines; 10% trichloroacetic acid in acetone with 2.0 mM tributylphosphine). The sequential extraction was performed in the presence and absence of protease inhibitors (aprotinin, trypsin inhibitors, phenylmethylsulfonyl fluoride, EDTA); protease inhibitors and inhibitor "cocktails" are described below. The following conclusions are posited:

- Sonication is more effective than "homogenization" with glass beads.
- Urea buffer extracted more protein but the Tris or TCA/acetone extracts provided samples with better resolution.
- Tris buffer extraction yielded more spots than TCA/acetone and, in addition, provide more specific protein expression patterns.
- Sequential extraction yields more data than independent extraction.

The majority of issues that might be considered somewhat unique to bacterial and plant systems have been discussed above. The general considerations are as valid for eukaryotic systems. A substantial amount of discussion regarding what might be considered issues unique to eukaryotic systems is provided in Chapter 5; as such, a limited amount of discussion will be included here. Borlak and coworkers[105] evaluated three extraction methods for obtaining membrane proteins for proteomic analysis. The methods were based on protein solubility (sequential extraction[70]), detergent solubiization (digitonin lysis, Triton X-100 solublization) and differential centrifugation. The detergent-based extraction provided the highest yield of protein but cleaner preparations of membrane proteins were obtained with either the sequential extraction or differential centrifugation.

TABLE 4.2
Common Protease Inhibitors

Protease Inhibitor	Reference
4-(2-aminoethyl)-benzenesulfonyl fluoride (AEBSF)	1, 2
Phenylmethyl sulfonyl fluoride (PMSF)	3, 4
Diisopropylphosphorofluoridate (DFP)	5, 6
3,4-Dichloroisocoumarin (3,4-DCI)	7, 8
Tosyl Phenylalanine Chloromethyl Ketone (TPCK)	9, 10
Tosyl Lysyl Chloromethyl Ketone (TLCK)	11, 12
Ethylenediaminetetraacetic acid (EDTA)	13, 14
Benzamidine	15, 16
Pepstatin	17, 18
Ortho-phenanthroline (*o*-phenanthroline, 1,10-phenanthroline)	19, 20
E-64	21, 22
Leupeptin (*N*-acetyl-leucyl-leucyl-argininal)	23–26

REFERENCES FOR TABLE 4.2

1. Diatchuk, V., Loten, O., Koshkin, V., Wikstrom, P., and Pick, E., Inhibition of NADPH oxidase activation by 4-(2-aminoethyl)-benzenesulfonyl fluoride, and related compounds, *J.Biol.Chem.* 272, 13292–13301, 1997.
2. Wechck, J.B., Goins,W.F., Glorioso, J.C., and Haai, M.M., Effect of protease inhibitors on the yield of HSV-1-based viral vectors, *Biotechnol.Progress.* 16, 493–496, 2000.
3. Fahrney, D.A. and Gold, A.M., Sulfonyl fluorides as inhibitors of esterases. I. Rates of reaction with acetylcholinesterase, α-chymotrypsin, and trypsin, *J.Amer.Chem.Soc.* 85, 997–1000, 1963.
4. Munger, W.E., Berrebi, G.A., and Henkart, P.A., Possible involvement of CTL granule proteases in target cell DNA breakdown, *Immunol.Res.* 103, 99–109, 1988.
5. Dano, K. and Reich, E., Serine enzymes released by cultured neoplastic cells, *J.Exptl.Med.* 147, 745–757, 1978.
6. Richards, P. and Lees, J., Functional proteomics using microchannel plate detectors, *Proteomics* 2, 256–261, 2002.
7. Harper, J.W., Hemmi, K., and Powers, J.C., Reaction of serine proteases with substituted isocoumarins: Discovery of 3,4-dichloroisocoumarin, a new general mechanism based serine protease inhibitor, *Biochemistry* 24, 1831–1841, 1985.
8. Ames, M. and Steven, F.S., Inhibition of a tumour protease with 3,4-dichloroisocoumarin, pentamidine-isethionate, and guanidino derivatives, *J. Enzyme Inhib.* 8, 213–221, 1994.
8a. Powers, J.C., Asgian, J.L., Ekici, O.D., et al., Irreversible inhibitors of serine, cysteine, and threonine proteases, *Chem.Rev.* 102, 4639–4750, 2002.
9. Schoelmann, G. and Shaw, E., Direct evidence for histidine at the active center of chymotrypsin, *Biochemistry* 2, 252–255, 1963.
10. Bennett, M.J., Van Leeuwen, J.E.M., and Kearse, K.P., Calnexin association is not sufficient to protect T cell receptor α proteins from rapid degradation in CD4[+]CD8[+] thymocytes, *J.Biol.Chem.* 273, 23674–23680, 1998.

11. Shaw, E. and Glover, G., Further observations on substrate-derived chloromethyl ketones that inactivate trypsin, *Archs.Biochem.Biophys.* 139, 398–305, 1970.

12. Anderle, P., Langguth, P., Rubas, W., and Merkle, H.P., *In vitro* assessment of intestinal IGF-1 stability, *J.Pharm.Sci.* 91, 290–300, 2002.

13. Varandani, P.T. and Natz, M.A., Insulin degradation. XVI. Evidence for sequential degradation pathway in isolated liver cells, *Diabetes* 25, 173–179, 1976.

14. Gotch, T., Miyazaki, Y., Kikuehi, K., and Bentley, W.E., Investigation of sequential behavior of carboxyl protease and cysteine protease activities in virus-infected Sf-9 insect cell cultures by inhibition assay, *Appl.Microbiol.Biotechnol.* 56, 742–749, 2001.

15. Supuran, C.T., Scozzafava, A., Briganti, F., and Clare, B.W., Protease inhibitors: synthesis and QSAR study of novel classes of nonbasis thrombin inhibitors incorporating sulfonylguanidine and *O*-methylsulfonylisourea, *J.Med.Chem.* 43, 1793–1806, 2000.

16. Biro, A., Herincs, Z., Fellinger, E., et al., Characterization of a trypsin-like serine protease of activated B cells, *Biochim.Biophys. Acta.* 1624, 60–69, 2003.

17. Roulston, J.E., Sanger, B., and Wathen, C.G., The stability of angiotensin I formed at room temperature in the presence of ethylenediaminetetraacetate to subsequent incubation at 37°C, *J.Clin.Chem.Clin. Biochem.* 21, 703–707, 1983.

17a. Lamcapere, J.J., Robert, J.C., and Thomas-Soumarmon, A., Efficient solubilization and purification of the gastric H+, K+-ATPase for functional and structural studies, *Biochem. J.* 345, 239–245, 2000.

18. Musial, A. and Eissa, N.T., Inducible nitric-oxide synthase is regulated by the proteosome degradation pathway, *J.Biol.Chem.* 276, 24268–24273, 2001.

18a. d-'Avila-Levy, C.M., Souza, R.F., Gomes, R.C., Vermelho, A.B., and Braquinha, M.H., A novel extracellular calcium-dependent cysteine proteinase from *Crithidia deanei*, *Arch.Biochem.Biophys.* 420, 1–8, 2003.

19. Wolz, R.L. and Zwilling, R., Kinetic evidence for cooperative binding of two ortho-phenanthroline molecules to *Astacus* protease during metal removal, *J. Inorganic Biochem.* 35, 157–167, 1989.

19a. Fernandez-Patron, C., Steward, K.G., Zhang, Y., Koivunen, E., Radomski, M.W., and Davidge, S.T., Vascular matrix metalloproteinase-2-dependent cleavage of calcitonin gene-related peptide promotes vasoconstriction, *Circ.Res.* 87, 670–676, 2000.

20. Motta, G., Shariat-Madara, Z., Mahdi, F., Sampaio, C.A., and Schmaier, A.H., Assembly of high molecular weight kininogen and activation of prekallikrein on cell matrix, *Thromb.Haemostas.* 86, 840–847, 2001.

21. Shaw, E., Cysteinyl proteinases and their selective inactivation, *Adv.Enzymol.Relat. AreasMol.Biol.* 63, 217–347, 1990.

22. Greenbaum, D., Medzihradszky, K.F., Burlingame, A.L., and Bogyo, M., Epoxide electrophiles as activity-dependent cysteine protease profiling and discovery tools, *Chem.Biol.* 7, 569–581, 2000.

23. Knowles, S.E. and Ballard, F.J., Selective control of the degradation of normal and aberrant proteins in Reuber H35 hepatoma cells, *Biochem.J.* 156, 609–617, 1976.

24. von Figura, K., Steckel, F., Conary, J., Hasilak, A., and Shaw, E., Heterogeneity in late-onset metachromatic leukodystrophy. Effect of inhibitors of cysteine proteinases, *Am.J.Hum.Genet.* 39, 371–382, 1986.

25. Yamada, T., Shinnoh, N., and Kobayashi, T., Proteases inhibitors suppress the degradation of mutant adrenoleukodystrophy proteins but do not correct impairment of very long chain fatty acid metabolism in adrenoleukodystrophy fibroblasts, *Neurochem. Res.* 22, 233–237, 1997.

26. Musial, A. and Eissa, N.T., Inducible nitric-oxide synthase is regulated by the proteosome degradation pathway, *J.Biol.Chem.* 276, 24268–24273, 2001.

SAMPLE STABILITY ISSUES

Degradation of the sample during processing can be issue. The major problem is proteolysis, although nucleases and phosphatases, for example, can also provide problems. Samples taken into trichloroacetic/acetone are unlikely to undergo degradation due to proteolysis. Protein modification with cyanate or thiocyanate derived from urea or thiourea is a possibility[106] but unlikely considering the purity of the reagents available today;[107] samples heated for a significant period of time before analysis can be an exception. Heterogeneity in the sample can also be induced by direct oxidation and reaction with oxidized lipids.[108,109]

Analyte stability in blood plasma and serum is of considerable importance because of the clinical use of these materials and the reader is directed to several studies which provide a general review of this area.[110–113] While these studies use plasma or serum, the considerations are useful for other biological fluids and cell/tissue extracts. Freezing a sample to provide stability is frequently assumed but in one situation,[114] it does not protect an analyte from degradation.

INHIBITION OF PROTEOLYSIS

While protease inhibitor cocktails have been in use for some time,[115] few rigorous studies exist examining the effect on proteolysis and very few are concerned with proteolytic degradation during sample preparation for proteomic analysis.[116] At the onset, proteolysis is assumed to be a problem and protease inhibitor cocktails are usually included as part of a protocol without the provision of justification. Few studies exist that actually demonstrate the necessity of the inclusion of proteases inhibitors. Several excellent review articles are found in this area that have been overlooked and thus profoundly ignored by the proteomic community. As one would suspect, this defect has not improved the quality of literature in this area. The first is by Salvesen and Nagase,[117] which discusses the inhibition of proteolytic enzymes in great detail. This article should be required reading for individuals using protease inhibitors in proteomic research. Of particular interest is the discussion of the relationship between inhibitor concentration, inhibitor/enzyme binding constants (association constants, binding constants, $t_{1/2}$, inhibition constants, etc.) and enzyme inhibition. For example, with a reversible enzyme inhibitor (such as benzamidine), if the K_i value is 100 nM, a 100 μM concentration of inhibitor would be required to decrease protease activity by 99.9%. Salvesen and Nagase[117] also note the well-known differences in the reaction rate of inhibitors such as DFP and PMSF (Figure 4.4) with the active site of serine proteases. DFP is much faster than PMSF with trypsin but equivalent rates are seen with chymotrypsin. PMSF is included in commercial protease inhibitor cocktails because of its lack of toxicity compared to DFP. Saveson and Nagase do observe that 3,4-dichloroisocoumarin (3,4-DCI, Figure 4.4) as described by Powers and coworkers[118] is faster than either

Phenylmethylsulfonyl fluoride (PMSF)
M.W. 174.2

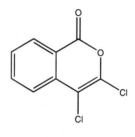

3,4-dichloroisocoumarin

4-(2-aminoethyl)benzenesulfonyl fluoride
Pefabloc SC'; M.W.

Diisopropylphosphorofluoridate

FIGURE 4.4 Some inhibitors of serine proteases.

DFP or PMSF. Also, the issue that enzyme inhibition is occurring in the presence of substrate (proteins) will influence the effectiveness of both irreversible and reversible enzyme inhibitors. In addition, some protease inhibitor cocktails include both PMSF and benzamidine. Benzamidine (Figure 4.5) is a competitive inhibitor of tryptic-like serine proteases and slows the rate of inactivation of such enzymes by reagents such as PMSF.[119]

The readers are also referred to an article by North and Benyon[120] that addresses the issue with respect to prevention of proteolysis by cellular factors. In addition to approaches with the addition of exogenous inhibitors, these investigators note the usefulness of acid pH (extraction with trichloroacetic acid such as that described above for plant tissue[95]) or heat (>100°C). Most proteases are far less active in solution below pH 5.0 but there are exceptions, such as the acid proteases pepsin and rennin. This rather nice article describes the range of protease inhibitors available and suggests concentrations for a mixture (cocktail) of inhibitors. These investigators also discuss some issues unique to individual tissue sources, such as high levels of aspartic acid proteases in yeast, as well as problems, such as the inclusion of a reducing agent (dithiothreitol), that might actually result in the activation of cysteine

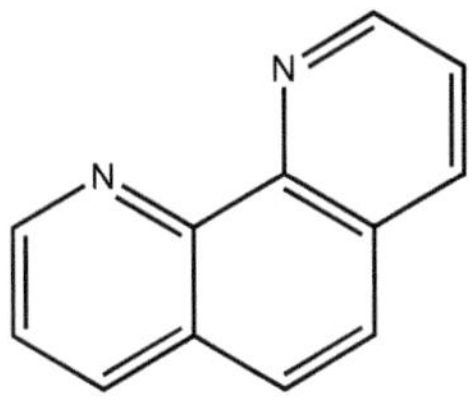

o-phenanthroline
M.W. 180.21

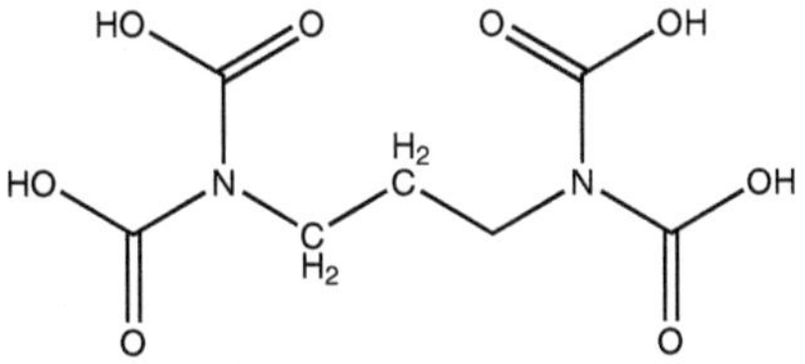

Edetic Acid; EDTA; ethylenediaminetetraacetic acid;
N,N'-1,2-ethanediaminediyl-bis-[N-(carboxymethylglycine
M.W. 292.24

H₂N   NH

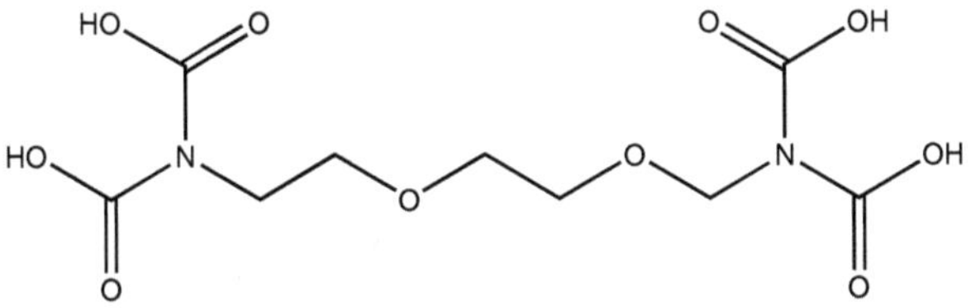

EGTA, 3,12-bis(carboxymethyl)-6,9-dioxa-3,12-diazatetradecanedoicic acid
M.W. 380.35

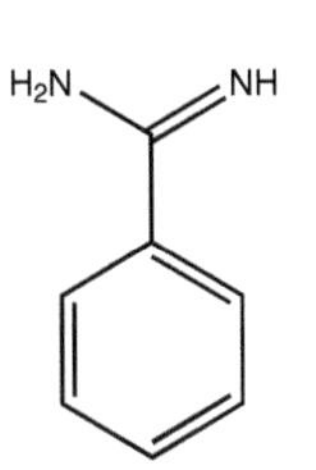

Benzamidine
M.W. 156.6 as
the hydrochloride

FIGURE 4.5 Benzamidine and some metal ion chelators used in proteomic research.

proteases resulting in the inactivation of specific enzymes. Issues such as the instability of inhibitors such as PMSF and DFP in aqueous solutions are also mentioned, although this is more of a problem when kinetics are the object of study.

Several articles are found on proteases that deserve attention. Pringle[120a] presented a review on problems with proteolytic activity in the study of yeast proteins. This article was published in 1975 and has been extensively cited. Pringle discusses the importance of solvent conditions (pH, buffer concentration) and temperature as well as the selective use of exogenous protease inhibitors such as PMSF, DFP, p-chloromercuribenzoate (pCMB, for sulfydryl proteases), pepstatin and chymostatin. Pringle notes the lack of selectivity of pCMB, which would influence the subsequent modification of sulfydryl groups with, for example, ICAT reagents; the same concern would hold for iodoacetamide, used by Weidner and coworkers[120b] for studies on periodontal pathogens such as *Porphyromonas gingivalis*. These latter investigators also studied TLCK and DFP as well as heat and pH. Pridmore and colleagues[120c] extended these observations on *Porphyromonas gingivalis*, comparing TCA extraction with TLCK treatment on two-dimensional gel electrophoresis. Either treatment was useful; failure to control proteolysis resulted in unacceptable analytical results. Hassel and coworkers[120d] studied proteolysis in hydra. These investigators used zymography[120e] to evaluate specific protease inhibitors in the development of a specific cocktail for use with this organism; the custom inhibitor mixture (cocktail) was more effective than a commercial preparation. The custom preparation did contain 4-(2-aminoethyl)-benzenesulfonyl fluoride which does present some problems, as discussed below. Lee and Goldberg[121] reviewed inhibitors of the proteosome with particular emphasis on the use of lactones, vinyl sulfones and boronate inhibitors (Figure 4.6). Inhibitors to capases (Figure 4.7) can also be considered in this group of proteosome inhibitors.

Castellanos-Serra and Paz-Lago[122] studied the effect of a protease (trypsin) and some enzymes thought to be contaminated with pancreatic protease (RNAse and DNAse) on a relatively sensitive substrate (streptokinase) under conditions used during the preparation of a sample for proteomic analysis. The enzymes were present at a relatively high concentration (trypsin to streptokinase ratio of 1:20) to provide a "worst case analysis." Proteolysis was evaluated by electrophoretic analysis. The DNAse had substantial protease activity while the RNAse did not have detectable

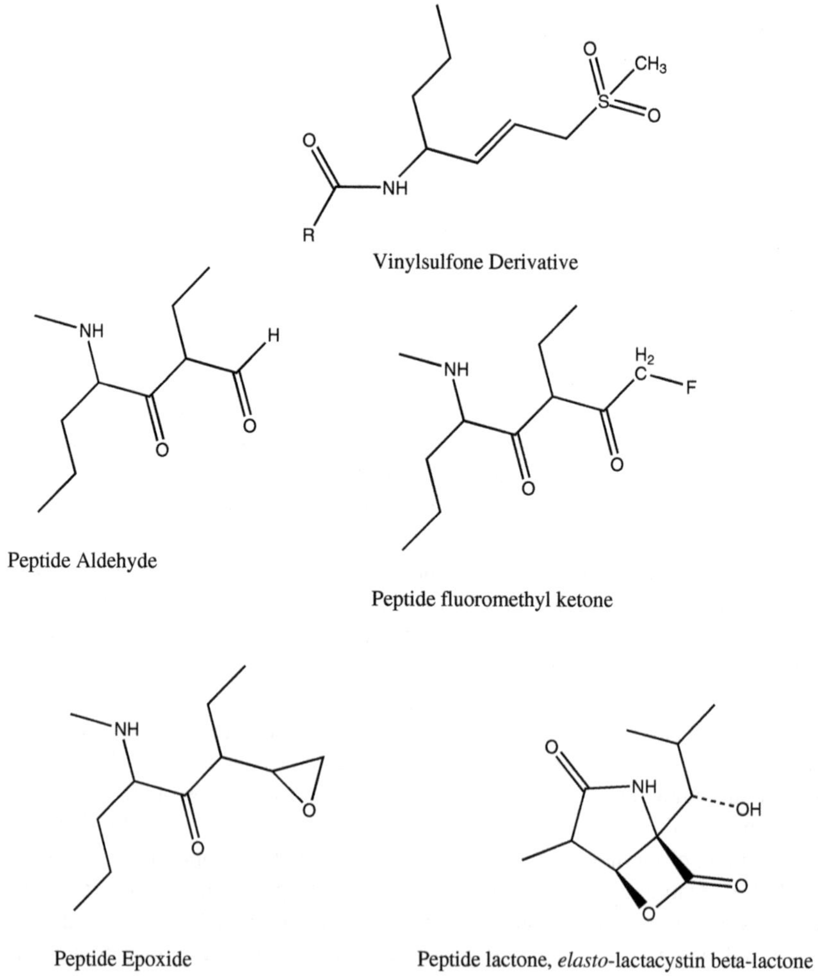

Vinylsulfone Derivative

Peptide Aldehyde

Peptide fluoromethyl ketone

Peptide Epoxide

Peptide lactone, *elasto*-lactacystin beta-lactone

FIGURE 4.6 Peptide vinyl sulfones, peptide aldehydes, peptide chloromethyl ketones, peptide epoxides and peptide lactones.

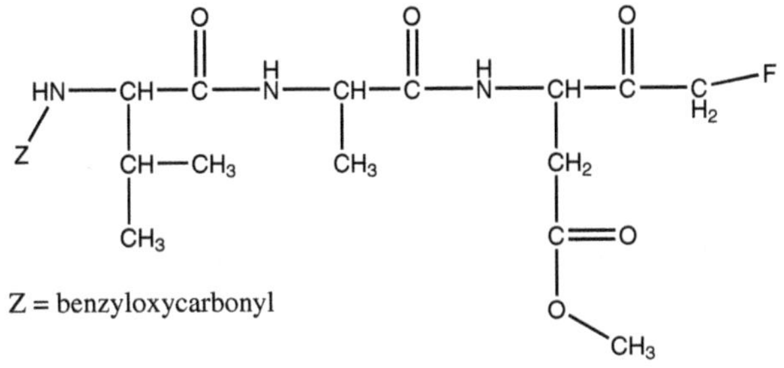

Z = benzyloxycarbonyl

Benzyloxcarbonyl-valylalanylasparate(beta methyl ester) fluoromethyl ketone (ZVAD.FMK)

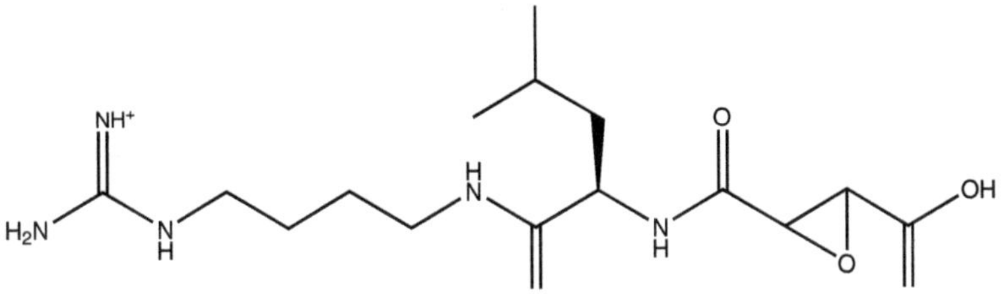

FIGURE 4.7 Some inhibitors of capases.

protease activity. Considerable proteolysis was observed in 40 mM Tris, pH 11.0 (40 mM Tris base). Heating the trypsin/streptokinase mixture in 1% SDS, 1% DTT (5 minutes at 95°C; 10 minutes at 95°C) reduced but did not abolish activity. Interestingly enough, the addition of protease inhibitors did not abolish proteolysis when added after heating (Roche complete proteinase inhibitor, composition not provided and not easily available at the Roche website; presumably contains PMSF, leuptin, EDTA; inhibitors to serine proteases, metalloproteases, cathepsins, Figure 4.4 to Figure 4.7). Inhibitors such as PMSF (phenylmethylsulfonyl fluoride) are relatively slow inhibitors of trypsin ($k_2 = 2.71 \times 10^2$ L·M^{-1}·min^{-1}); it is much more rapid with chymotrypsin ($k_2 = 1.49 \times 10^4$ L·M^{-1}·min^{-1});[123] reaction with both proteases is inhibited by the presence of substrate. The presence of 7.0 M urea did not abolish proteolytic activity but the addition of 2.0 thiourea and CHAPS resulted in the inhibition of proteolytic activity. Thiourea is suggested as a potent inhibitor of trypsin and likely has some resonance forms where there would be a guanidine function with an ionized thiol function (Figure 4.2). The guanidine would be a potent inhibitor of trypsin as the inhibition of proteolysis is observed with guanidine.[123a]

Some suggestion exists that proteolysis may be a common occurrence during two-dimensional electrophoresis.[124] However, protease inhibitors can cause problems as has been demonstrated for 4-(2-aminoethyl)-benzenesulfonyl fluoride (AEBSF, Figure 4.4).[125,126] Other examples exist and the reader is directed to a recent review.[127] Thus, the inclusion of protease inhibitor cocktails as a routine issue may not always be useful. Some samples are found where the inclusion of protease inhibitors is critical, such as for the study of pancreatic juice and pancreatic tissue.[128] In these situations, designing a custom inhibitor formulation is probably more effective (and likely less expensive). Finally, while proteolysis during sample preparation may be a problem, a "branch" of proteomics, protease degradomics, exists that proposes to identify proteases and their substrates within the proteome.[129]

QUALITY ATTRIBUTES FOR A PROTEOMIC SAMPLE

What are the quality attributes of a sample for proteomic analysis? The most frequently used measure seems to be protein concentration and it is infrequently cited with respect to method and standard. Since many of the studies cited herein are hoped to eventually form the basis of a new diagnostic, the fact that many lack adequate experimental information is indeed unfortunate, a fact glaringly obvious with the determination of the protein concentration of the sample. Not only does this determination allow comparison with other samples, it provides information critical for sample loading as well as for pre-electrophoretic chemical modification with fluorescent dyes. Omission of the technique used is common and, when cited, is inaccurate. The reference standard is usually not given. Another omission is an accurate reference to temperature; room temperature or ambient temperature does not provide information of quality for a regulatory submission.

DETERMINATION OF PROTEIN CONCENTRATION IN PROTEOMIC SAMPLES

The determination of protein concentration is a challenge since the samples are frequently prepared in the presence of urea, thiourea and reducing agents that interfere with most commonly used assays. A large number of papers on the use of proteomic techniques for the discovery of biomarkers for the measurement of protein concentration cite a modification[130] of the Coomassie Blue G250 dye-binding method developed by Bradford.[131] This technique is sensitive, robust and reproducible but not necessarily accurate.[132,133]

The Coomassie Blue dye-binding assay depends on a chromogenic response to the binding of the dye to the protein. Coomassie Brilliant Blue R250, a textile dye (magenta family, acid blue 83, Color Index 43660, a triphenylmethane dye) was introduced by St. Groth and colleagues in 1973[134] for protein staining on solid electrophoretic media. These investigators demonstrated quantitative staining with both the Coomassie dye and Procion Brilliant Blue RS. Subsequently, Diezel and coworkers[135] introduced Coomassie Brilliant Blue G250 (Colour Index 42655) which had improved staining properties. These early observations were extended by Reisner and coworkers[136] in 1973 who observed a color change with the G250 dyes in dilute (3.5% w/v) perchloric acid when the dye bound to protein. The dye is brown-orange in color (λ_{max} = 595 nm) which changes to blue (λ_{max} = 695 nm) upon binding to a protein. Despite considerable work as reviewed by Sapan and others[132] and by Lundblad and Price,[133] the structural features on a protein responsible for this blue shift are still not identified. While the assay is reasonably accurate with "average" proteins (MW 30 to 80 kDa, "normal" amino acid composition), it will be grossly inaccurate with smaller proteins and proteins with "abnormal" amino acid compositions such as histones. The Coomassie Blue dye-binding assay is relatively insensitive to interference by common reagents such as EDTA, NaCl and ammonium sulfate; an exception is provided by detergents, phenol and tris buffer.[131]

Ramagli and Rodriguez[130] modified the original Coomassie Blue G250 dye binding procedure by acidification of the reaction mixture and the use of ovalbumin as the standard protein. Ovalbumin is said to be a more accurate standard for the modified Coomassie Blue G250 dye binding assay but the assay is still markedly dependent on protein quality. The modified assay still shows nonlinearity when A595 is plotted vs. protein concentration but is useful as the nonlinearity is not as pronounced as that observed with the commercial Coomassie Brilliant Blue G250 preparations. In addition to the Ramagli and Rodriquez study,[130] a number of modifications are found of the original Coomassie Blue dye-binding assay to improve accuracy and reproducibility.[132,137–141] None of these modifications have been totally satisfactory but deserve consideration for use in proteomic assays.

The message from the above is that the Coomassie Brilliant Blue G250 assay has the potential to be useful for the determination of samples for proteomic analysis. The assay system should be optimized for each sample type and must be validated for each experimental system. Alternatively, the overall assay system (in this case, protein modification followed by two-dimensional electrophoresis) can possibly be demonstrated insensitive to protein concentration over a reasonable range of sample concentration.

The standard protein must be identified such that studies can be compared. Most studies use bovine serum albumin as a standard.[132,133] Immunoglobulin G (IgG) is also used as a standard, as is ovalbumin as discussed above. The best standard is a material that is similar to the test item and the extent to which this can be a problem is provided by a study on the determination of protein concentration in saliva[142] where protein concentration varied from 0.67 mg/mL (Coomassie Blue G250 in phosphoric acid) to 2.37 mg/mL (Biuret) using bovine serum albumin as the standard; when polylysine was used as the standard, the values varied from 2.13 mg/ml. (Biuret) to 65.5 (Coomassie Blue G250 with HCl). Given these results, the choice of assay and standard deserves consideration and preparation of a standard composed of cell lysates might be useful for some of the studies described below. If the samples do not contain urea or thiourea, use of a ninhydrin-based assay as described by Starcher[143] might be possible. Finally, diluting a portion of the sample for the measurement of protein concentration at 205 nm as described by Scopes[144] might also be possible. A fluorescence-based solid phase protein assay has recently been reported[145] which looks to be a promising technique. This assay is a solid-phase assay where protein is applied to the assay paper. A proprietary fluorescent reagent is added and protein concentration is determined by fluorescence using a fluorescence microplate reader. A dynamic range of 10 ng to 5 µg is claimed by these investigators; protein-to-protein variability is said to be comparable to the bicinchoninic acid assay or the Lowry assay. The assay can be performed in the presence of components of IEF sample buffer such as urea, detergents and reducing agents.

The above approaches are certainly worth considering. However, somewhat remarkable is that more attention has not been given to the amido black (Amidoschwarz 10B) assay developed by Schaeffner and Weissman.[146] This assay is based on the quantitative precipitation of protein from solution with trichloroacetic acid, the capture of the precipitated material by filtration followed by quantitative measurement of captured protein with amidoschwarz dye (amido black). This study has been cited 2356 times (ISI) since its publication in 1973. The original assay used the addition of an equal amount of 60% trichloroacetic acid to a final concentration of 10%. The precipitate is captured by filtration and stained with amidoschwarz dye in methanol/glacial acetic acid/H_2O. The protein is visualized as a blue spot on an almost colorless background. The spot is excised from the filter, eluted with 25 mM NaOH, 0.05 mM EDTA in 50% aqueous ethanol. The absorbance of the eluate at 630 nm is determined and concentration is determined by comparison with the results obtained with a standard protein. This assay has been used for the determination of protein concentration in grape juices and wines,[147] low concentration of protein in phospholipids[148] and the *Escherichia coli* multidrug transporter EmrE[149] in the presence of detergents. Of direct relevance to the various proteomic analyses described herein are the studies of Elaine and coworkers[150] on the determination of protein concentration of a *Medicago truncatula* root microsomal fraction with the amidoblack assay in a solution composed of 7.0 M urea, 2.0 M thiourea, 4% CHAPS (w/v), 0.1%(w/v) Triton X-100, 2 mM tributylphosphine, 2% ampholines. The reader is also recommended the study by Tate and coworkers[149] who validated the amidoblack assay with quantitative amino acid analysis, a recommend method of choice for accurate determination of protein concentration.[133] The study with lactose permease[151] used the bicinchoninic acid assay[152] for the determination of protein concentration as this assay can be used with protein bound to a surface.[132,133]

SAMPLE QUALITY

The above discussion addresses the measurement of the quantity of sample in terms of protein content. The second issue concerns sample quality. Is the extract an accurate measure of the protein content of the cell or tissue and is the extraction method reproducible?

The question of reproducibility is addressed by repeating the sample preparation process. The number of times required should be determined in consultation with a biostatistician. The issue in sample quality is a decision on what measurements will be used to monitor sample quality. This issue is of particular importance when multiple samples are being compared, such as would be the situation in the identification of a biomarker. Ideally, a biomarker is one or several "unique" factors whose expression would be indicative of a pathology. Such a determination depends on the use of other markers that can be regarded as constant. Protein concentration is an appropriate normalizing factor but issues persist with the use of this measurement as discussed above. The use of other markers of the biological landscape that could be used as guides in the identification of biomarkers is then useful to consider.

Stewart and coworkers[153] used the relative variation of the intensity of endogenous peptide species and the presence of one or more trans-membrane domains.[154,155] An approach can use the expression of "housekeeping" proteins as an internal control in proteomic studies. Housekeeping genes and their protein products are defined as those genes or proteins that are responsible for basic cell functions which are not likely to change as result of external changes; examples include glyceraldehydes-3-phosphate dehydrogenase, glutamate dehydrogenase, cyclophilin and tubulin.[156–159] While potentially useful, some marker proteins such as glutamate dehydrogenase exist as both a housekeeping form (heat-resistant) and tissue-specific form (heat-labile)[160] which can present some difficulty. Banks and colleagues[160] discuss this in further detail with respect to proteomics.

CONCLUSIONS

This chapter has attempted to provide an overview of some of the issues associated with the reproducible preparations of samples for proteomic analysis. Each tissue and fluid will provide some unique challenges but there are common issues that will depend to some extent on the analytical technique. For example, with shotgun proteomics, somewhat less concern exists about proteolytic degradation of the sample than there might be with surface enhanced laser desorption/ionization-mass spectrometry (SELDI-MS). Fourier transform ion cyclotron resonance (FTICR) mass spectrometry is exquisitely sensitive but somewhat less forgiving of errors in sample preparation. The use of standard operating procedures (SOP) for the preparation of sample will be regarded by some as being more arduous than necessary; however, the benefits in reproducible sample preparation are worth the effort.

REFERENCES

1. Molloy, M.P., Brzezinski, E.E., Hang, J. et al., Overcoming technical variations and biological variations in quantitative proteomics, *Proteomics*, 3, 1912, 2003.
2. Downey, T., Using statistics to improve the quality of genomic and proteomic data, *Sci. Comp*, 30, August, 2004.
3. Blomberg, A., Blomberg, L., Norbeck, J. et al., Intralaboratory reproducibility of yeast protein patterns analyzed by immobilized pH gradient two-dimensional gel electrophoresis, *Electrophoresis*, 16, 1935, 1995.
4. Williams, G.Z., Young, D.S., Stein, M.R. and Cotlove, E., Biological and analytic components of variation in long-term studies of serum constituents in normal subjects, *Clin. Chem.*, 16, 1016, 1970.
5. *World Proteomics Sample Preparation Market*, Frost & Sullivan, Williamstown, MA, 2004.
6. Bodovitz, S., Dunphy, J. and Gentile, F.M., The proteomics markets, in *Protein Arrays, Biochips, and Proteomics*, J.S. Albala and J. Humphrey-Smith, Eds., Marcel Dekker, New York, NY, Ch. 14, 337, 2003.
7. Bodovitz, S. and Joos, T., The proteomics bottleneck: Strategies of preliminary validation of potential biomarkers and drug targets, *Trends Biotech.*, 22, 4, 2004.

186 The Evolution from Protein Chemistry to Proteomics

8. Boenish, O., Ehmke, K.D., Heddergott, A. et al., C-Reactive protein and cytokine plasma levels in hemodialysis patients, *J. Neph.*, 15, 547, 2002.
9. Molls, R.R., Ahluwalia, N., Fick, T. et al., Inter- and intra-individual variation in test of cell-mediated immunity in young and old women, *Mech. Aging Dev.*, 124, 619, 2003.
10. Anderle, M., Ray, S., Lin, H., Becker, C. and Joho, K., Quantifying reproducibility for differential proteomics: Noise analysis for protein liquid chromatography-mass spectrometry of human serum, *Bioinformatics*, 20, 3575, 2004.
11. McCormack, A.L., Schieltz, D.M., Goode, B. et al., Direct analysis and identification of proteins in mixtures by LC/MS/MS and database searching at the low-femtomole level, *Anal. Chem.*, 69, 767, 1997.
12. Choe, L.H. and Lee, K.H., Quantitative and qualitative measure of intralaboratory two-dimensional protein gel reproducibility and the effects of sample preparation, sample load, and image analysis, *Electrophoresis*, 24, 3500, 2003.
13. Jiang, L., He, L. and Fountoulakis, M., Comparison of protein precipitation methods for sample preparation prior to proteomic analysis, *J. Chrom. A*, 1023, 317, 2004.
14. Adkins, J.N., Varnum, S.M., Auberry, K.J. et al., Toward a human blood serum proteome. Analysis by multidimensional separation coupled with mass spectrometry, *Mol. Cell. Proteomics*, 1, 947, 2002.
15. Wilmarth, P.A., Riviere, M.A., Rustvold, D.L. et al., Two-dimensional liquid chromatography study of the human whole saliva proteome, *J. Proteome Res.*, 3, 1017, 2004.
16. Ahmed, N., Barker, G., Oliva, K.T. et al., Proteomic-based identification of haptoglobin-1 precursor as a novel circulating biomarker of ovarian cancer, *Brit. J. Cancer*, 91, 129, 2004.
17. Wadsworth, J.T., Somers, K.D., Cazares, L.H. et al., Serum protein profiles to identify head and neck cancer, *Clin. Cancer Res.*, 10, 1625, 2004.
18. Petricoin, E.F., III., Ardekani, A.M., Hitt, B.A. et al., Use of proteomic patterns in serum to identify ovarian cancer, *Lancet*, 359, 572, 2002.
19. Huang, R.-P., Detection of multiple proteins in an antibody-based protein microarray system, *J. Immunol. Meth.*, 255, 1, 2001.
20. Wolters, D.A., Washburn, M.P. and Yates, J.R., III, An automated multidimensional protein identification technology for shotgun proteomics, *Anal. Chem.*, 73, 5683, 2001.
21. Ui, N., Isoelectric points and conformation of proteins. I Effect of urea on the behavior of some proteins in isoelectric points, *Biochim. Biophys. Acta*, 229, 567, 1971.
22. O'Farrell, P.H., High resolution two-dimensional electrophoresis of proteins, *J. Biol. Chem.*, 250, 4007, 1975.
23. Richard, P., Protein electrophoresis, in *Molecular Biomethods Handbook*, R. Rapley and J.M. Walker, Eds., Humana Press, Towata, NJ, Ch. 32, 413–424, 1998.
24. Patel, D. and Rickwood, D., Electrophoresis methods, in *Protein Labfax*, N.C. Price, Ed., BIOS Scientific Publishers, Oxford, United Kingdom, Ch. 9, 95–99, 1999.
25. Perdew, G.H., Schaup, H.W. and Selivonchick, D.P., The use of a zwitterionic detergent in two-dimensional gel electrophoresis of trout liver microsomes, *Anal. Biochem.*, 135, 453, 1983.
26. Tostet, C., Chamml, S., Chevallet, M. et al., Structure-efficiency relationships in zwitterionic detergents as protein solubilizers in two-dimensional electrophoresis, *Proteomics*, 3, 111, 2003.
27. Fountoulakis, M. and Takács, B., Effect of strong detergents and chaotropes on the detection of proteins in two-dimensional gels, *Electrophoresis*, 22, 1593, 2001.

28. Stanley, B.A., Neverova, I., Brown, H.A. and Van Eyk, J.E., Optimizing protein solubility for two-dimensional gel electrophoresis analysis of human myocardium, *Proteomics*, 3, 815, 2003.

29. Chevalet, M., Santoni, V., Poinas, A. et al., New zwitterionic detergents improve the analysis of membrane proteins by two-dimensional electrophoresis, *Electrophoresis*, 19, 1901, 1998.

30. Hjelmeland, L.M. and Chrambach, A., Solubilization of functional membrane proteins, *Meth. Enzymology*, 104, 305, 1984.

31. Furth, A.J., Bolton, H., Potter, J. and Priddle, J.D., Separating detergents from proteins, *Meth. Enzymology*, 104, 318, 1984.

32. Lichtenberg, D., Robson, R.J. and Dennis, E.A., Solubilization of phospholipids by detergents. Structural and kinetic aspects, *Biochim. Biophys. Acta.*, 737, 285, 1983.

33. Racker, E., Resolution and reconstitution of biological pathways from 1919 to 1984, *Fed. Proc.*, 42, 2899, 1983.

34. Reynolds, J.A. and McCaslin, D.R., The role of detergents in membrane reconstitution, *Subcell. Biochem.*, 14, 1, 1989.

35. Rosebusch, J.P., The critical role of detergents in the crystallization of membrane proteins, *J. Struct. Biol.*, 104, 134, 1990.

36. Henry, G.D. and Sykes, B.D., Methods to study membrane protein structure in solution, *Meth. Enzymology*, 239, 515, 1994.

37. Fujimoto, K., SDS-digested freeze-fracture replica labeling electron microscopy to study the two-dimensional distribution of integral membrane proteins and phospholipids in biomembranes: Practical procedure, interpretation and application, *Histochem. Cell Biol.*, 107, 87, 1997.

38. Jones, M.N., Surfactants in membrane solubilisation, *Intl. J. Pharm.*, 177, 137, 1999.

39. le Maire, M., Champiel, P. and Moller, J.V., Interaction of membrane proteins and lipids with solubilizing detergents, *Biochim. Biophys. Acta*, 1508, 86, 2000.

40. Rigaud, J., Chami, M., Lambert, O., Levy, D. and Ranck, J., Use of detergents in two-dimensional crystallization of membrane proteins, *Biochim. Biophys. Acta*, 1608, 112, 2000.

41. Shogomori, H. and Brown, D.A., Use of detergents to study membrane rafts: The good, the bad, and the ugly, *Biol. Chem.*, 384, 1259, 2003.

42. Laemmli, U.K., Cleavage of structural proteins during the assembly of the head of bacteriophage T4, *Nature*, 227, 680, 1970.

43. Wilson, D.L., Hall, M.E., Stone, G.C. and Rubin, R.W., Some improvements in two-dimensional gel electrophoresis of proteins. Protein mapping of eukaryotic tissue extracts, *Anal. Biochem.*, 83, 33, 1977.

44. Ames, G.F.-L. and Nikaido, K., Two-dimensional electrophoresis of membrane proteins, *Biochemistry*, 15, 616, 1976.

45. Rubin, R.W. and Milikowski, C., Over two hundred polypeptides resolved from human erythrocyte membranes, *Biochim. Biophys. Acta*, 509, 100, 1978.

46. Rubin, R.W. and Leonardi, C.L., Two-dimensional polyacrylamide gel electrophoresis of membrane proteins, *Meth. Enzymology*, 96, 184–192, 1983.

47. Herbert, B., Galvani, M., Hamdan, M. et al., Reduction and alkylation of proteins in preparation of two-dimensional map analysis: Why, when, and how? *Electrophoresis*, 22, 2046, 2001.

48. Brunet, S., Thibault, P., Gagnon, E. et al., Organelle proteomics: Looking at less to see more, *Trends Cell Biol.*, 13, 629, 2003.

49. Melle, C., Ernst, G., Schimmel, B. et al., Biomarker discovery and identification in laser microdissected head and neck squamous cell carcinoma with ProteinChip® technology, two-dimensional gel electrophoresis, tandem mass spectrometry, and immunohistochemistry, *Mol. Cell. Proteomics*, 2, 443, 2003.

50. Guillemin, I., Becker, M., Ociepka, K., Friauf, E. and Northwang, H.G., A subcellular prefractionation protocol for minute amounts of mammalian cell cultures and tissues, *Proteomics*, 5, 35, 2004.

51. Görg, A., Obermaier, C., Boguth, G. et al., The current status of two-dimensional electrophoresis with immobilized pH gradients, *Electrophoresis*, 21, 1037, 2000.

52. Rabilloud, T., Aolessi, C., Giraudel, A. and Lunardi, J., Improvement of the solubilization of proteins in two-dimensional electrophoresis with immobilized pH gradients, *Electrophoresis*, 18, 307, 1997.

53. Lanne, B., Potthast, F. and Höglund, Å., Thiourea enhances mapping of the proteome from murine white adipose tissue, *Proteomics*, 1, 819, 2001.

54. Herbert, B., Advances in protein solubilization for two-dimensional electrophoresis, *Electrophoresis*, 20, 660, 1999.

55. Galvani, M., Rovatti, L., Hamdan, M., Herbert, B. and Righetti, P.G., Protein alkylation in the presence/absence of thiourea in proteome analysis: A matrix assisted laser desorption/ionization-time-of-flight-mass spectrometry investigation, *Electrophoresis*, 22, 2066, 2001.

56. Tyagarajan, K., Pretzer, E. and Wiktorowicz, J.E., Thiol-reactive dyes for fluorescence labeling of proteomic samples, *Electrophoresis*, 24, 2348, 2003.

57. Britto, P.J., Knipling, L. and Wolff, J., The local electrostatic environment determines cysteine reactivity of tubulin, *J. Biol. Chem.*, 277, 29018, 2002.

58. Cleland, W.W., Dithiothreitol, a new protective reagents for SH groups, *Biochemistry*, 2, 480, 1962.

59. Herbert, B.R., Molloy, M.P., Gooley, A.A., Walsh, G.J., Bryson, W.G. and Williams, K.L, Improved protein solubility in two-dimensional electrophoresis using tributylphosphine as reducing agent, *Electrophoresis*, 19, 845, 1998.

60. Shaw, M.M. and Riderer, B.M., Sample preparation for two-dimensional electrophoresis, *Proteomics*, 3, 1408, 2003.

61. Toyo'oka, T. and Imai, K., High performance liquid chromatography and fluorometric detection of biological important thiols, derivatized with ammonium-7-fluorobenzenzo-2-oxa-1,3-diazole-4-sulphonate (SBD-Fl), *J. Chrom.*, 282, 495, 1981.

62. Feussner, A., Ralinsko, B., Weiss, N. et al., Determination of total homocysteine in human plasma by isocratic high-performance liquid chromatography, *Eur. J. Clin. Chem. Biochem.*, 35, 687, 1997.

63. Zinello, A., Carro, C., Galistu, F. et al., N-Methyl-D-glucosamine improves the laser-induced fluorescence capillary electrophoresis performance in total plasma thiols measurement, *Electrophoresis*, 24, 2796, 2003.

64. Rüegg, U.T. and Rudinger, J., Reductive cleavage of cysteine disulfides with tributylphosphine, *Meth. Enzymology*, 47, 111, 1977.

65. Burns, J.A., Butler, J.C, Moran, J. and Whitesides, G.M., Selective reduction of disulfides by tris(2-carboxyethyl)phosphine, *J. Org. Chem.*, 56, 2648, 1991.

66. Krijt, J., Vackova, M. and Kožich, V., Measurement of homocysteine and other amino thiols in plasma: Advantages of using tris-(2-carboxyethyl) phosphine as reductant compound with tri-n-butylphosphine, *Clin. Chem.*, 47, 1821, 2001.

67. Amons, R., Vapor-phase modification of sulfydryl groups in proteins, *FEBS Letters*, 212, 68, 1987.

68. Kirley, T.L., Reduction and fluorescent labeling of cysteine-containing proteins for subsequent structural analysis, *Anal. Biochem.*, 180, 231, 1989.

69. Cobb, K.A. and Novatny, M.V., Peptide mapping of complex proteins at low-picomole level with capillary electrophoresis separation, *Anal. Chem.*, 64, 879, 1992.

70. Chin, C.C. and Wold, F., The use of tri-*n*-butylphosphine and 4-(aminosulfonyl)-7-fluoro-2,1,3-benzoxadiazole in the study of protein sulfydryls and disulfides, *Anal. Biochem.*, 214, 128, 1993.

71. Hou, X.-L., Fan, R.-H. and Dai, L.-X., Tributylphosphine: A remarkable promoting reagent for the ring-opening reaction of aziridines, *J. Org. Chem.*, 67, 5295, 2002.

72. Fan, R.-H. and Hou, X.-L., Efficient ring-opening reaction of expoxides and aziridines promoted by tributylphosphine in water, *J. Org. Chem.*, 68, 726, 2003.

73. Liu, B., Davis, R., Joshi, B. and Reynolds, D.W., Phosphine-catalyzed annulation of thioamides and 2-alkynoates: A new synthesis of thiazolines, *J. Org. Chem.*, 67, 4595, 2002.

74. Krel, A., Latajka, R., Bujacz, G.D. and Bal, W., Coordination properties of tris(2-carboxyethyl)phosphine, *Inorg. Chem.*, 42, 1994–2003, 2003.

75. Chan, L.L., Lo, S.C.-L. and Hodgkiss, I.J., Proteomic study of a model causative agent of harmful red tide, *Prorocentrum triestinum* I: Optimization of sample preparation methodologies for analyzing with two-dimensional electrophoresis, *Proteomics*, 2, 1169, 2002.

76. Anderson, M.T., Turdell, J.R., Voehringer, D.W. et al, An improved monobromobimane assay for glutathione utilizing tris(2-carboxyethyl)phosphine as reductant, *Anal. Biochem.*, 277, 107, 1999.

77. Gray, W.P., Echistatin disulfide bridges: Selective reduction and linkage assignment, *Protein Sci.*, 2, 1749, 1993.

78. Wu, J., Gage, D.A. and Watson, J.T., A strategy to locate cysteine residues in proteins by specific chemical cleavage followed by matrix-assisted laser desorption/ionization time-of-flight mass spectrometry, *Anal. Biochem.*, 235, 161, 1996.

79. Lundell, N. and Schreitmuller, T., Sample preparation for peptide mapping — A pharmaceutical quality-control perspective, *Anal. Biochem.*, 266, 31, 1999.

80. Tetenbaum, J. and Muller, L.M., A new spectroscopic approach to examining the role of disulfide bonds in the structure and unfolding of soybean trypsin inhibitor, *Biochemistry*, 40, 12215, 2001.

81. Bergendahl, V., Anthony, L.C., Heyduk, T. and Burgess, R.R., On-column tris-(2-carboxyethyl)phosphine reduction and IC5-maleimide labeling during purification of an RpoC fragment on a nickel-nitrilotriacetic acid column, *Anal. Biochem.*, 307, 368, 2002.

82. Gray, W.R., Disulfide structures of highly bridged peptides: A new strategy for analysis, *Protein Sci.*, 2, 1732, 1993.

83. Qi, J., Wu, J., Somkuti, G.A. and Watson, J.T., Determination of the disulfide structure of silucin, a highly knotted, cysteine-rich peptide by cyanylation/cleavage mass mapping, *Biochemistry*, 40, 4531, 2001.

84. Kelleher, N.L., Hong, Y., Valaskovic, G.A. et al., Top down versus bottom up protein characterization by tandem high-resolution mass spectrometry, *J. Am. Chem. Soc.*, 121, 806, 1999.

85. Pfeiffer, C.M., Huff, D.L. and Gunter, E.W., Rapid and accurate HPLC assay for plasma total homocysteine and cysteine in a clinical laboratory setting, *Clin. Chem.*, 45, 290, 1999.

86. Garcia, J.J. and Apitz-Castro, R., Plasma homocysteine quantification. An improvement of the classical high-performance liquid chromatographic method with fluorescence detectors for the thiol-SBD derivatives, *J. Chrom. B*, 779, 359, 2002.

87. Shaw, J., Rowlinson, R., Nickson, J. et al., Evaluation of saturation labelling two-dimensional difference gel electrophoresis fluorescent dyes, *Proteomics*, 3, 1181, 2003.

88. Crestfield, A.M., Moore, S. and Stein, W.H., The preparation and enzymatic hydrolysis of reduced and *S*-carboxymethylated proteins, *J. Biol. Chem.*, 238, 622, 1963.

89. Hirs, C.H.W., Reduction and *S*-carboxymethylation of proteins, *Meth. Enzymology*, 11, 199, 1967.

90. Molloy, M.P., Herbert, B.R., Walsh, B.J. et al., Extraction of membrane proteins by differential solubilization for separation using two-dimensional gel electrophoresis, *Electrophoresis*, 19, 837, 1998.

91. Ferro, M., Seigneurin-Berny, D., Rolland, N. et al., Organic solvent extraction as a versatile procedure to identify hydrophobic chloroplast membrane proteins, *Electrophoresis*, 21, 3517, 2000.

92. Chua, N.-H., Electrophoretic analysis of chloroplast-protein, *Meth. Enzymology*, 69, 434, 1980.

93. Marmagne, A., Rouet, M.-A., Ferro, M. et al., Identification of new intrinsic proteins in *Arabidopsis* plasma membrane proteome, *Mol. Cell. Proteomics*, 3, 675, 2004.

93a. Meinke, D.W., Cheng, J.M., Dean, C., Rounsley, S.D. and Koorneef, M., *Arabidopsis thaliana*: A model plant for genomic analysis, *Science*, 282, 662, 1998.

94. Saravanan, R.S. and Rose, J.K.C., A critical evaluation of sample extraction techniques for enhanced protein analysis of recalcitrant plant tissues, *Proteomics*, 4, 2522, 2004.

95. Molloy, M.P., Herbert, B.R., Williams, K.L. and Gooely, A.A., Extraction of *Escherichia coli* proteins with organic solvents prior to two-dimensional electrophoresis, *Electrophoresis*, 20, 701, 1999.

96. Granier, F., Extraction of plant proteins for two-dimensional electrophoresis, *Electrophoresis*, 9, 712, 1988.

97. Damerval, C., Quantification of silver-stained proteins resolved by two-dimensional electrophoresis: Genetic variability as related to abundance and solubility in two maize lines, *Electrophoresis*, 15, 1573, 1994.

98. Barraclough, D., Obenland, D., Laing, W. and Carroll, T., A general method for two-dimensional protein electrophoresis of fruit samples, *Postharvest Biol. Tech.*, 32, 175, 2004.

99. Wang, W., Scali, M., Vignani, R. et al., Protein extraction for two-dimensional electrophoresis from olive leaf, a plant tissue containing high levels of interfering compounds, *Electrophoresis*, 24, 2369, 2003.

100. Jacobs, D.T., van Rijssen, M.S., van der Heijden, R. and Verpoorte, R., Sequential solubilization of proteins precipitated with trichloroacetic acid in acetone from cultured *Catharanthus roseus* cells yield 52% more spots after two-dimensional electrophoresis, *Proteomics*, 1, 1345, 2001.

101. Islam, N., Lonsdale, M., Upadehyayan, N.M. et al., Protein extraction for mature rice leaves for two-dimensional gel electrophoresis and its application in proteome analysis, *Proteomics*, 4, 1903, 2004.

102. Trisirroj, A., Jeyachok, N. and Chen, S.T., Proteomics characterization of different bran proteins between aromatic and nonaromatic rice, *Proteomics*, 4, 2047, 2004.

103. Borderles, G., Jamet, E., Lafitte, C. et al., Proteomics of loosely bound cell wall proteins of *Arabidopsis thaliana* cell suspension cultures: A critical analysis, *Electrophoresis*, 24, 3421, 2003.

104. Chan, L.L., Lo, S.C.-L. and Hodgkiss, I.J., Proteomic analysis of a model causative agent of harmful red tide, *Prorocentrum triestinum* I: Optimization of sample preparation methodologies for analyzing with two-dimensional electrophoresis, *Proteomics*, 2, 1169, 2002.

105. Lehner, I., Niehof, M. and Borlak, J., An optimized method for the isolation and identification of membrane proteins, *Electrophoresis*, 24, 1795, 2003.

106. Lippincott, J. and Apostol, I., Carbamylation of cysteine: A potential artifact in peptide mapping of hemoglobins in the presence of urea, *Anal. Biochem.*, 267, 57, 1999.

107. McCarthy, J., Hopwood, F., Oxley, D. et al., Carbamylation of proteins in 2-D electrophoresis — Myth or reality? *J. Proteomc Res.*, 2, 239, 2003.

108. Reubsaet, J.L.E., Beijnen, J.H., Bult, A. et al., Analytical techniques used to study the degradation of proteins and peptides: Chemical instability, *J. Pharm. Biomed. Anal.*, 17, 955, 1998.

109. Reubsaet, J.L.E., Beijnen, J.H., Bult, A. et al., Analytical techniques used to study the degradation of proteins and peptides: Physical instability, *J. Pharm. Biomed. Anal.*, 17, 979, 1998.

110. José, M., Curtu, S., Gajardo, R. and Jorquera, J.I., The effect of storage at different temperatures on the stability of hepatitis C virus RNA in plasma samples, *Biologicals*, 31, 1, 2003.

111. Fura, A., Harper, T.W., Zhang, H. et al., Shift in pH of biological fluids during storage and processing: Effect on bioanalysis, *J. Pharm. Biomed. Anal.*, 32, 513, 2003.

112. Bel-Tal, O., Zwang, E., Eichel, R. et al., Vitamin K-dependent coagulation factors and fibrinogen levels in FPP remain stable upon freezing and thawing, *Transfusion*, 43, 873, 2003.

113. Lewis, M.R., Callas, P.W., Jenny, N.S. and Tracy, R.P., Longitudinal stability of coagulation, fibrinolysis, and inflammation factors in stored plasma samples, *Thromb. Haemostas.*, 86, 1495, 2001.

114. Rouy, D., Ernens, I., Jeanty, C. and Wagner, D.R., Plasma storage at 80°C does not protect matrix metalloproteinase-9 from degradation, *Anal. Biochem.*, 338, 294, 2005.

115. Takei, Y., Marzi, I., Kauffman, F.C. et al., Increase in survival time of liver transplants by protease inhibitors and a calcium channel blocker, nisolidpine, *Transplantation*, 50, 14, 1990.

116. Pyle, L.E., Barton, P., Fujiwara, Y., Mitchell, A. and Fidge, N., Secretion of biologically active human proapolipoprotein A-1 in a baculovirus-insect cell system: Protection from degradation by protease inhibitors, *J. Lipid Res.*, 36, 2355, 1995.

117. Salvesen, G. and Nagase, H., Inhibition of proteolytic enzymes, in *Proteolytic Enyzymes: Practical Approaches*, 2nd Edition, R. Benyon and J.S. Bond, Eds., Oxford University Press, Oxford, Ch. 5, 105, 2001.

118. Harper, J.W., Hemmi, K. and Powers, J.C., Reaction of serine proteases with substituted isocoumarins: Discovery of 3,4-dichloroisocoumarin, a new general mechanism based serine protease inhibitor, *Biochemistry*, 24, 1831, 1985.

119. Lundblad, R.L., A rapid method for the purification of bovine thrombin and the inhibition of the purified enzyme with phenylmethylsulfonyl fluoride, *Biochemistry*, 10, 2501, 1971.

120. North, M.J. and Benyon, R.J., Prevention of unwanted proteolysis, in *Proteolytic Enyzymes: Practical Approaches*, 2nd Edition, R. Benyon and J.S. Bond, Eds., Oxford University Press, Oxford, United Kingdom, Ch. 9, 211, 2001.

120a. Pringle, J.R., Methods for avoiding proteolytic artifacts in studies with enzymes and other proteins from yeasts, *Meth. Cell Biol.*, 12, 149, 1975.

120b. Weidner, M.-F., Grenier, D. and Mayrand, D., Proteolytic artifacts in SDS-PAGE analysis of selected periodontal pathogens, *Oral Microbiol. Immunol.*, 11, 103, 1996.

120c. Pridmore, A.M., Devine, D.A., Bonass, W.A. et al., Influence of sample preparation technique on two-dimensional gel electrophoresis of proteins form *Porphyromonas gingivalis*, *Letters Appl. Microbiol.*, 28, 245, 1999.

120d. Hassel, M., Klenk, G. and Frohme, M., Prevention of unwanted proteolysis during extraction of proteins from protease-rich tissue, *Anal. Biochem.*, 242, 274, 1996.

120e. Kleiner, D.E. and Stetler-Stevenson, W.G., Quantitative zymography: Detection of pictogram quantities of gelatinase, *Anal. Biochem.*, 218, 325, 1994.

121. Lee, D.H. and Goldberg, A.L., The proteome inhibitors and their uses, in *Proteosomes: The World of Regulatory Proteolysis*, W. Hilt and D.H. Wolf, Eds., Landis Bioscience, Georgetown, TX, Ch. 10, 154, 2000.

122. Castellanos-Serra, L. and Paz-Lago, D., Inhibition of unwanted proteolysis during sample preparation: Evaluation of its efficiency in challenge experiments, *Electrophoresis*, 23, 1745, 2002.

123. Fahrney, D.E. and Gold, A.M., Sulfonyl fluorides as inhibitors of esterases. I. Rates of reaction with acetylcholinesterase, α-chymotrypsin, and trypsin, *J. Am. Chem. Soc.*, 85, 997, 1963.

123a. Hirabayashi, T., Tamura, R., Mitsui, I. and Watanabe, Y., Investigation of actin in *Tetrahymena* cells. A comparison with skeletal muscle actin by a devised two-dimensional gel electrophoresis method, *J. Biochem.*, 93, 461, 1983.

124. Finnie, C. and Svensson, B., Proteolysis during the isoelectric focusing step of two-dimensional gel electrophoresis may be a common problem, *Anal. Biochem.*, 311, 182, 2002.

125. Wedchuck, J.B., Goins, W.F., Glorioso, J.C. and Ataai, M.M., Effect of protease inhibitors on yield of HSV-1 based viral vectors, *Biotech. Prog.*, 16, 493, 2000.

126. Bergen, H.R., III, Klug, M.G., Bolander, M.E. and Muddiman, D.C., Informed use of proteolytic inhibitors in biomarker discovery, *Rapid Comm. Mass Spec.*, 18, 1001, 2004.

127. Lundblad, R.L., Chemical Reagents for Protein Modification, 3rd Edition, CRC Press, Boca Raton, FL, Ch. 1, 2004.

128. Wandeschneider, S., Fehring, V., Jacobs-Emeis, S. et al., Autoimmune pancreatic disease: Preparation of pancreatic juice for proteome analysis, *Electrophoresis*, 22, 4383, 2001.

129. López-Otín, C. and Overall, C.M., Protease degradomics: A new challenge for proteomics, *Nat. Rev. Mol. Cell Biol.*, 3, 509, 2002.

130. Ramagli, L.S. and Rodriguez, L.V., Quantitation of microgram quantities of protein in two-dimensional electrophoresis sample buffer, *Electrophoresis*, 6, 559, 1985.

131. Bradford, M.M., A rapid and sensitive method for the quantitation of microgram quantities of protein utilizing the principle of protein-dye binding, *Anal. Biochem.*, 72, 248, 1976.

132. Sapan, C.V., Lundblad, R.L. and Price, N.C., Colorimetric protein assay techniques, *Biotech. Appl. Biochem.*, 29, 99, 1999.

133. Lundblad, R.L. and Price, N.C., Protein concentration determination. The Achilles' heel of cGMP, *Bioprocess Intl.*, 1, January, 2004.

134. de St. Groth, S.F., Webster, R.G. and Datyner, A., Two new staining procedures for quantitative estimation of proteins on electrophoretic strips, *Biochim. Biophys. Acta*, 71, 377, 1963.

135. Diezel, W., Kopperschläger, G. and Hofmann, E., An improved procedure for protein staining in polyacrylamide gels with a new type of Coomassie Brilliant Blue, *Anal. Biochem.*, 48, 617, 1972.

136. Reisner, A.H., Nemes, P. and Bucholtz, C., The use of Coomassie Brilliant Blue G250 perchloric acid solution for staining in electrophoresis and isoelectric focusing on polyacrylamide gels, *Anal. Biochem.*, 64, 509, 1975.

137. Sedmak, J.J. and Grossberg, S.E., A rapid, sensitive, and versatile assay for protein using Coomassie Brilliant Blue G250, *Anal. Biochem.*, 79, 544, 1977.

138. Spector, T., Refinement of the Coomassie Blue method of protein quantitation. A simple and linear spectrometric assay for less than or equal to 0.5 to 50 microgram of protein, *Anal. Biochem.*, 86, 142, 1978.

139. Read, S.M. and Northcote, D.H., Minimization of variation in the response to different proteins of the Coomassie Blue G dye-binding assay for protein, *Anal. Biochem.*, 116, 53, 1981.

140. Lea, M.A., Grasso, S.V., Hu, J. and Seidler, N., Factors affecting the assay of histone H1 and polylysine by binding of Coomassie Blue G, *Anal. Biochem.*, 141, 390, 1984.

141. Marshall, T. and Williams, K.M., Phenol addition to the Bradford dye binding assay improves sensitivity and gives a characteristic response with different proteins, *J. Biochem. Biophys. Meth.*, 13, 145, 1986.

141a. Zor, T. and Selinger, Z., Linearization of the Bradford assay increases its sensitivity: Theoretical and experimental studies, *Anal. Biochem.*, 236, 302, 1996.

142. Jenzano, J.W., Hogan, S.L., Noyes, C.M., Featherstone, G.L. and Lundblad, R.L., Comparison of five techniques for the determination of protein content in mixed human saliva, *Anal. Biochem.*, 159, 370, 1986.

143. Starcher, B., A ninhydrin-based assay to quantitate the total protein content of tissue samples, *Anal. Biochem.*, 292, 125, 2001.

144. Scopes, R.K., Measurement of protein by spectrometry at 205 nm, *Anal. Biochem.*, 59, 277, 1974.

145. Agnew, B.J., Murray, D. and Patton W.F., A rapid solid-phase fluorescence-based protein assay for quantitation of protein electrophoresis samples containing detergents, chaotropic dyes, and reducing agents, *Electrophoresis*, 25, 2478, 2004.

146. Schaeffner, W. and Weissman, C., A rapid, sensitive, and specific method for the determination of protein in dilute solutions, *Anal. Biochem.*, 56, 502, 1973.

147. Weiss, K.C. and Bisson, L.F., Optimisation of the Amido Black assay for the determination of protein content of grape juices and wines, *J. Sci. Food Agr.*, 81, 583, 2001.

148. Bergo, H.O. and Christiansen, C., Determination of low levels of protein impurities in phospholipids samples, *Anal. Biochem.*, 288, 225, 2001.

149. Butler, P.J.G., Ubarretxena-Belandia, I., Warne, T. and Tate, C.G., The *Escherichia coli* multidrug transporter EmrE is a dimer in the detergent solubilized state, *J. Mol. Biol.*, 340, 797, 2004.

150. Valot, B., Gianinazzi, S. and Elaine, D.-G., Sub-cellular proteomic analysis of a *Medicago truncatula* root microsomal fraction, *Phytochemistry*, 65, 1721, 2004.

151. Engel, C.K., Chen, L. and Privé, G.C., Stability of the lactose permease in detergent solutions, *Biochim. Biophys. Acta*, 1564, 47, 2002.

152. Smith, P.K., Krohn, R.I., Hermanson, G.T. et al., Measurement of protein using bicinchoninic acid, *Anal. Biochem.*, 150, 76, 1985.

153. Stewart, I.I., Zhao, L., Le Bihan, T. et al., The reproducible acquisition of comparative liquid chromatography/tandem mass spectrometry data from complex biological samples, *Rapid Comm. Mass Spec.*, 18, 1697, 2004.

154. Krogh, A., Larsson, B., von Heijne, G. and Sonnhammer, E.L.L., Predicting transmembrane protein topology with a hidden Markov model: Application to complete genome, *J. Mol. Biol.*, 305, 567, 2001.

155. Arai, M., Okumura, K., Satake, K. and Shimizu, T., Proteome-wide functional classification and identification of prokaryotic transmembrane proteins by transmembrane topology similarity comparisons, *Protein Sci.*, 13, 2170, 2004.

156. Garrett, S.H., Samji, S., Todd, J.H. et al., Differential expression of human metallothionein isoforms I mRNA in human proximal tubule cells exposed to metals, *Env. Health Pers.*, 106, 825, 1998.

157. Drinkwater, S.L., Burnand, K.G., Ding, R. and Smith, A., Increased but ineffective angiogenic drive in nonhealing venous leg ulcers, *J. Vasc. Surg.*, 38, 1106, 2003.

158. Gehrmann, M., Brunnen, M., Pfister, K. et al., Differential up-regulation of cytolsolic and membrane-bound heat shock protein 70 in tumor cells by anti-inflammatory drugs, *Clin. Cancer Res.*, 4, 3084, 2004.

159. Piterch, A., Abian, J., Carrascal, M., Sánchez, M., Nombela, C. and Gil, C., Proteomics-based identificataion of novel *Candida albicans* antigens for diagnosis of system candidiasis in patients with underlying hematological malignancies, *Proteomics*, 4, 3084, 2004.

160. Ferguson, R.E., Carroll, H.P., Harris, A., Maher, E.R., Selby, P.J. and Banks, R.F., Housekeeping proteins: A preliminary study illustrating some limitations as useful references in protein expression studies, *Proteomics*, 5, 556, 2005.

5 Prefractionation

CONTENTS

INTRODUCTION

The goal of proteomic analysis is the identification of the constituents of the proteome together with their functional interactions. This task is a formidable and impractical undertaking requiring the identification of an absolute minimum of 30,000 proteins.[1] Future technologies might make this possible but would likely require a minimum of a three-dimensional orthogonal format that would require the definition of a third general protein characteristic in addition to size and charge. For the present, we are presented with the luxury of having a superb analytical tool available in mass spectrometry; our challenge is to prepare a sample worthy of the analytical instrumentation. For the purpose of the following discussion, our focus is on prefractionation, defined as the processes of molecular separations prior to analytical procedures, such as mass spectrometry or Western blotting, which result in the identification of proteins or other materials that can be used as biomarkers. Prefractionation is not an extensively referenced concept; a total of 77 citations were obtained from a PUBMED search in November 2004 with the first citation[2] dating to 1972; if "prefractionation" is used as the search term, 723 citations are obtained. Prefractionation could not be found as an entry in a current dictionary; the word is assumed to be derived from the Latin *prae* (in front of, before) and fractionation (the process of separation as a mixture) into different portions. As such, this method is not anywhere near a new or novel concept. For this reason, the reader is referred to some early volumes of *Methods in Enzymology* as well other early tomes[3-10] that will be particularly useful for investigators who are starting with tissue or cellular samples, as much of the early work on subcellular fractionation to obtain sample materials rich in mitochondria, plasma membranes and nuclei are these early materials. Obtaining information about marker enzymes/proteins that would be of value in normalizing samples would likely also be possible.

Chapter 4 discusses sample preparation for proteomic analysis while the current chapter discusses sample prefractionation and conceptual overlap between the two chapters is recognized. An alternative approach would have been one long unwieldy chapter, so this approach was chosen to emphasize the utility of protein chemistry, protein purification and subcellular fractionation in proteomic research. As such, most of proteomic prefractionation is concerned with the application of these well-established techniques in protein purification including ion-exchange chromatography and affinity chromatography.

The typical sample for proteomic analysis is derived from blood serum/plasma or tissue/cell extracts. These samples are both qualitatively and quantitatively heterogeneous. The number of proteins in a proteome is likely in the middle to high five figures (30,000 to 90,000) which includes proteins modified by various post-translational modifications, whose numbers are legion; some estimate that millions of discrete species exist in a proteome.[11] This compositional heterogeneity is qualitative heterogeneity. The concentrations of proteins in a proteomic sample easily vary over four orders of magnitude and most likely more.[12–14] Variation in proteome protein concentration can be referred to as quantitative heterogeneity. These problems are major issues for proteomic analysis and are somewhat complicated by the fact that most studies are not hypothesis-based and thus do not have a target analyte in the first analysis. The effective resolving range (dynamic range) for most current analytical technologies used in proteomics is at best three orders of magnitude and more likely two orders of magnitude for routine investigations. For the sake of argument, let us assume that there are 10^5 separate species (qualitative variations) with a quantitative range of 10^5; a simple x-y plot will demonstrate the problem. This analogy is not completely valid as concentration is not a separation variable but rather a confounding factor as with albumin in the study of blood plasma. Until we can move to a third dimension of separation (at present we can separate by charge and size, giving us two dimensions of separation), this requires the absolute necessity of a prefractionation step to obtain fractions with a workable number of analyte species prior to two-dimensional gel electrophoresis,[15–17] multidimensional protein identification technology (MuDPIT)[18–23] or surface-enhanced laser desorption/ ionization mass spectrometry (affinity mass spectrometry).[24–26] The key factor here is to obtain a reproducible technique that will yield a series of fractions containing 200 to 500 protein species each and the collective series of fractions is an adequate representation of the starting material. Furthermore, it would be quite useful, if not essential, that such a prefractionation technique be orthogonal to the subsequent analytical process. In other words, the prefractionation technique to obtain a sample for two-dimensional electrophoresis ideally would use molecular characteristics other than charge and size, providing a considerable challenge. One useful approach is the use of isocratic ion-exchange chromatography[27–30] to obtain discrete charge species that could then be expanded over of a narrow range isoelectric focusing.[31–33] The reader is directed to an excellent recent review by Lescuyer and collegues[34] on the use of chromatography in the prefractionation of samples for proteomic analysis by two-dimensional gel electrophoresis. This review discusses the use of ion-exchange chromatography, affinity chromatography, hydrophobic interaction chromatography (HIC, hydrophobic affinity chromatography) and size-exclusion chromatography (gel filtration). The various approaches are useful but the issue exists of the loss of material from irreversible and frequently nonspecific binding to the separation matrices. A summary of some approaches for the prefractionation of samples for proteomic analysis is presented in Table 5.1.

BLOOD AND SERUM

Blood is an extremely popular source for biological samples. Samples are reasonably easy to obtain, the samples are technically and psychologically easy to process (e.g., feces, while easy to obtain, generally is considered difficult to process because of cultural reasons) and samples are mostly considered homogeneous when compared to saliva or urine, both of which are somewhat compositionally dependent on fluid flow rates.

The initial blood sample is removed from the circulatory system via venipuncture.[35,36] If the sample is withdrawn in the presence of an anticoagulant and centrifuged to remove cellular elements, a plasma sample is obtained; in the absence of anticoagulant, the blood clots and can be centrifuged immediately to remove the fibrin clot and cellular elements or allowed to stand for period of time and clot retraction occurs.[37,38] As noted below, drawing blood through a resin-containing device that depletes the blood of calcium and precludes coagulation is possible.[39] The addition of corn trypsin inhibitor to the blood as withdrawn,[40,41] greatly slowing the process of blood coagulation, permitted removal of the cellular elements and transfer,

---

TABLE 5.1
Analysis of Methods of Biological Fluid Fractionation

Method	Advantages	Disadvantages
Precipitation[a]	–Minimal sample transfer –Quantitative Recovery –Reproducible	–Incomplete resolution between fractions –May need to clean precipitate
Ultrafiltration	–Minimum sample transfer –Reproducible	–Incomplete resolution –Non-specific binding to filter material
Isocratic Chromatography[b]	–Minimum sample transfer –Reproducible –Small Sample Volume	–Incomplete resolution between fractions
Gradient Chromatography[b]	–Minimun sample transfer –Purification technique –Small sample volume	–Reproducibility an issue –Higher loss of analytes than with isocratic

a This includes such techniques as ammonium sulfate fractionation, ethanol fractionation, acetone precipitation, trichloroacetic acid precipitation/extraction.

b This includes all chromatography matrices which would include affinity, immunoaffinity, anion-exchange, cation-exchange and is not system-dependent with the caveat that while it is easy to run a batch absorption/elution process (isocratic) in a microplate format, it is a little more difficult to visualize a gradient format.

for example, into urea-detergent prior to two-dimensional gel electrophoresis. Multiple anticoagulants are available for the collection of blood plasma, including citrate and heparin; the choice of anticoagulant and temperature can influence storage stability of the analyte.[42–47] The material nature of the process and storage containers also influences stability.[48] Personal experience suggests that the time between venipuncture and freezing, process/storage containers, centrifugation speed and the temperature of storage are the most critical variables for plasma.

Blood serum is technically easier to work with than plasma. However, serum is, of course, not plasma or whole blood and a clear distinction needs to be made in experimental design and reporting of data. The bulk of the fibrinogen has been removed along with the platelets, which have either been physically bound in the fibrin matrix or activated to form aggregates or both. Varying amounts of other proteins are removed into the fibrin clot either by specific or nonspecific interactions. A number of other changes exist[49] including the formation of protease-serpin complex and protein fragments such as D-dimer and prothrombin fragment 1. Contrary to at least one statement in the literature,[50] many of the coagulation factors such as factor IX, factor X and factor XIa are retained in serum with some of the original names such as serum prothrombin conversion accelerator (now known as factor VIIa); as noted below, substantial changes exist in the transition from whole blood to serum that render qualitative and quantitative differences between blood plasma and blood serum, and a generalized loss of coagulation factors is not one of such changes. As a result of the clotting of fibrinogen, the protein concentration of serum is less than that of plasma.[51,52] In the process of whole blood coagulation, the cellular elements (erythrocytes, leukocytes, platelets) can secrete components. In particular, platelets contribute a variety of components to blood serum.[53–55] Vascular endothelial growth factor (VEGF) is an excellent example.[56,57] In one study,[56] normal individuals had a serum concentration of 250 pg/mL with a plasma concentration of 30 pg/mL; breast cancer patients with thrombocytosis had a median VEGF concentration of 833 pg/mL compared to 249 pg/mL in other patients. Both studies suggest that platelets can contribute to VEGF levels in both plasma and serum but more markedly so in serum. The immediate separation of plasma or serum from the cellular elements is suggested to provide optimal analyte stability.[58] Critical process variables for serum are process/storage containers, time of clot retraction/removal of the fibrin clot with associated platelets and other cellular elements, centrifugation speed and temperature of storage.

An alternative for sample preparation that has not been explored for proteomic research is the defibrination of plasma with *Bothrops jararraca venom* (reptilase).[59–61] One study is found[61] on the use of reptilase for the removal of fibrinogen from plasma for clinical chemistry studies that appears to be useful for the preparation of samples for protein electrophoresis. Note that defibrinated plasma is not equivalent to serum.

Many factors other than the underlying biology exist that can influence a blood sample. Elimination of these factors is difficult, if not impossible; the best that one can do is to very carefully document the conditions of blood processing. Establishment of a standard operating procedure (SOP) is absolutely critical for the process of obtaining a blood sample. Only by doing this is one able to assure reproducibility of samples and to allow some rationale comparison of data from various laboratories.

REMOVAL OF HIGH- ABUNDANCE PROTEINS FROM PLASMA AND SERUM

The next problem with the use of blood as a sample for proteomic analysis is the dynamic qualitative and quantitative range of proteins. In addition to these differences as described above, substantial size heterogeneity exists in plasma with smaller proteins such as the vitamin K-dependent proteins, which have molecular weights in 50 kDa range, fibrinogen, which is approximately 340 kDa, to von Willebrand factor in the 10 to 20 million dalton range.[62] Considering this heterogeneity, prefractionation is essential for the proteomic analysis of blood. At the onset, free boundary electrophoresis[63–68] is recommended for plasma/serum prefractionation. While this technique has been used extensively for the study of biological fluids, it can be used for tissue and cellular extracts. The bulk of the plasma proteins are comprised of several proteins and their isoforms, albumin, immunoglobulins, alpha-1-antitrypsin, fibrinogen and haptoglobin. The presence of large amounts of these proteins creates technical difficulties in identifying minor components with current technology. One of these proteins, fibrinogen, is absent from serum creating an advantage for serum as compared to plasma. Most efforts have been directed at methods for the removal of albumin and immunoglobulin as they are present in the highest concentrations (Table 5.2).[62] These approaches use affinity matrices that remove a good portion but not all of the proteins; more problematic is that these columns also remove other components of plasma/serum by "nonspecific" binding. Since most studies do not have a specific target protein, knowing whether a biomarker of interest is lost on the removal of albumin or immunoglobulin is not possible. Components would be less likely lost during the free boundary electrophoresis prefractionation step. Again, the issue is that if one are looking for an "unknown" biomarker, one would not know if the material has been lost in the process of sample preparation, which would include the prefractionation step. To this

TABLE 5.2
Characteristics of Major Human Plasma Proteins

Protein	Concentration (mg/mL)	M.W. (kDa)
Albumin	50	67
Immunoglobulin G	15	150
Fibrinogen	3	340
Haptoglobin	2.0	100
α_1-Antitrypsin	2.7	54
Immunoglobulin A	2.6	160
α_2-Macroglobulin	2.7	800
Blood coagulation factor VII	0.0005	50

end, the fact that more investigators have not used multiple analytical approaches such SEDLI and two-dimensional gel electrophoresis as complementary approaches for obtaining samples for mass spectrometry analysis is somewhat remarkable.

Affinity binding to Cibacron blue dyes has been widely used for the depletion of albumin in biological fluids including plasma, serum and cerebrospinal fluid samples prior to proteomic analysis.[69–78] The binding of albumin to Cibacron blue dyes is nonspecific compared to the specific binding of dehydrogenases and phosphatases via the dinucleotide fold.[79–81] More recent work in this area has used immunoaffinity columns for the removal of albumin.[82–87] Phage display has been used to develop a specific peptide sequence for binding to albumin and subsequent removal from serum.[88]

Removal of IgG is accomplished with either Protein A column or Protein G columns.[83,84,86,87] As a final note, these approaches are all based on "negative selection affinity" where "contaminating proteins" are removed[89–91] and with analysis for the removal of other proteins, the process cannot necessarily be considered a validated process in preparing samples for subsequent analysis. The use of free-boundary electrophoresis approaches has been discussed above. Other approaches have been suggested for the prefractionation of serum and several groups have used multiple prefractionation columns.[92,93] Pieper and coworkers used immunoaffinity columns with antibodies for albumin, transferrin, haptoglobin and other proteins (Immunoaffinity subtraction chromatography, IMAC).[92] The antibodies used in the immunoaffinity columns were obtained by purification from commercial polyclonal sera or IgG-enriched fractions from polyclonal sera. The polyclonal antibodies were of goat or rabbit origin. These crude preparations were purified by affinity chromatography using the appropriate antigen coupled to an aldehyde matrix (POROS® AL, Applied Biosystems). Mixed bed immunoaffinity resins were prepared by combining portions of the individual immunoaffinity matrices; approximately 50% of the mixed bed resin was anti-albumin resin with small amounts of the other resins.

Zhang and coworkers[93] used ion-exchange columns for prefractionation of serum samples from ovarian cancer patients for analysis by SELDI-MS technology using multiple chip matrices including IMAC-Cu, strong anion exchange, weak cation exchange and hydrophobic affinity. This practice permitted the identification of some potential biomarkers. Candidate biomarkers were then purified by a combination of anion-exchange chromatography, size-exclusion chromatography and hydrophobic affinity chromatography. The fractionation process is monitored by SELDI-MS analysis. The proteins are identified from SEQUEST after proteolysis and peptide analysis by mass spectrometry. Immunoassays were obtained via available antibodies permitting conventional ELISA assays. A model was then developed using a combination of CA125 and three biomarkers permitting an increase in sensitivity and specificity. Another group[94] used multiple matrices including anion-exchange, cation-exchange, lectin-affinity and metal chelate affinity in a micro-column format for the prefractionation of serum prior to surface-enhanced laser desorption/ionization mass spectrometry. A batch format was used for this procedure with binding at low salt and elution at high salt; the concanavalin A column was eluted with D-methylglucoside. These investigators also used a variety of ProteinChip® matrices including SAX (strong anion exchange), (IMAC3-Cu) Copper ion IMAC, H50 (hydrophobic) and WCX2 (weak cation exchange) for analysis of the various fractions. These investigators also discuss the issue of the removal of potential biomarkers when high-abundance proteins such as albumin or IgG are removed from serum or plasma samples. Potential biomarkers were identified in sera from patients with Alzheimer's disease, type-II diabetes (insulin-resistant diabetes) and congestive heart failure. This study included negative-expression biomarkers (biomarkers at lower concentration in control sera) and positive-expression biomarkers (higher expression in test sera).

An alternative approach for the prefractionation of biological fluids would be differential precipitation with salts such as ammonium sulfate or organic solvents. Both of these approaches are examples of classic methods of protein fractionation. In general, the methods are little used today although ethanol fractionation still forms the basis the human blood plasma fractionation business, useful in that this should be applicable to a small volume of materials with minimal sample transfer. Fountoulakis and coworkers[95] evaluated several different methods for the precipitation of plasma/serum proteins prior to proteomic analysis. These investigators found that ultrafiltration, precipitation with trichloroacetic acid or acetone was useful for the preparation of samples. They also observed that ammonium sulfate precipitation could be useful as it could separate albumin; however, other proteins are also likely removed.

TISSUE AND CELL SAMPLES — LASER CAPTURE MICRODISSECTION

Processing of tissue and cell samples for proteomics analysis presents a different challenge to the investigator. In approaching this problem, the focus is on the use of tissue as starting material for the study of biomarkers in cancer as this appears to the most popular target, based on a PUBMED search. Other pathologies are discussed in the chapter on clinical proteomics.

The problem of detection of low-abundance proteins has been discussed above and addressed for blood proteins where the protein extract (blood plasma or serum) is separated from cellular elements such as erythrocytes, neutrophils and platelets by centrifugation or clot retraction. In the case of tissues, morphologically defined cell populations can be obtained from complex primary tissue by laser capture microdissection[96–98] of tissue sections.[99] Recent advances in tissue fixation technology such as the HEPES-glutamic acid buffer mediated organic solvent protection effect (HOPE)[100–102] is proving useful. This fixation technique combined with embedding in a paraffin matrix (paraffin-embedding) preserves soft tissue morphology as effectively as formalin-fixation. In the HOPE approach, acetone is the only dehydrating agent used in the processing of tissue. This approach preserves biological structure (proteins) for a considerable period of time (up to five years) as well as preserving nucleic acids for PCR and DNA microarray analysis. The mechanism by which this solvent combination (HEPES/glutamic acid/acetone) preserves structure is not clear. Acetone removal of water will decrease degradative activity; acetone can also form Schiff bases with amino groups. Avoidance of formalin fixation results in increased retention of biological activity. Other investigators have also shown the advantage of ethanol fixation over formalin fixations.[103,104] Gillespie and coworkers[103] used 70% ethanol to fix tissue samples with success in preserving protein/nucleic acids for molecular profiling. Phenoxyethanol has been suggested to be a biochemical preservative[105,106] but has not been used in proteomics.

Laser capture microdissection is a relatively new[107] technology but has seen very quick acceptance as a useful technique for both genomics, epigenomics, transcriptomics and proteomics;[108–116] more work is found on genomics and transcriptomics than on proteomics. However, increasing use of laser capture microdissection is seen in proteomics. Early work by Banks and coworkers[109] demonstrated that laser capture microdissection could be used to study proteins from renal tumors. This study is seminal in the use of laser capture microdissection for proteomic research. Serial sections were taken from snap-frozen tissue samples and either directly taken into lysis buffer or placed on a glass slide, fixed with ethanol and stained with hematoxylin and eosin. The hematoxylin and eosin solutions contained a protease inhibitor "cocktail." The stained sections were then either scraped into lysis buffer or subjected to laser capture microdissection. Comparison of the two-dimensional gel electrophoretic analysis of the three tissue samples — unstained tissue, stained tissue and laser capture microdissected tissue — did not demonstrate any gross differences. The antigenicity of two proteins, HSP-60 and β_2-microglobulin, was not effected by the laser treatment as evaluated by Western blot analysis. This is an excellent study that has been extensively cited in the short time since its publication.

The traditional hemotoxylin and eosin (H&E) stain technology[99] can be effectively used to process tissues prior to laser capture microdissection.[117–119] Some use of immunohistochemistry[120,121] for guidance exists but requires more prior knowledge of the target protein than is required for H&E staining. Immunohistochemistry is used increasingly in surgical pathology[122] and has been used to demonstrate increased expression of S100 proteins in protkeratosis,[123] which correlated with increased mRNA expression. von Egglining and coworkers[124] have used immunohistochemistry to validate observations of increased protein expression in tumor tissue as assessed with ProteinChip® technology. Samples were obtained from head and neck tumor tissue and normal mucosal tissue after H&E staining by laser capture microdissection, extracted with lysis buffer and assayed with ProteinChip® technology using a strong anion exchange chip (SAX2). Increased expression of annexin was observed and subsequently confirmed by immunohistochemistry using immunodepletion. These investigators suggest that successful biomarker identification requires a relatively pure microdissected tumor tissue sample.

While there has been some concern about the use of H&E stains for subsequent proteomic analysis,[125–127] subsequent studies[128,129] have suggested that this procedure does not create any substantial analytical issues for two-dimensional electrophoresis and subsequent protein identification by mass spectrometry. Some issues may exist with respect to immunological reactivity that stem from tissue processing but these likely need to be pursued on an individual study basis; the same consideration is required for evaluation of various fixation technologies including formalin fixation. Although both of the above studies[128,129] are useful, the reader is specifically directed to the excellent study by Rose and coworkers.[129] These investigators evaluated the effect of four different tissue processing procedures (no staining; ethanol fixation,

H&E staining; ethanol fixation, H&E staining, xylene dehydration; acetone fixation, H&E staining, xylene dehydration) on subsequent proteomic analysis after laser capture microdissection. While some qualitative differences were found, no significant difference emerged in the number of spots obtained on the two-dimensional gel electrophoretograms (silver staining). A quantity of 171 matched spots were selected and the intensities measured and used to examine the correlations between various processed tissue protocols. Evaluation was accomplished with scatter plots and subsequent calculation of Spearman correlation coefficients; the lowest R value was 0.726 with $P < 0.0001$ for all groups. A difference in immunological reactivity (antibody to smooth muscle α-actin) was observed; acetone treatment appears to result in the loss of immunological reactive spots on the two-dimensional electrophoretograms.

An approach using laser capture microdissection that may provide increased sensitivity and accuracy has been reported by Hirohashi and coworkers.[130] The power of laser capture microdissection was coupled with the use of Cy dyes[131] in differential gel electrophoresis (DIGE).[132] The extracts containing fluorescent-labeled proteins were subjected to two-dimensional gel electrophoresis. Difference spots were excised and digested *in situ* with trypsin and the digests analyzed by MALDI-MS. Proteins were also identified by Western blotting after initial identification by mass spectrometry; upregulated proteins included prohibitin, 14-3-3 zeta, tropomyosin 3 and Hsp84. Zhou and coworkers[133] previous reported the use of Cy dyes in DIGE in the study of esophageal cancer. Laser microdissection was used to capture samples (approximately 250,000 cells/sample) from tumor tissue and normal epithelial tissue. These investigators identified 58 spots which were upregulated and 107 which were downregulated. There are other examples of the use of laser capture microdissection in the chapter on clinical proteomics.

Other approaches exist that can be used enrich tissue samples for specific cell types. Kellner and coworkers[127] used immunobeads (antibody coupled to magnetic beads) to isolate epithelial cells from the pancreas. Tumor or nontumor tissue was taken to cells by passing through cell mesh followed by centrifugation. Bef-Ep beads (Dynabeads®, Epithelial Enrich, Dynal®, Norway) were used to capture the epithelial cells from pancreatic tissue cell homogenates obtained from suspension of the pellets obtained above. This process was performed in the presence of protease and RNAse inhibitors. The Dynabeads® with the attached epithelial cells were obtained by centrifugation, washed and frozen at $-80°C$. The purified epithelial cells were placed into a lysis/denaturing buffer prior to two-dimensional gel electrophoresis.

Performance of mass spectrometric analyses have been possible on intact tissues.[134,135] While this procedure is in the process of development, it has immense potential in the search for useful biomarkers in that potential candidates will not be lost during the process of tissue extraction and subsequent sample processing.

SUBPROTEOMICS

Since a true total analysis of a proteome, as defined, is not possible with current technology, the effective analysis of discrete and well-characterized portion of the proteome is essential. This approach is a lead-in to the type of experimental method referred to as subproteomics.[136–140] Two examples will be cited briefly: Navarre and coworkers[137] purified yeast cell membranes by removing cytosolic proteins by treatment with deoxycholate and ribosomal proteins by sucrose density flotation. Subsequent analysis of the isolated membrane by two-dimensional gel electrophoresis demonstrated the presence of known membrane proteins and one new membrane protein with a single transmembrane segment. Shiozaki and coworkers[138] obtained different fractions by differential tissue extraction. Extraction with urea and Nonidet P-40 yielded 1300 proteins spots on two-dimensional gel electrophoresis; extraction with urea/thiourea/SB3-10/CHAPS yielded 500 protein spots.

Other work exists on the application of proteomic analysis to subcellular fractions[141–143] including membrane fractions.[144–147] Some examples include studies on prostasomes,[148] aging mitochondria[149] and proteosomes.[150]

AFFINITY SELECTION

Affinity selection procedures can use a matrix such as an agarose matrix or a magnetic bead. Any matrix that would selectively bind the analyte and then could be removed from bulk solution would prove satisfactory. The basic concept is also used in affinity electrophoresis[151–153] and affinity partitioning.[154–156]

Positive affinity selection has not been extensively used in proteomics as it does select a group of proteins rather than the entire proteome. The reader is directed to recent reviews[157,158] for a more global discussion of affinity purification and proteomics. The discussion does not include a consideration of tandem affinity purification,[159–161] a valuable approach for the study of *in vivo* protein complexes. Affinity purification is generally considered to be a highly specific technology[162–164] that would not appear to be useful for a more global protein analysis. However, with a broad enough affinity approach, useful information can be obtained by proteomic analysis. An example is provided by the use of dye affinity columns to concentrated low abundance proteins in *Escherichia coli* lysates.[165] Six different dyes were used in this study including Reactive Green, Reactive Brown, Reactive Red, Cibacron blue, Reactive Blue and Reactive Yellow bound to an agarose matrix (Sigma Chemical Corporation, St. Louis, MO; http://www.sigma-aldrich.com). Lysates were passed over these columns (0.01 M Tris-Cl, pH 7.5) and bound materials eluted with 1.5 M NaCl. Depending on the dye, 0.38 to 4.42% of the proteins were eluted and enriched by this process. For example, with Reactive Green, 278 proteins were detected by subsequent two-dimensional gel electrophoresis with 128 (46%) enriched over the parent lysates; with Reactive Yellow, 175 proteins were bound with 109 (62%) enriched. Isacchi and coworkers[166] used inhibitor affinity chromatography to isolate kinases from HTC-116 cells and pancreatic acinar cell lysates. A Cdk2 inhibitor (PHA-539236 cyclopropylpyrazole) was bound to an agarose matrix and used as an affinity purification matrix. Other investigators have also presented affinity chromatography approaches to the study of kinases and kinase substrates.[167–169] Glycoproteins can be isolated by affinity chromatography on lectins.[170] A large number of potential matrices are available[171–176] that could be explored. The goal is population capture and enrichment for subsequent analysis. Thus, one looks for a relatively simple isocratic elution that can be managed either in a chromatographic column, disc or "spin-tube" that is reproducible and, as such, amenable to validation. The caveat, of course, is that the extensive purification of the proteomic sample defeats the study of the proteome.

CONCLUSIONS

The current level of our separation and analytical technology requires prefractionation of the sample. A wealth of information is found in the literature on the purification of proteins. Much of this information has direct application to proteomics. The recommended approach is a batchwise process using either ion-exchange or affinity. The capture process is limited only by diffusion in the case of ion-exchange; with affinity matrices, defined on-off rates are encountered that must be considered in sample capture and usually are not. Matrix effects also must be considered in the development of a robust procedure. Nonspecific absorption effects must be considered and any prefractionation process must be understood to the extent that it could be validated.

REFERENCES

1. Brunet, S., Thibault, P., Gagnon, E. et al., Organelle proteomics: looking at less to see more, *Trends Cell Biol.*, 13, 629, 2003.
2. Schwabe, E., Ribosomal proteins. XLIV. A simple procedure for prefractionation of proteins from *Escherichia coli* 70S ribosomes and from 30S and 50S subunits, *Hoppe Seylers Z. Physiol. Chem.*, 353, 1899, 1972.
3. Colowick, S.P. and Kaplan, N.O. Eds., *Meth. Enzymology*, 1, 1955.
4. Green, A.A. and Hughes, W.L., Protein fractionation on the basis of solubility in aqueous solutions of salts and organic solvents, *Meth. Enzymology*, 1, 67, 1955.
5. Colowick, S.P. and Kaplan, N.O. Eds., *Meth. Enzymology*, 5, 1962.
6. Bier, M., Preparative electrophoresis, *Meth. Enyzmology*, 5, 33, 1962.
7. Affinity Techniques Enzyme Purification, W.B. Jacoby and M. Wilchek, Eds., *Meth. Enzymology*, Part B, 34, 1974.
8. Immobilized Enzymes, *Meth. Enzymology*, K. Mosbach, Ed., 44, 1976.
9. Enzyme Purification and Related Techniques, W.B. Jacoby, Ed., *Meth. Enzymology*, Part C, 104, 1984.
10. Sundarm, P.V. and Eckstein, F., *Theory and Practice in Affinity Techniques*, Academic Press, London, United Kingdom, 1978.
11. Henschen-Edman, A.H., Fibrinogen non-inherited heterogeneity and its relationship to function in health and disease, *Ann. N.Y. Acad. Sci.*, 936, 580, 2001.
12. Corthals, G.L., Wasinger, V.C., Hochstrasser, D.F. and Sanchez, J.-C., The dynamic range of protein expression: A challenge for proteomic research, *Electrophoresis*, 21, 1104, 2000.
13. Rabilloud, T., Two-dimensional gel electrophoresis in proteomics: Old, old fashioned, but it still climbs up the mountain, *Proteomics*, 2, 3, 2002.
14. Greennough, C., Jenkins, R.E., Kitteringham, N.R. et al., A method for the rapid depletion of albumin and immunoglobulin from human plasma, *Proteomics*, 4, 3107, 2004.

15. Görg, A., Obermaier, C., Boguth, G. et al., The current state of two-dimensional electrophoresis with immobilized pH gradients, *Electrophoresis*, 21, 1037, 2000.

16. Challapalli, K.K., Zabel, C., Schuchhardt, J. et al., High reproducibility of large-gel two-dimensional electrophoresis, *Electrophoresis*, 25, 3040, 2004.

17. Rais, I., Karas, M. and Schäger, H., Two-dimensional electrophoresis for the isolation of integral membrane proteins and mass spectrometric identification, *Proteomics*, 4, 2567, 2004.

18. Wolters, D.A., Washburn, M.P. and Yates, J.R., III, An automated multidimensional protein identification technology for shotgun proteomics, *Anal. Chem.*, 73, 5683, 2001.

19. Washburn, M.P., Wolters, D. and Yates, J.R., III, Large-scale analysis of the yeast proteome by multidimensional protein identification technology, *Nat. Biotech.*, 1, 242, 2001.

20. Washburn, M.P., Ulaszek, R.R. and Yates, J.R., III, Reproducibility of quantitative proteomic analyses of complex biological mixtures by multidimensional protein iden- tification technology, *Anal. Chem.*, 75, 5054, 2003.

21. Wienkoop, S., Glinski, M., Tanaka, N. et al., Linking protein fractionation with multidimensional monolithic reversed-phase peptide chromatography/mass spectro- metry enhances protein identification from complex mixtures even in the presence of abundant proteins, *Rapid Comm. Mass Spec.*, 18, 643, 2004.

22. Kaiser, T., Wittka, S., Just, I. et al., Capillary electrophoresis coupled to mass spec- trometer for automated and robust polypeptide determination in body fluids for clinical use, *Electrophoresis*, 25, 2044, 2004.

23. Li, Y., Xiang, R., Wilkins, J.A. and Horváth, C., Capillary electrochromatography of peptides and proteins, *Electrophoresis*, 25, 2242, 2004.

24. Van de Water, J. and Gershwin, M.E., Detection of molecular determinants in complex biological systems using MALDI-TOF affinity mass spectrometry, *Meth. Mol. Biol.*, 146, 453, 2000.

25. Dick, L.W., Jr. and McGown, L.B., Aptamer-enhanced laser desorption/ionization for affinity mass spectrometry, *Anal. Chem.*, 76, 3037, 2004.

26. Tang, N., Tornatore, P. and Weinberger, S.R., Current developments in SELDI affinity technology, *Mass Spec. Rev.*, 23, 34, 2004.

27. Yao, K. and Hjerten, S., Gradient and isocratic high-performance liquid chromato- graphy of proteins on a new agarose-based anion exchanger, *J. Chrom.*, 385, 87, 1987.

28. DePhillips, P. and Lenoff, A.M., Determinants of protein retention characteristics on cation-exchange adsorbents, *J. Chrom. A*, 933, 57, 2001.

29. Kato, Y., Nakamura, K., Kitamura, T. et al., Separation of proteins by hydrophobic interaction chromatography at low salt concentration, *J. Chrom. A*, 971, 143, 2002.

30. Shi, Q., Zhou, Y. and Sun, Y., Influence of pH and ionic strength on the steric mass- action model parameters around the isoelectric point of protein, *Biotech. Prog.*, 21, 516, 2005.

31. Solassol, J., Marin, P., Demettre, E. et al., Proteomic detection of prostate-specific anitgen using a new serum fractionation procedure: Potential implication for new low-abundance cancer biomarkers detection, *Anal. Biochem.*, 338, 26, 2005.

32. Hoving, S., Gerrits, B., Voshol, H. et al., Preparative two-dimensional gel electro- phoresis at alkaline pH using narrow range immobilized pH gradients, *Proteomics*, 2, 127, 2002.

33. Cohen, A.M., Rumpel, K., Coombs, G.H. et al., Characterization of global protein expression by two-dimensional electrophoresis and mass spectrometry: Proteomics of *Toxoplasma gondii*, *Intl. J. Parisitol.*, 32, 39, 2002.

34. Lescuyer, P., Hochstrasser, D.F. and Sanchez, J.-C., Comprehensive proteome analysis by chromatographic protein prefractionation, *Electrophoresis*, 25, 1125, 2004.

35. Koepke, J.A., Specimen collection — Cellular hematology, in *Laboratory Hematology*, J.A. Koepke and C. Livingstone, Eds., Vol. 2, Ch. 32, 821, 1984.

36. Thompson, J.M., Specimen collection for blood coagulation testing, in *Laboratory Hematology*, J.A. Keopke and C. Livingstone, Eds., Vol 2, Ch. 33, 846, 1984.

37. Budtz-Olsen, O.E., *Clot Retraction*, C.C. Thomas, Springfield, IL, 1951.

38. MacFarlane, R.G., A single method for maximum clot retraction, *Lancet*, I, 236, 1199, 1939.

39. Kingdon, H.S. and Lundblad, R.L., Factors affecting the evolution of factor XIa during blood coagulation, *J. Lab. Clin. Med.*, 85, 826, 1976.

40. Rand, M.D., Lock, J.B., van 't Veer, C., Gaffney, D.P. and Mann, K.G., Blood clotting in minimally altered blood, *Blood*, 88, 3432, 1996.

41. Schneider, D.J., Tracy, P.B., Mann, K.G. and Sobel, B.E., Differential effects of anticoagulants on the activation of platelets *ex vivo*, *Circulation*, 96, 2877, 1997.

42. Evans, M.J., Livesey, J.H., Ellis, M.J. and Yandle, T.O., Effect of anticoagulants and storage temperatures on stability of plasma and serum hormones, *Clin. Biochem.*, 34, 107, 2001.

43. Prisco, D., Panicca, R., Bandinelli, B. et al., Euglobulin lysis time in fresh and stored samples, *Am. J. Clin. Path.*, 102, 794, 1994.

44. Guder, W.G., Who cares about the stability of analytes? *Eur. J. Clin. Chem. Biochem.*, 33, 177, 1995.

45. Heins, M., Heil, W. and Withold, W., Storage of serum or whole blood samples? Effects of time and temperature on 22 serum analytes, *Eur. J. Clin. Chem. Biochem.*, 33, 231, 1995.

46. Qvist, P., Munk, M., Hoyle, N. and Christianen, C., Serum and plasma fragments of C-telopeptides of type I collagen (CTX) are stable during storage at low temperatures for 3 years, *Clin. Chim. Acta*, 350, 167, 2004.

47. Kioukia-Fouglia, N., Christofidis, I. and Strantzalis, N., Physicochemical conditions affecting the formation/stability of serum complexes and the determination of prostate-specific antigen (PSA), *Anticancer Res.*, 19, 3315, 1999.

48. Preissner, C.M., Reilly, W.M., Cyr, R.C. et al., Plastic versus glass tubes: Effects on analytical performance of selected serum and plasma hormone assays, *Clin. Chem.*, 50, 1245, 2004.

49. Faulk, W.P., Torny, D.S. and McIntyre, J.A., Effects of serum versus plasma on agglutination of antibody-coated indicator cells by human rheumatoid factor, *Clin. Immunol. Immunopath.*, 46, 169, 1988.

50. Adkins, J.N., Varnum, S.M., Auberry, K.J. et al., Toward a human blood serum proteome. Analysis by multidimensional separation coupled with mass spectrometry, *Mol. Cell. Proteomics*, 1, 947, 2002.

51. Lum, G. and Gambino, S.R., A comparison of serum versus heparinized plasma for routine clinical tests, *Am. J. Clin. Path.*, 61, 108, 1974.

52. Ladenson, J.H., Tsai, L.-M., Michael, J.M. et al., Serum versus heparinized plasma for eighteen common chemistry tests, *Am. J. Clin. Path.*, 62, 545, 1974.

53. George, J.N., Thai, L.L., McManus, L.M. and Reiman, T.A., Isolation of human platelet membrane microparticles from plasma serum, *Blood*, 60, 834–840, 1982.

54. Levine, R.B. and Rebellino, E.M., Platelet glycoprotein-IIb and glycoprotein-IIIa associated with blood monocytes are derived from platelets, *Blood*, 67, 207, 1986.

55. Lindemann, S., Tolley, N.D., Dixon, P.A. et al., Activated platelets mediate inflammatory signaling by regulated interleukin 1 beta synthesis, *J. Cell. Biol.*, 154, 485, 2001.

56. Benoy, I., Salgado, R., Colpaert, C. et al., Serum interleukin 6, plasma VEGF, serum VEGF, and VEGF platelet load in breast cancer patients, *Clin. Breast Cancer*, 2, 311, 2002.

57. Spence, G.M., Graham, A.N., Mulholland, K. et al., Vascular endothelial growth factor levels in serum and platelets following esophageal cancer resection — Relationship to platelet count, *Intl. J. Biol. Markers*, 17, 119, 2002.

58. Boyanton, B.L., Jr. and Blick, K.E., Stability studies of twenty-four analytes in human plasma and serum, *Clin. Chem.*, 48, 2242, 2002.

59. Blomback, G., Blomback, M. and Nilsson, I.M., Coagulation studies on reptilase, an extract of the venom from *Bothrops jaraca*, *Thromb. Diath. Haemor.*, 15, 76, 1958.

60. Funk, C., Gmur, J., Herold, R. and Straub, P.W., Reptilase® — A new reagent in blood coagulation, *Brit. J. Haematol.*, 21, 43, 1971.

61. Ibrahim, Y., Volkmann, M., Hassoun, R., Fiehn, W. and Rossman, H., Serum protein electrophoresis: Reptilase treatment is superior to ethanol precipition for specific removal of fibrinogen from heparinized plasma samples, *Clin. Chem.*, 50, 1100, 2004.

62. Heide, K., Haupt, H. and Schwick, H.G., Plasma protein fractionation, in *The Plasma Proteins*, F.W. Putnam, Ed., Vol. 3, Ch. 8, 545, Academic Press, New York, NY, 1997.

63. Burggraf, D., Weber, C. and Lottspeich, F., Free flow-isoelectric focusing of human cellular lysates as sample preparation of protein analysis, *Electrophoresis*, 16, 1010, 1995.

64. Herbert, B. and Righetti, P.G., A turning point in proteome analysis: Sample prefractionation via multicompartment electrolyzers with isoelectric membranes, *Electrophoresis*, 21, 3639, 2000.

65. Moritz, R.L., Ji, H., Schültz, F. et al., A proteome strategy for fractionating proteins and peptides using continuous free-flow electrophoresis coupled off-line to reversed-phase high-performance liquid chromatography, *Anal. Chem.*, 76, 4811, 2004.

66. Marshall, J., Jankowski, A., Furesz, S. et al., Human serum proteins preseparated by electrophoresis or chromatography followed by tandem mass spectrometry, *J. Proteome Res.*, 3, 264, 2004.

67. Pang, L., Fryksdale, B.G., Chow, N. et al., Impact of prefractionation using Gradiflow on two-dimensional gel electrophoresis and protein identification by matrix assisted laser desorption/ionization-time-of-flight mass spectrometry, *Electrophoresis*, 24, 3484, 2004.

68. Hoffman, P., Ji, H., Moritz, R.L. et al., Continuous free-flow electrophoresis separation of cytosolic proteins from the human colon carcinoma cell line LIM 1215: A non two-dimensional gel electrophoresis-based proteome analysis strategy, *Proteomics*, 1, 807, 2001.

69. Travis, J. and Pannell, R., Selective removal of albumin from plasma by affinity chromatography, *Clin. Chim. Acta*, 49, 49, 1973.

70. Travis, J., Bower, J., Tewksbury, D., Johnson, D. and Pannell, R., Isolation of albumin from whole plasma and fractionation of albumin-depleted plasma, *Biochem. J.*, 157, 301, 1976.

71. Leatherbarrow, R.J. and Dean, P.D., Studies on the mechanism of binding of serum albumin to immobilized Cibacron Blue F3GA, *Biochem. J.*, 189, 27, 1980.

72. Shaw, M.M. and Riederer, B.M., Sample preparation of two-dimensional electrophoresis, *Proteomics*, 3, 1408, 2003.

73. Bharti, A., Ma, P.A., Maulik, G. et al., Haptoglobin -subunit and hepatocyte growth factor can potentially serve as serum tumor markers in small cell lung carcinoma, *Anticancer Res.*, 24, 1031, 2004.

74. Kubo, K., Honda, E., Imoto, M. and Morishima, Y., Capillary zone electrophoresis of albumin-depleted human serum using a linear polyacrylamide-coated capillary: Separation of serum α- and β-globulins into individual components, *Electrophoresis*, 21, 396, 2000.

75. Li, C. and Lee, K.H., Affinity depletion of albumin from human cerebrospinal fluid using Cibacron-blue-3GA-derivatized photopatterned copolymer in a microfluidic device, *Anal. Biochem.*, 333, 381, 2004.

76. Ahmed, N., Barker, G., Oliva, K. et al., An approach to remove albumin for the proteomic analysis of low abundance biomarkers in human serum, *Proteomics*, 3, 1980, 2003.

77. Hammack, B.N., Owens, G.P., Burgoon, M.P. and Gilden, D.H., Improved resolution of human cerebrospinal fluid proteins on two-dimensional gels, *Multiple Sclerosis*, 9, 472, 2003.

78. Ahmed, N., Barker, G., Oliva, K.T. et al., Proteomic-based identification of hapto-globin-1 precursor as a novel circulating biomarker of ovarian cancer, *Brit. J. Cancer*, 91, 129, 2004.

79. Thompson, S.T., Cass, K.H. and Stellwagen, E., Blue dextran-sepharose: an affinity column for the dinucleotide fold in proteins, *Proc. Natl. Acad. Sci. USA*, 72, 669, 1975.

80. Reuter, R., Naumann, M. and Kopperschlager, G., New aspects of dye-ligand affinity-chromatography of lactate-dehydrogenase applying spacer-mediated beaded cellulose, *J. Chrom.*, 510, 189, 1990.

81. Labrou, N.E., Eliopoulos, E. and Clonis, Y.D., Dye-affinity labeling of bovine heart mitochondrial malate dehydrogenase and study of the NADH-binding site, *Biochem. J.*, 315, 687, 1996.

82. Breski, H., Katenhusen, R.A., Sullivan, A.G. et al., Albumin depletion method of improved plasma glycoprotein analysis by two-dimensional difference gel electrophoresis, *BioTechniques*, 35, 1128, 2003.

83. Govarukhina, N.T., Keizer-Gunnick, A., van der Zee, A.G. et al., Sample preparation of human serum for the analysis of tumor markers. Comparison of different approaches for albumin and gamma-globulin depletion, *J. Chrom. A*, 1009, 171, 2003.

84. Gennough, C., Jenkins, R.E., Heringham, N.R., Dirmohamed, M., Palk, B.K. and Pennington, S.R., A method for the rapid depletion of albumin and immunoglobulin from human plasma, *Proteomics*, 4, 3107, 2004.

85. Quero, C., Colomó, N., Prieto, M.R. et al., Determination of protein markers in human serum: Analysis of protein expression in toxic oil syndrome studies, *Proteomics*, 4, 303, 2004.

86. Wang, Y.Y., Cheng, P. and Chan, D.W., A simple affinity spin tube filter method for removing high-abundant common proteins or enriching low-abundant biomarkers for serum proteomic analysis, *Proteomics*, 3, 243, 2003.

87. Wenner, B.R., Lovell, M.A. and Lynn, B.C., Proteomic analysis of human ventricular cerebrospinal fluid from neurologically normal, elderly subjects using two-dimensional LC-MS/MS, *J. Proteome Res.*, 3, 97, 2004.

88. Sato, A.K., Sexton, D.J., Morganelli, L.A. et al., Development of mammalian serum albumin affinity purification media by peptide phage display, *Biotech. Prog.*, 18, 182, 2002.

89. Ehle, H. and Horn, A., Immunoaffinity chromatography of enzymes, *Bioseparation*, 1, 97, 1990.

90. Kleine-Tebbe, J., Hamilton, R.G., Roebber, M. et al., Purification of immunoglobulin E (IgE) antibodies from sera with high IgE titers, *J. Immunol. Meth.*, 179, 153, 1995.

91. Benet, C. and Van Cutsem, P., Negative purification method of the selection of specific antibodies from polyclonal antisera, *BioTechniques*, 33, 1051, 2002.

92. Pieper, R., Su, Q., Gatlin, C.L. et al., Multi-component immunoaffinity subtraction chromatography: An innovative step towards a comprehensive survey of the human plasma proteome, *Proteomics*, 3, 422, 2003.

93. Zhang, Z., Bast, R.C., Jr., Yu, Y. et al., Three biomarkers identified from serum proteomic analysis for the detection of early stage ovarian cancer, *Cancer Res.*, 64, 5882, 2004.

94. Zhang, R., Barker, L., Pinchev, D. et al., Mining biomarkers in human sera using proteomic tools, *Proteomics*, 4, 244, 2004.

95. Jiang, L., He, L. and Fountoulakis, M., Comparison of protein precipitation methods for sample preparation prior to proteomic analysis, *J. Chrom. A*, 1023, 317, 2004.

96. Fend, F., Kremer, M. and Quintanilla-Martinez, L., Laser capture microdissection: Methodological aspects and applications with emphasis on immuno-laser capture microdissection, *Pathobiology*, 68, 209, 2000.

97. Craven, R.A. and Banks, R.E., Use of laser capture microdissection to selectively obtain distinct populations of cells for proteomic analysis, *Meth. Enzymology*, 356, 33, 2002.

98. Laser capture microscopy and microdissection, P.M. Coan, Ed., *Meth. Enzmology*, 355, 2002.

99. Roche, P.E. and His, E.D., Immunohistochemistry — Principles and advances, in *Manual of Clinical Laboratory Immunology*, N.R. Rose, R.G. Hamilton and B. Detrick, Eds., ASM Press, Washington, DC, Ch. 41, 380, 2002.

100. Olert, J., Wiedorn, K.-H., Goldmann, T. et al., HOPE fixation: A novel fixing method and paraffin-embedding technique for human soft tissues, *Path. Res. Prac.*, 197, 823, 2001.

101. Wiedorn, K.H., Olert, J., Stacy, R.A. et al., HOPE-a new fixing technique enables preservation and extraction of high molecular weight DNA and RNA of >20 kb from paraffin-embedded tissues, Hepes glutamic acid buffer mediated organic solvent protection effect, *Path. Res. Prac.*, 198, 735, 2002.

102. Goldman, T., Vollmer, E. and Gerdes, J., What's cooking? Detection of important biomarkers in HOPE-fixed, paraffin-embedded tissues eliminates the need for antigen retrieval, *Am. J. Path.*, 163, 2683, 2003.

103. Gillespie, J.W., Best, C.J., Bichsel, V.E. et al., Evaluation of non-formalin tissue fixation for molecular profiling studies, *Am. J. Path.*, 160, 449, 2002.

104. Permutter, M.A., Best, C.J., Gillepie, J.W. et al., Comparison of snap freezing versus ethanol fixation for gene expression profiling of tissue samples, *J. Mol. Diag.*, 6, 371, 2004.

105. Nakanishi, M., Wilson, A.C., Nolan, R.A. et al., Phenoxyethanol: Protein preservative for taxonomists, *Science*, 163, 681, 1969.

106. Wineski, L.E. and English, A.W., Phenoxyethanol as a nontoxic preservative in the dissection laboratory, *Acta Anat. (Basel)*, 136, 155, 1989.

107. Emmert-Burk, M.R., Banner, R.F., Smith, P.P. et al., Laser capture microdissection, *Science*, 274, 998, 1996.

108. Simone, N.L., Benner, R.F., Gillespie, J.W. et al., Laser-capture microdissection opening the microscopic frontier to molecular analysis, *Trends Gen.*, 14, 272, 1998.

109. Banks, R.E., Dunn, M.J., Forbes, M.A. et al., The potential use of laser capture microdissection to selectively obtain distinct populations of cells for proteomic analysis — Preliminary findings, *Electrophoresis*, 20, 689, 1999.

110. Fend, F. and Raffeld, M., Laser capture microdissection in pathology, *J. Clin. Path.*, 53, 666, 2000.
111. Eltoum, I.A., Siegel, G.P. and Frost, A.R., Microdissection of histologic sections: Past, present, and future, *Adv. Anat. Path.*, 9, 316, 2002.
112. Heinmöller, E., Schlacke, G., Rente, B. et al., Microdissection and molecular analysis of single cell clusters in pathology and diagnosis — Significance and challenges, *Anal. Cell. Path.*, 24, 125, 2002.
113. Fuller, A.P., Palmer-Toy, D., Erlander, M.G., Jr. and Sgroi, D.C., Laser capture microdissection and advanced molecular analysis of human breast cancer, *J. Mamm. Gland Biol. Neoplasia*, 8, 335, 2003.
114. Zieziolewicz, T.J., Unfricht, D.W., Hadjout, N. et al., Shrinking the biologic world — Nanobiotechnologies for toxicology, *Tox. Sci.*, 74, 235, 2003.
115. Fend, F., Kremer, M., Specht, K. and Quintanille-Martinez, L., Laser microdissection in hematopathology, *Path. Res. Prac.*, 199, 425, 2003.
116. Heergenhahn, M., Kenzelmann, M. and Gröne, H.-J., Laser-controlled microdissection of tissues opens a window of new opportunities, *Path. Res. Prac.*, 419, 2003.
117. Tachikawa, T. and Irié, T., A new molecular biology approach in morphology: basic method and application of laser microdissection, *Med. Elec. Micro.*, 37, 82, 2004.
118. Kondo, T., Seike, M., Mori, Y. et al., Application of sensitive fluorescent dyes in linkage of laser microdissection and two-dimensional gel electrophoresis as a cancer proteomic study tool, *Proteomics*, 3, 1758, 2003.
119. Ross, M.H. and Romrell, L.J., *Histology: A Text and Atlas*, 2nd Edition, Williams & Wilkins, Baltimore, MD, Ch. 1, 1–15, 1989.
120. Asen, E., Progress in focus: Recent advances in histochemistry and cell biology, *Histochem. Cell Biol.*, 118, 507, 2002.
121. Angeletti, C., Applications of proteomic technologies to cytologic specimens: A review, *Acta Cytologica*, 47, 535, 2003.
122. Jaffer, S. and Bleiweiss, I.J., Beyond hematoxylin and eosin — The role of immunohistocytochemistry in surgical pathology, *Cancer Invest.*, 22, 445, 2004.
123. Hivnor, C., Williams, N., Singh, F. et al., Gene expression profiling of porokeratosis demonstrates similarities with psoriasis, *J. Cut. Path.*, 31, 657, 2004.
124. Melle, C., Ernst, G., Schimmil, B. et al., Biomarker discovery and identification in laser microdissected head and neck tumor squamous cell carcinoma with Protein-Chip® technology, two-dimensional electrophoresis, tandem mass spectrometry and immunohistochemistry, *Mol. Cell. Proteomics*, 3, 443, 2003.
125. Ahram, M., Flaig, M.J., Gillespie, J.W. et al., Evaluation of ethanol-fixed, paraffin-embedded tissues for proteomic applications, *Proteomics*, 3, 413, 2003.
126. Mojsilovic-Petrovic, J., Nesic, M., Pen, A. et al., Development of rapid staining protocols for laser-capture microdissection of brain vessels from human and rat coupled to gene expression analyses, *J. Neuro. Meth.*, 133, 39, 2004.
127. Kellner, U., Steinert, R., Siebert, V. et al., Epithelial cell preparations for proteomic and transcriptomic analysis in human pancreatic tissue, *Path. Res. Prac.*, 200, 155, 2004.
128. Tadros, Y., Ruiz-Deya, G., Crawford, B.E. et al., *In vivo* proteomic analysis of cytokine expression in laser capture-microdissected urothelial cells of obstructed ureteropelvic junction procured by laparoscopic dismembered pyeloplasty, *J. Endourology*, 17, 333, 2003.
129. De Souza, A.J., McGregor, E., Dunn, M.J. and Rose, M.L., Preparation of human heart for laser microdissection and proteomics, *Proteomics*, 4, 578, 2004.

130. Kondo, T., Seike, M., Mori, Y. et al., Application of sensitive fluorescent dyes in linkage of lasere microdissection and two-dimensional electrophoresis as a cancer proteomic study tool, *Proteomics*, 3, 1758, 2003.

131. Ernst, L.A., Gupta, R.K., Mujumblar, R.B. and Waggoner, A.S., Cyanine dye labeling reagents for sulfydryl groups, *Cytometry*, 10, 3, 1989.

132. Unlu, M., Morgan, M.E. and Minden, J.S., Difference gel electrophoresis: A single gel method for detecting changes in protein extracts, *Electrophoresis*, 18, 2071, 1997.

133. Zhou, G., Li, H., DeCoup, D. et al., 2D Differential in-gel electrophoresis for the identification of esophageal scans cells cancer-specific protein markers, *Mol. Cell. Proteomics*, 1, 117, 2002.

134. Cordwell, S.J., Nouwans, A.S., Verrills, N.M. et al., Subproteomics based upon protein cellular location and relative solubilities in conjunction with composite two-dimensional electrophoresis gels, *Electrophoresis*, 21, 1094, 2000.

135. Chaurand, P., Sander, M.E., Jensen, R.A. and Caprioli, R.M., Proteomics in diagnostic pathology. Profiling and imaging proteins directly in tissue sections, *Am. J. Path.*, 165, 1057, 2004.

136. Imam-Sghiouar, N., Laude-Lemaire, I., Labas, I. et al., Subproteomics analysis of phosphorylated proteins: Application to the study of B-lymphoblasts from a patient with Scott syndrome, *Proteomics*, 2, 828, 2002.

137. Navarre, C., Degand, H., Bennett, K.L. et al., Subproteomics: Identification of plasma membrane proteins from the yeast *Saccharomyces cerevisiae*, *Proteomics*, 2, 1706, 2002.

138. Shiozaki, A., Tsuji, T., Kohno, R. et al., Proteome analysis of brain proteins in Alzheimer's disease: Subproteomics following sequentially extracted protein preparation, *J. Alzheimers Dis.*, 6, 257, 2004.

139. Journet, A. and Ferro, M., The potentials of MS-based subproteomics approaches in medical science: The case of lysosomes and breast cancer, *Mass Spec. Rev.*, 23, 393, 2004.

140. Doran, P., Dowling, P., Lohan, J. et al., Subproteomics analysis of Ca+-binding proteins demonstrates decreased calsequestrin expression in dystrophic mouse skeletal muscle, *Eur. J. Biochem.*, 271, 3943, 2004.

141. Huber, L.A., Pfallen, K. and Victor, I., Organelle proteomics. Implications for subcellular fractionation in proteomics. *Circulation Res.*, 92, 962, 2003.

142. Dreyer, M., Subcellular proteomics, *Mass Spec. Rev.*, 22, 27, 2003.

143. Toyln, S.W., Warnock, D.E., Glenn, G.M. et al., An alternative strategy to determine the mitochondrial proteome using sucrose density gradient fractionation, *J. Proteome Res.*, 1, 451, 2002.

144. Wu, C.C., MacCoss, M.J., Howell, K.E. and Yates, J.R., III, A method for the comprehensive proteomic analysis of membrane proteins, *Nat. Biotech.*, 21, 532, 2003.

145. Schirmer, E.C., Florens, L., Guan, T. et al., Nuclear membrane proteins with potential disease links found by subtractive proteomics, *Science*, 301, 1380, 2003.

146. Sprenger, R.R., Speijer, D. and Back, J.W., Comparative proteomics of human endothelial cell cavaeolae and rafts using two-dimensional gel electrophoresis and mass spectrometry, *Electrophoresis*, 25, 156, 2004.

147. Bard, M.R., Hegmans, J.P., Hemmes, A. et al., Proteomic analysis of exosomes isolated from human malignant pleural effusions, *Am. J. Respir. Cell. Mol. Biol.*, 31, 114, 2004.

148. Utleg, A.G., Yi, E.C., Xie, T. et al., Proteomic analysis of human prostasomes, *The Prostate*, 56, 150, 2003.

149. Chang, J., Van Remmen, H., Cornell, J. et al., Comparative proteomics: Characterization of a two-dimensional gel electrophoresis system to study the effect of aging on mitochondrial proteins, *Mech. Aging Dev.*, 124, 33, 2003.

150. Claverol, S., Burlet-Schiltz, O., Girbal-Neuhauser, E., Gariin, J.E. and Monsarrat, B., Mapping and structural dissection of human 20 S proteosome using proteomic approaches, *Mol. Cell. Proteomics*, 1, 561, 2002.

151. Horejsi, V., Affinity electrophoresis, *Anal. Biochem.*, 112, 1, 1981.

152. Nakamura, K. and Takeo, K., Affinity electrophoresis and its applications to studies of immune response, *J. Chrom. B Biomed. Sci. Appl.*, 715, 125, 1998.

153. Lee, B.S., Gupta, S., Krisnanchettier, S. et al., Catching and separating protein ligands by functional affinity electrophoresis, *Anal. Biochem.*, 334, 106, 2004.

154. Kopperschlager, G., Effects of specific binding reactions on the partitioning behavior of biomaterials, *Intl. Rev. Cytol.*, 192, 61, 2000.

155. Johansson, G., Affinity partitioning of proteins, *Meth. Mol. Biol.*, 147, 105, 2000.

156. Barinaga-Rementeria Ramirez, I., Mebrahtu, S. and Jergil, B., Affinity partitioning for membrane purification exploiting the biotin-NeutrAvidin interaction. Model study of mixed liposomes and membranes, *J. Chrom. A*, 971, 117–127, 2002.

157. Lowe, C.R., Combinatorial approaches to affinity chromatography, *Curr. Opin. Chem. Biol.*, 5, 248–256, 2001.

158. Lee, W.-C. and Lee, K.H., Applications of affinity chromatography in proteomics, *Anal. Biochem.*, 324, 1–10, 2004.

159. Puig, O., Caspary, F., Rigaut, G. et al., The tandem affinity purification (TAP) method: A general procedure of protein complex purification, *Methods*, 24, 218–229, 2001.

160. Forler, D., Kocher, T., Rode, M., Gentzel, M., Isaurralde, E. and Wilm, M., An efficient protein complex purification for functional proteomics in higher eukaryotes, *Nat. Biotech.*, 21, 89–92, 2003.

161. Agaton, C., Uhlén, M. and Hober, S., Genome-based proteomics, *Electrophoresis*, 25, 1280–1288, 2004.

162. Cuatrecasas, P., Affinity chromatography of macromolecules, *Adv. Enzymol. Rel. Areas Mol. Biol.*, 36, 29–89, 1972.

163. March, S.C., Parikh, I. and Cuatrecasas, P., Affinity chromatography — Old problems and new approaches, *Adv. Expt. Med. Biol.*, 42, 3–14, 1974.

164. Wilchek, M. and Hexter, C.S., The purification of biologically active compounds by affinity chromatography, *Meth. Biochem. Anal.*, 23, 347–385, 1976.

165. Birch, R.M., O'Bryrne, C., Booth, I.R. and Cash, P., Enrichment of *Escherichia coli* proteins by column chromatography on reactive dye columns, *Proteomics*, 3, 764–776, 2003.

166. Lolli, G., Thaler, F., Valsasina, B. et al., Inhibitor affinity chromatography: Profiling the specific reactivity of the proteome with immobilized molecules, *Proteomics*, 3, 1287–1298, 2003.

167. Brill, L.M., Salomon, A.R., Ficarro, S.B. et al., Robust phosphoproteomic profiling of tyrosine phosphorylation sites from human T cells using immobilized metal affinity chromatography and tandem mass spectrometry, *Anal. Chem.*, 76, 2763–2772, 2004.

168. Daub, H., Godl, K., Brehmer, D., Klebl, B. and Muller, G., Evaluation of kinase inhibitor selectivity by chemical proteomics, *Assay Drug Dev. Tech*, 2, 215–224, 2004.

169. Archambault, V., Chang, E.J., Drapkin, B.J. et al., Targeted proteomic study of the cyclin-cdk module, *Mol. Cell.*, 14, 699–711, 2004.

170. Xiong, L., Andrews, D. and Regnier, F., Comparative proteomics of glycoproteins based on lectin selection and isotope coding, *J. Proteome Res.*, 2, 618–625, 2003.

171. Lowe, C.R., *An Introduction to Affinity Chromatography*, Elsevier-North Holland, Amersterdam, 1979.

172. Scouten, W.H., *Affinity Chromatography*, Wiley, New York, NY, 1980.

173. *Affinity Chromatography — A Practical Approach*, P.D.G. Dean, W.S. Johnson and F.A. Middle, Eds., IRL Press, Oxford, United Kingdom, 1985.

174. *Analytical Affinity Chromatography*, I.M. Chaiken, Ed., CRC Press, Boca Raton, FL, 1987.

175. *Handbook of Affinity Chromatography*, T. Kline, Ed., Dekker, New York, NY, 1993.

176. *Affinity Chromatography: Methods and Protocols*, P. Bailon, et al., Eds., Humana Press, Totowa, NJ, 2000.

6 An Overview of Analytical Technologies Used in Proteomic Research

CONTENTS

INTRODUCTION

The success of proteomics is driven not by concept or hypothesis but by discovery, which is driven by marked advances in analytical technologies including microarray analysis (including surface-plasmon resonance technology) and mass spectrometry. Associated developments exist in separation science including electrophoresis and chromatography necessary for the preparation of samples for analysis by mass spectrometry. Protein microarray technology has not been of significant use as of the time of this writing. Unlike DNA microarray technology where the construction of multiple oligonucleotide arrays with specificity is comparatively easy because of knowledge derived from the genomic sequence,[1-4] the use of protein microarray requires prior knowledge of the analyte for preparation of the microarray and in most cases, antibodies are used to capture analyte. While specific base-pairing is well-understood and is the basis for oligonucleotide/polynucleotide interaction, similar specific phenomena is not available for protein–protein or protein–peptide interactions. Such interactions are inherently more complex because of the number of different amino acids involved compared the number of purine and pyrimidine bases as well as the number of interaction modalities including hydrogen-bonding, van der Waals forces and hydrophobic interactions. Thus, while using combinatorial chemistry or phage display is possible to obtain "protein" microarrays similar to the DNA microarrays in binding site numbers, the rationale for site design is missing. One possibility that has not been explored is the use of antisense peptides. Antisense peptides are derived from the "transcription/translation" of the antisense or non-coding strand of DNA. The original hypothesis stated there would be a specific interaction between a sense protein/peptide and an antisense protein/peptide. While the hypothesis has not been validated, enough results exist[5-9] to suggest there are specific interactions that could be useful for microarray technology.

The use of a protein microarray does not clearly provide any intrinsic advantage other than scale over current immunoassay technology. However, also clear is that immunoassay technology cannot be effectively used to identify the number of proteins possible with mass spectrometry/mass fingerprinting.[10] The ability to multiplex sandwich assays in a microarray technology does appear to have some potential in providing an advantage over current immunoassay technology.[11] Significant advances in protein microarray technology will likely require the development of new paradigms. One such approach is peptidomics,[12] where the heterogeneous protein sample is digested to yield peptides that can be used as samples. This approach provides a more physically homogeneous sample that has the potential for less promiscuous binding. The direct analysis of samples from the matrix provides an advantage over the usual fluorescence-based detection. Another approach use aptamers as the affinity matrix.[13] Several recent reviews are found in this area[14,15] and little additional discussion of this technology is necessary.

Tissue microarray technology is of great potential value to proteomic research. The ability to simultaneously process a large number of tissue samples under identical conditions is a great advantage[16-18] as the use of laser capture microscopy and direct mass spectrometric analysis of tissue samples is developed.[19,20]

The point from the above microarray discussion is that protein microarrays analogous to DNA microarrays may be available at a later date. The lack of sufficiently diverse target design precludes true utility but does not preclude confusion in the literature and care must be taken in the evaluation of information. ProteinChip® technology such as that described below for surface-enhanced laser desorption/ionization (SELDI) mass spectrometry are not the same as DNA chips, which are DNA microarray chips.

DEVELOPMENT IN PROTEIN ANALYTICAL TECHNOLOGY

Emphasis of one critical point is necessary in the discussion of the several analytical techniques used in proteomics. Proteomics is based on discovery, as opposed to being hypothesis-based, meaning (1) the investigator will not know what has been missed since it is not known that it is present in the starting material, and (2) at least in initial studies, it is not known whether the amount of protein detected is, in fact, a true measure of the content in the starting material. This problem can be solved in later studies when, for example, an immunoassay can be used to follow the fractionation/assay pathway.

The chromatographic and electrophoretic fractionation of proteins and peptides is a major part of proteomic analysis. Thus, the fact that little time is devoted to these technologies in this chapter might be strange. Consideration of chromatography is scattered through various parts of this work as required for a specific task. Much of the basis work was done years ago as part of classical protein chemistry. The reader is directed to a collection of extremely useful review articles on the use of chromatography and electrophoresis in proteomic analysis.[21–28] Righetti[21] has provided an excellent review of the development of separation technology since the work of Svedberg and Tiselius. Analytical ultracentrifugation, developed from the early work of Svedberg, has not seen as much use as the electrophoretic techniques but can be valuable for the characterization of proteins and protein–protein interactions, as can techniques such as light scattering.[29] Application of these biophysical techniques are somewhat hampered by the sample size requirements as compared to other analytical techniques such as electrophoresis and mass spectrometry; these techniques are also somewhat difficult. Free boundary electrophoresis as developed by Tiselius was supplanted by electrophoretic techniques based on matrices such as starch gel and, later, polyacrylamide gels. Free boundary electrophoresis has seen somewhat of a renaissance for prefractionation of proteins for proteomic analysis (see Chapter 5) and for large scale protein fractionation.[30]

MASS SPECTROMETRY

The major technology advances responsible for proteomics have been in mass spectrometry[31–41] and the associated development of data analysis for protein identification.[42–49] Mass spectrometry continues to evolve at a breakneck pace with the development of Fourier transform ion cyclotron resonance (FITCR) in proteomic analysis.[50–55] As the analytical techniques increase in accuracy, an equal increase in the sophistication of experimental design will also be required.

The developments in mass spectrometry as applied to proteomics have been driven by a number of investigators.[56–62] Advances have included both instrument development and, in the case of proteins, "soft ionization" methods such as matrix-assisted laser desorption/ionization (MALDI) and electrospray ionization (ESI), which permit the volatilization of proteins and peptides. Caprioli and Suter[63] have compared the attributes of ESI, MALDI and plasma desorption mass spectrometry. These advances in sample preparation, sensitivity and resolution, when combined with the completion of the Genome Project and increased sophistication of protein sequence databases, have permitted the identification of individual proteins in complex mixtures. The novice is directed to the book by Herbert and Johnstone[36] and the review by Glish and Vachet[62] to obtain a strong basis in mass spectrometry. Both of these sources provide a basic understanding of the physical and chemical concepts underlying the analysis of peptides and proteins by mass spectrometry.

One of the more interesting developments in mass spectrometry is the emergence of FITCR mass spectrometry. This technique is extremely accurate, sensitive and expensive. It is based on electron capture dissociation (ECD), where positively charged ions capture electrons, resulting in neutralization and fragmentation.[53] While this concept is not completely understood, the technique is gaining considerable use.[52,54,55] Some post-translational modifications are interesting, such as γ-carboxyl groups in vitamin K-dependent proteins that are stable to ECD. The use of accurate mass tags (AMT) permits the facile identification of proteins. An accurate mass tag is a peptide of reasonably unique mass and elution time from liquid chromatography[52] that has seen practical use.[64–66]

Mass spectrometry has been a powerful tool not only for protein identification in proteome analysis but also for the study of the post-translational modification of proteins.[67–77] Wold has compiled a list of the mass charge changes associated with post-translational modifications[78,79] but this compilation is likely outdated. Post-translational modification increases the challenge for proteome analysis as the number of analytes increase. Most significant post-translational modifications are specific, enzyme-catalyzed reactions such as phosphorylation, sulfation, farnesylation and gamma-glutamic acid carboxylation. Other post-translational modifications, such as the reaction of nitric oxide with sulfydryl groups and tyrosine residues in proteins while "enzyme-catalyzed" in the sense the nitric oxide synthetase is involved, are not site-specific modifications; the same caveat holds for the modification of proteins such as the formation of carbonyl groups in lysine via oxidation.

Mass spectrometry has also proved useful for the study of the chemical modification of proteins.[80–86] Chemical modification can be useful for improving the accuracy of mass spectrometric measurements. Modification of lysine residues in peptides with O-methylisourea results in the formation of homoarginine (Figure 6.1) and results in more intense signals on MALDI analysis.[87,88] Further modification of the homoarginine peptides with chlorosulfonylacetyl chloride resulting in sulfonylation of the amino-terminal residues (Figure 6.2) permitted more facile sequencing by mass spectrometry.[88,89] Specificity of N-terminal modification is provided by prior modification of the terminal lysine residue to form homoarginine precluding reaction

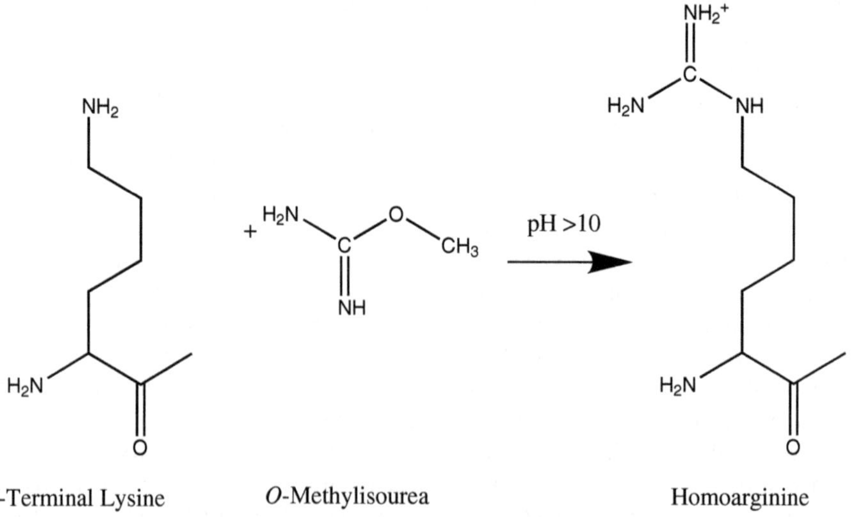

N-Terminal Lysine   O-Methylisourea   Homoarginine

FIGURE 6.1 The chemical mutation of lysine to homoarginine.

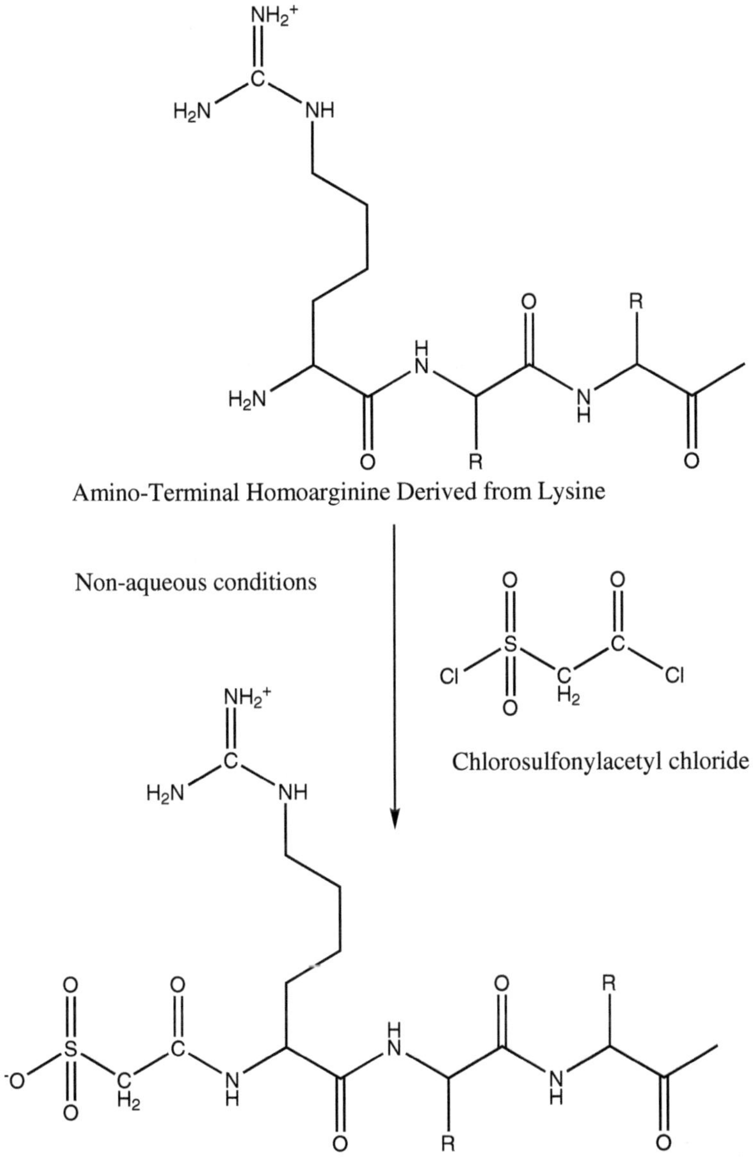

FIGURE 6.2 The modification of *N*-terminal homoarginine with chlorosulfonylacetyl chloride.

at the ε-amino group. More recently, Samyn and colleagues[90] prepared an *N*-terminal sulfonyl derivative with 2-sulfobenzoic acid cyclic anhydride (Figure 6.3). The sulfonylation reaction adds a strong negative charge to the peptide. Modification with labeled (deuterium) propionic anhydride and unlabeled propionic anhydride has provided a sequence-independent approach for ICAT technology since modification occurs at the amino-terminal homoarginine residue.[91] An approach designated as

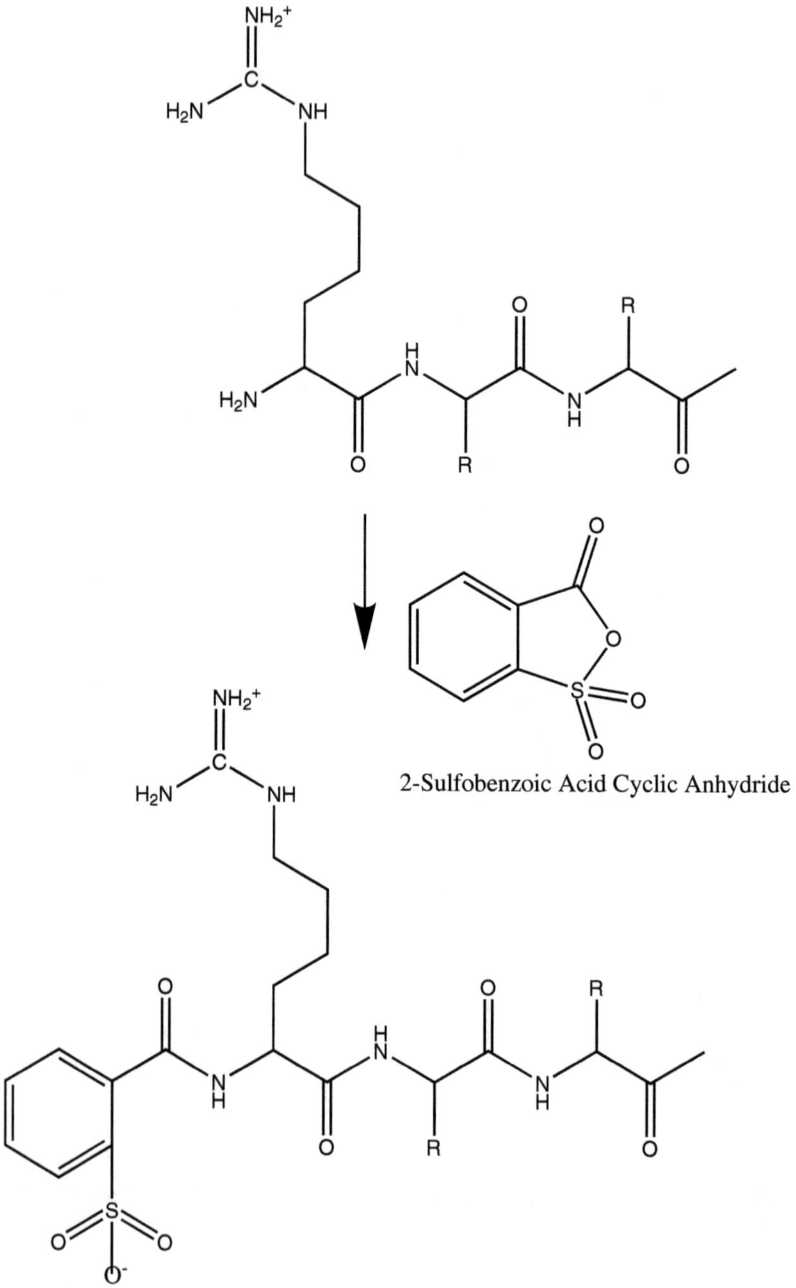

FIGURE 6.3 The modification of *N*-terminal homoarginine with 2-sulfobenzoic acid cyclic anhydride.

quantitation using enhanced signal tags (QUEST) uses the modification of peptides with S-methylthioacetamidate and S-methylthiopropioamidate (Figure 6.4).[92,93] These reagents would react with both the ε-amino group of lysine and the peptide α-amino group. Improvement of mass spectral resolution of tryptic peptides is also seen with oxidation with performic acid,[94] which converts cysteine to cysteic acid, methionine to methionine sulfoxide and tryptophan to the oxindole derivatives. Selective reductive methylation (Figure 6.5) of the N-terminal amino groups also appears to be a useful approach for enhancing mass spectrometric signal.[95,96]

The above brief discussion is not intended to provide anything close to a global discussion of mass spectrometry and the reader is directed to the above cited references for more useful technical information. For our purpose, mass spectrometry and related bioinformatics tools are analytical techniques for the identification of proteins. As noted, the past decade has seen truly spectacular technical advances in mass spectrometry. However, such technology does not substitute for either experimental design or sample quality/preparation.[97] Review of a comprehensive treatise[98,99] on analytical chemistry is useful before engaging in proteomic research.

TWO-HYBRID SYSTEMS AND TANDEM-AFFINITY PURIFICATION — STUDY OF PROTEIN–PROTEIN INTERACTIONS

We acknowledge the value of two-hybrid systems[100–104] and tandem-affinity chromatography (tandem-affinity purification, TAP)[105–109] in proteomics as analytical techniques to probe *in vivo* protein–protein interactions. Both of these approaches require the use of protein engineering to express "bait" proteins. These systems are of great value in the study of the effect of stimuli on cells in culture and likely will be of great benefit to the development of therapeutic products. Another approach uses chemical cross-linking followed by mass spectrometry to measure intramolecular and intermolecular interactions.[110,111] The chemistry of cross-linking is discussed in Chapter 2. Other approaches to the study of *in vivo* protein–protein interactions include purification of protein complexes under native conditions,[112–115] co-immunoprecipitation,[116–119] and GST-pulldowns. GST-pulldowns involve the expression of a GST-fusion protein with the "bait" protein, binding the GST-fusion protein to a glutathione matrix and eluting the protein complexes by cleavage of the fusion protein.[120–122] Other fusion partners such as hexahistidine could be used for the purpose of complex purification.[123,124] Figeys[125] has reviewed several methods of mapping protein interactions. One approach is based on the preparation of a fusion protein composed of a lactamase fragment and a "bait" protein. Potential target proteins are expressed as fusion proteins with a lactamase fragment that complements the fragment in the bait fusion protein. Binding of the target protein to the bait protein results in enzyme activity that can be measured as an indication of the interaction.

Tandem affinity purification (TAP) has been mostly used for bacterial and yeast systems. Technical difficulties have precluded extensive use in plant and mammalian systems but this is changing.[126–130] Baserga and colleagues present an example of

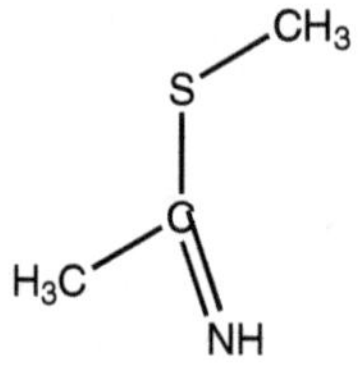

S-Methyl thioacetimidate

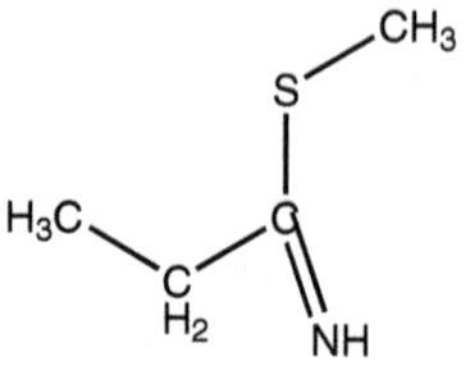

S-Methyl thiopropioimidiate

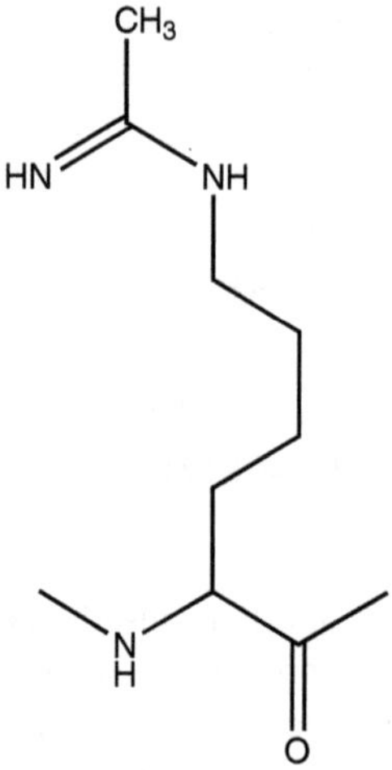

N-acetimidolysine

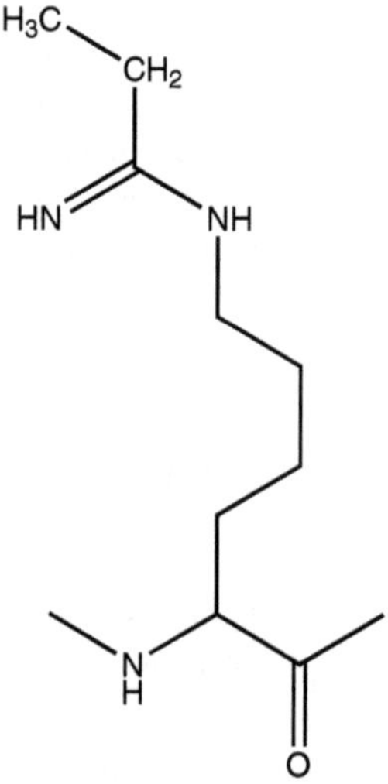

N-propionimidolysine

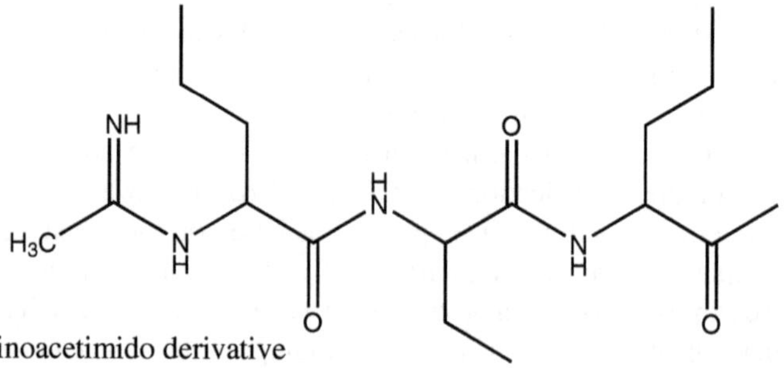

alpha-aminoacetimido derivative

FIGURE 6.4 The modification of cysteine with S-methylthioacetamidate — the QUEST strategy.

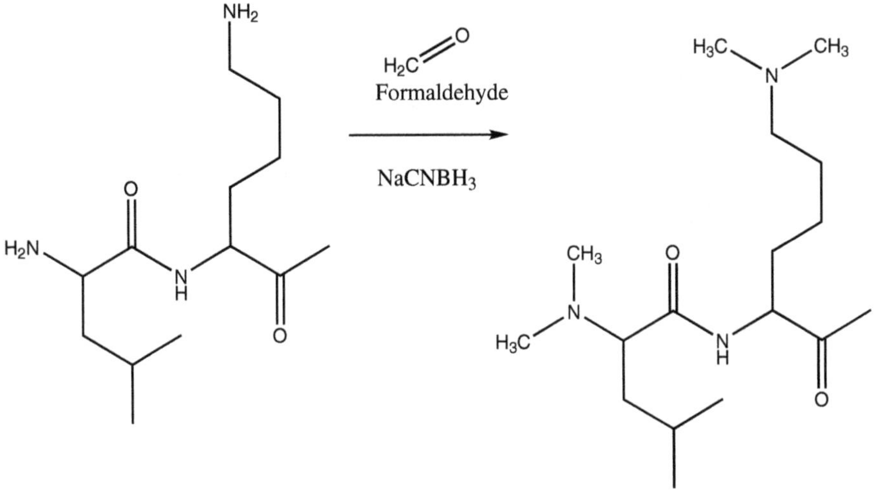

FIGURE 6.5 The reductive alkylation of lysine residues in proteins and peptides.

the application of TAP in mammalian systems. The affinity tag for these studies is a construct of two IgG binding units from Protein A from *Staphylococcus aureus* together with a cleavage site for tobacco etch virus (TEV) protease and calmodulin binding peptide. The fusion protein is bound to Protein A-agarose; proteins inter-acting with the bait domain are released by digestion with TEV protease. The protein complex in the supernatant fraction is captured with monomeric avidin resin and the bound proteins released at low pH (glycine, pH 2.8) and analyzed by mass spectrometry. The objective of this study was to analyze proteins bound to IRS-1 (insulin-substrate receptor-1). A number of proteins were identified including histone h1/h5, histone h1, prohibitin, beta-catenin and pp2A alpha; the presence of these proteins was confirmed by coimmunoprecipitation.

TWO-DIMENSIONAL GEL ELECTROPHORESIS

The remainder of the discussion will focus on the three most commonly used techniques for proteomics research: two-dimensional gel electrophoresis/mass spec-trometry, multi-dimensional protein identification technologies (MuDPIT)/"shotgun proteomics" and surface-enhanced laser desorption/ionization (SELDI) mass spec-trometry. These methods are quite different technologies with individual strengths and limitations. All technologies rely on mass spectrometry for analysis and, in general, identification of the proteins is based on mass fingerprinting.[131–133] The concept of "fingerprinting" is thought to have originated with Ingram in 1958[134] on the comparison of tryptic peptides obtained from normal and sickle cell hemoglobin by two-dimension paper chromatography. These early studies were extended by Anfinsen and coworkers[135] using a two-dimensional approach combining electro-phoresis and chromatography. The term "fingerprint" is thought to have originated from the similarity of the "spots" to images obtained by a fingerprint on the media detected by ninhydrin staining. Another thought is the pattern would be a unique identifier of the protein, analogous to a person's fingerprints.

The oldest of the techniques and likely the most frequently used is two-dimensional gel electrophoresis,[136-140] which would include the traditional two-dimensional gel electrophoresis (2D-GEL) and difference gel electrophoresis (DIGE). The difference between the two techniques is that the sample is modified with a fluorescent dye prior to electrophoresis in the DIGE system. The first dimension is isoelectric focusing with an immobilized pH gradient and separation is based on the charge of the protein (isoelectric point); the second dimension is performed in the presence of sodium dodecyl sulfate and separation is based on molecular size. This system is complex but has been demonstrated to be reproducible.[141,142] Depending on the skill of the investigator and the detection system, the number of protein "spots" can number more than 5000 with a sensitivity of 1 ng protein/spot. Prefractionation of the sample is critical for success[143] and Chapter 5 is devoted to this activity. One recent approach to improving the resolution of two-dimensional gel electrophoresis, described as microscale solution isoelectric focusing,[144] has significant potential in improving resolution.

Two-dimensional gel electrophoresis does have limitations. The tendency of this system is to "miss" low molecular weight proteins, proteins with a high pI and high molecular weight proteins. The molecular weight issue is a problem with the polyacrylamide gel in that low molecular weight proteins tend to "leech out" of the gel before and during the staining procedure and high molecular weight proteins tend to have a problem entering the gel as well as a tendency to aggregate under denaturing conditions; the high isoelectric point issue is an inherent shortcoming of the IEF system. Oh-Ishi and Maeda have presented a discussion of separation techniques for high molecular weight proteins.[145] These investigators describe a variety of approaches and present a listing of issues associated with specific tissue samples. The major difference is the use of agarose gel systems (agarose gel in the first dimension, the IEF dimension). Other specific issues include optimal concentration of chaotropic agents such as thiourea and urea. Detergents such as SDS, which are quite useful for the solubilization of high molecular weigh proteins, cannot be used because of incompatibility with the IEF step. Triton X-100 and CHAPS were evaluated but neither appeared optimal for solubilization. Detergent concentration should not increase above 3%. Sequential sample extraction as discussed in Chapter 5 might be extremely useful in the processing of tissue (plant or animal) samples for high molecular weight proteins. Proteolytic degradation of high molecular weight proteins is also an issue and protease inhibitor cocktails are useful. Protease inhibitors and protease inhibition are discussed in greater detail in Chapter 4.

Identification of biomarkers by 2D GEL is based on the comparison of diseased tissue versus normal tissue. In most studies, the protein "spots" are identified on the electrophoretograms by staining with Coomassie Blue, ammoniacal silver or with a fluorescent dye such as SYPRO, the gels compared, the difference "spots" excised either manually or with the aid of a robotic process and the proteins either digested *in situ* with proteolytic enzymes or analyzed directly by mass spectrometry.[146-151] Another option is the derivatization of the protein sample with fluorescent dyes prior to electrophoresis; selection of different fluorophores allows the more facile comparison of several electrophoretograms.[152–154] The chemistry of the fluorescent dyes is discussed elsewhere (Chapter 3 and Chapter 6). Somewhat remarkable is that the staining issues of considerable concern during the development of SDS-PAGE seem to be totally ignored in the current discussion.

Advances in the use of soft-ionization techniques such as MALDI or ESI have permitted the analysis of high molecular weight materials including intact proteins; however, analysis of peptides derived from the digestion of an intact protein is the most frequently used approach for the identification of proteins. A useful example of this approach is provided from the study of Tomonaga and coworkers on altered protein expression in primary colorectal cancer.[155] This study is noteworthy in that a two-dimensional agarose gel was used for sample fractionation rather than the conventional acrylamide gel system, permitting the fractionation of a broader size range of proteins. The authors did report a higher number of spots on the agarose gel as well as an increase in the number of differentially expressed proteins as assessed by Coomassie dye staining.

SHOTGUN PROTEOMICS/MULTIDIMENSIONAL PROTEIN IDENTIFICATION TECHNOLOGY (MuDPIT)

The use of a separation technology that would be a direct alternative to two-dimensional gel electrophoresis for protein identification in mixtures can be traced to the work of John Yates and colleagues in 1997.[156] This approach has been described as "shotgun proteomics," an analytical technique that provides the composition of a protein mixture without a prior fractionation procedure. The success of this approach is driven by the success in identifying a large number of peptides by mass spectrometry and the "reassembly" of these peptides in a parent protein using databases derived from the human genome structure.[157–162] While different separation modalities are used, the general concept of this multidimensional protein identification technology (MuDPIT)[163–167] is the use of orthogonal techniques; in the large majority of studies, an ion-exchange column is the first separation phase and a reversed-phase column would be the second phase. A number of other potential combinations exist such as capillary electrophoresis coupled with a reverse phase column.

Shotgun proteomics/MuDPIT technology[168,169] should be and is increasing in use in proteomic research. First, major prefractionation of the sample is usually required; the coupling of multiple systems such as free boundary electrophoresis/chromatography/liquid isoelectric focusing (capillary isoelectric focusing)/capillary electrophoresis with variants such as affinity chromatography and affinity electrophoresis should provide adequate and reproducible separation of the components of most samples for analysis by mass spectrometry. Second, low-abundance proteins are less likely to be missed. Third, the technique should be able to cover the dynamic range of protein content in most samples. Fourth, low molecular weight proteins are less likely to be excluded from analysis and loss from matrix should not be an issue. High molecular weight proteins such as von Willebrand factor, fibronectin and fibrinogen would still pose an analytical problem requiring the use of a separate sample separation matrix. A batch size exclusion step such as filtration or, in the case of blood plasma, cryoprecipitation seems reasonable.

Shotgun proteomics is a global approach to the analysis of a protein mixture as that derived from the lysis of a cell. No one single technical approach exists to the product used for analysis, a tryptic digest or a Lys-C/trypsin digest. Presumably, other fractionation approaches such as cyanogen bromide cleavage could be used, providing a "top-down" approach to complement the "bottom-up" approach. A protein mixture is reduced in the presence or absence of a denaturing agent; alternatively, a protein mixture might be denatured without reduction. The protein mixture in the presence of urea is digested with endoproteinase Lys-C, which may be followed by hydrolysis with trypsin. If trypsin is used either alone or after digestion with endoproteinase Lys-C, the denatured protein (usually in 6 to 9 M urea) is diluted (three to four fold) prior to hydrolysis.[170] The combined use of endoproteinase Lys-C (which cleaves at the carboxyl group of lysine residues) and trypsin appears to be useful in obtaining a complete set of tryptic peptides.[171,172] The reduced protein is usually carboxamidomethylated (reaction with iodoacetamide) or carboxymethylated (reaction with iodoacetic acid) prior to digestion with the proteases.[169] Alternatively, the digest can be alkylated after digestion.[173]

The protein digest is then subjected to at least two successive orthogonal separations. For example, the first separation is usually ion-exchange chromatography with either a strong cation-exchange (SCX) matrix with a sulfoethyl or sulfopropyl group or with a strong anion-exchange (SAX) matrix with a quaternary amine function. In earlier days, the cation exchange matrix would have been referred to as Dowex®-1 and the anion-exchange matrix as Dowex®-50. The second column is usually a reverse phase (C_{18}) column (RP-HPLC). The isolated peptide fragments are identified by mass spectrometry and reassembled using the above databases.

The study by Washburn, Wolters, and Yates[169] is probably the seminal paper in MuDPIT/shotgun proteomics, having been cited 562 times since its publication in 2001. In the Washburn study,[169] soluble and insoluble fractions were prepared from *Saccharomyces cerevisiae* following the method of Link and coworkers[170] (cell lysis in 300 mM NaF, 3.45 mM NaVO$_3$, 50 mM Tris, 12 mM EDTA, 250 mM NaCl, 140 mM sodium phosphate, pH 7.4 and glass beads) in the presence of a protease inhibitor cocktail.[174] The soluble fraction was reduced with dithiothreitol (1 mM) in the presence of 8.0 M urea (pH 8.5 with 1.0 M ammonium bicarbonate), carboxamidomethylated with iodoacetamide (10 mM) and digested with endoproteinase Lys-C (enzyme/substrate ratio of 1:100; 15 hours at 37°C). After dilution to 2.0 M urea with 100 mM NH$_4$HCO$_3$, pH 8.5 and CaCl$_2$ added to a final concentration of 1 mM. Immobilized trypsin was added to complete the digestion. Endoproteinase Lys-C and trypsin are frequently combined to produce a complete yield of "tryptic" peptides. This digest was chromatographed on a column first packed with a C_{18} reverse phase matrix followed by a sulfoethyl cation exchange matrix (SCX), allowing an orthogonal separation of peptides within the same chromatographic step. Effluent fractions are analyzed by mass spectrometry and the identified peptides are "reassembled" into the parent proteins using a database developed from the human genome structure. This process is truly analogous to the process of putting a puzzle together; you are, however, blessed by the facts that (1) the puzzle is, in fact, linear and (2) you do not need all the pieces; usually three or four tryptic peptides will be adequate for identification.

SURFACE-ENHANCED LASER DESORPTION/IONIZATION — MASS SPECTROMETRY (SELDI-MS)

The third most extensively used procedure has properties of both of the above procedures. Surface-enhanced laser desorption/ionization-time of flight mass spectrometry (SELDI-TOF-MS) was developed as affinity mass spectrometry.[175–179] SELDI technology is an attractive approach to many investigators because of its simplicity and the "black box" approach. Some issues exist with respect to the reproducibility of matrix manufacture and interlaboratory variability. SELDI can be described as a type of MALDI mass spectrometry where the sample matrix has an active role in sample purification as well as the desorption/ionization step.

The approach has been commercialized by Cipergen (Fremont, CA) as Protein-Chip® technology.[180–186] While the ProteinChip® technology is analogous to a microarray system, it is not necessarily the same as a protein chip[187] and thus care must be taken in literature searches. The basis of the ProteinChip® technology is the specific adsorption of the analyte from a mixture.[186] Current surface technologies, which seem to be most commonly used, include ion-exchange (anion exchange, SAX; cation-exchange, SCX) where protein charge is the determinant, immobilized metal affinity chromatography matrix (IMAC)[188] and hydrophobic matrices. In addition, a variety of affinity matrices may be prepared including DNA aptamers[178] and antibodies.[189] This use of more specific affinity matrices is referred to as surface-enhanced affinity capture (SEAC). Borchers and colleagues[190] have developed a variant technology based on the capture of peptides rather than proteins for use in an immunoassay for protein expression.

The original studies using SELDI technology did not use sample prefractionation. More recent studies have suggested that the use of ion-exchange chromatography prior to SELDI analysis yields more useful samples.[188,191] Most work has used isocratic chromatographic systems. Isocratic systems are, in general, more reproducible and easier to validate for regulatory purposes. However, isocratic systems do not have the resolving power of gradient systems and unique gradient systems are being developed for use in proteomic research.[192] In this study,[192] Timperman and coworkers developed a "saw-tooth" gradient for the reverse-phase HPLC fractionation of human serum. This technique combines features of an isocratic elution with gradient elution, permitting the separation of albumin from low-abundance proteins and the resolution of low molecular weight proteins and the approach appears to have great potential. A variety of other approaches exist and the reader is directed to a discussion of prefractionation in Chapter 5.

After washing, a suitable matrix additive such as sinapinic acid or α-cyano-4-hydroxy-cinnamic acid is added, the sample dried[193] and mass spectrometric analysis performed. The protein sample may be analyzed directly[194,195] or after proteolytic digestion of the adsorbed material.[196] Proteolysis of a protein after adsorption can be used to identify binding sites such as antibody epitope sites.[197]

CONCLUSIONS

Advances in mass spectrometry technique and data analysis (bioinformatics) have been responsible for the development of proteomics as a discipline. The challenge is in the reproducible preparation of samples for mass spectrometric analysis. Several technical approaches are available for the investigator. The SELDI-MS approach is the least challenging for the novice investigator but would also appear to be the system that is most likely to miss low-abundance biomarkers. MuDPIT and 2D-DIGE approaches are more technically challenging but also more time-consuming and expensive. Adherence to the basic principles of analytical chemistry will be essential for success in proteomic research.

REFERENCES

1. Lee, P.S. and Lee, K.H., Genomic analysis, *Curr. Opin. Biotech.*, 11, 171, 2000.
2. Blohm, D.H. and Guiseppi-Elie, A., New developments in microarray technology, *Curr. Opin. Biotech.*, 12, 41, 2001.
3. Olson, J.A., Application of microarray profiling to clinical trials in cancer, *Surgery*, 136, 519, 2004.
4. Geschwind, D.H., DNA Microarrays: Translation of the genome from laboratory to clinic, *Lancet Neurology*, 2, 275, 2003.
5. Shai, Y., Flashner, M., and Chaikin, I.M., Anti-sense peptide recognition of sense peptides: Direct quantitative characterization with the ribonuclease S-peptide system using analytical high-performance affinity chromatography, *Biochemistry*, 26, 669, 1987.
6. Tropsha, A., Kizer, J.S. and Chaiken, I.M., Making sense from antisense: A review of experimental data and developing ideas on sense — Antisense peptide recognition, *J. Mol. Recog.*, 5, 43, 1992.
7. Madhusudanan, K.P., Katti, S.B., Haq, W. and Misra, P.K., Antisense peptide interactions studied by electrospray ionization mass spectrometry, *J. Mass Spec.*, 35, 237, 2000.
8. Zhao, R., Xu, X., Liu, H. et al., Studies on the degeneracy of antisense peptides using affinity chromatography, *J. Chrom. A*, 913, 421, 2001.
9. Chen, J., He, Q., Zhang, R. et al., Allogenic donor splenocytes pretreated with antisense peptide against B7 prolong cardiac allograft survival, *Clin. Exp. Immunol.*, 138, 245, 2004.
10. Hancock, W.S., Proteomics versus immunoassays, *J. Proteome Res.*, 1, 393, 2002.
11. Nielsen, U.B. and Geierstanger, B.H., Multiplexed sandwich assays in microarray format, *J. Immunol. Meth.*, 290, 107, 2004.
12. Scrivener, E., Barry, R., Platt, A. et al., Peptidomics: A new approach to affinity protein microarrays, *Proteomics*, 3, 122, 2003.
13. McCauley, T.G., Hamaguchi, N. and Stanton, M., Aptamer-based biosensor arrays for detection and quantification of biological macromolecules, *Anal. Biochem.*, 319, 244, 2003.

14. *Protein Microarrays*, N. Schena, Ed., Jones & Bartlett, Sudbury, MA, 2005.
15. *Protein Microarray Technology*, D. Kambpampatil, Ed., John Wiley & Sons, Hoboken, NJ, 2004.
16. Shaknovich, R., Celestine, A., Yang, L. and Cattoretti, G., Novel relational database for tissue microarray analysis, *Arch. Path. Lab. Med.*, 127, 492, 2003.
17. Krenn, V., Petersen, I., Häupl, T. et al., Array technology and proteomics in autoimmune disease, *Path. Res. Prac.*, 200, 95, 2004.
18. Mobasheri, A., Airley, R., Foster, C.S. et al., Post-genomic applications of tissue microarrays: Basic research, prognostic oncology, clinical genomics, and drug discovery, *Histol. Histopath.*, 19, 325, 2004.
19. Chaurand, P., Saunders, M.E., Jensen, R.A. and Caprioli, R.M., Proteomics in diagnostic pathology: Profiling and imaging proteins directly on tissue sections, *Am. J. Path.*, 165, 1057, 2004.
20. Caldwell, R.L. and Caprioli, R.M., Tissue profiling by mass spectrometry: A review of methodology and applications, *Mol. Cell. Proteomics*, 4, 393, 2005.
21. Righetti, P.G., Bioanalysis: Its past, present, and the future, *Electrophoresis*, 25, 2111, 2004.
22. Weber, P.L., Capillary electrophoresis in peptide and protein analysis, Detection methods for, in *Encyclopedia of Analytical Chemistry. Applications, Theory and Instrumetnation*, R.A. Meyers, Ed., John Wiley & Sons, Chichester, United Kingdom, Vol. 7, 5614, 2000.
23. Tomlinson, A.J. and Naylor, S., Capillary electrophoresis/mass spectrometry in peptide and protein analysis, in *Encyclopedia of Analytical Chemistry. Applications, Theory and Instrumentation*, R.A. Meyers, Ed., John Wiley & Sons, Chichester, United Kingdom, Vol. 7, 5674, 2000.
24. Shen, T.L., High-performance liquid chromatography/mass spectrometry in peptide and protein analysis, in *Encyclopedia of Analytical Chemistry. Applications, Theory and Instrumentation*, R.A. Meyers, Ed., John Wiley & Sons, Chichester, United Kingdom, Vol. 7, 5845, 2000.
25. Manabe, T., Analysis of complex protein-polypeptide systems for proteomic studies, *J. Chrom. B*, 787, 29, 2003.
26. Shen, Y. and Smith, R.D., Proteomics based on high-efficiency capillary separations, *Electrophoresis*, 23, 3116, 2004.
27. Wang, H. and Hanash, S., Multi-dimensional liquid phase based separations in proteomics, *J. Chrom. B*, 787, 11, 2003.
28. Li, Y., Xiong, R., Wilkins, J.A. and Horváth, C., Capillary chromatography of peptides proteins, *Electrophoresis*, 25, 2242, 2004.
29. Neet, K.E. and Lee, J.C., Biophysical characterization of proteins in the post-genomic era of proteomics, *Mol. Cell. Proteomics*, 1, 415, 2002.
30. Evtushenko, M., Wang, K., Stokcs, H.W. and Nair, H., Blood protein purification and simultaneous removal of noneveloped viruses, using tangential-flow preparative electrophoresis, *Electrophoresis*, 26, 28, 2005.
31. http://masspec.scripps.edu/hist.html.
32. Suzdak, G., *Mass Spectrometry for Biotechnology*, Academic Press, San Diego, CA, 1996.
33. Johnstone, R.A.W., *Mass Spectrometry for Chemists and Biochemists*, 2nd Edition, Cambridge University Press, Cambridge, United Kingdom, 1996.
34. *Mass Spectrometry in Biology and Medicine*, A.L. Burlingame, S.A. Carr and M.A. Baldwin, Eds., Humana Press, Towata, NJ, 2000.
35. Throck, W.J., *Introduction to Mass Spectrometry*, Lippincott-Raven, Philadelphia, PA, 1997.

36. Herbert, J.C. and Johnstone, R.A.W., *Mass Spectrometry Basics*, CRC Press, Boca Raton, FL, 2003.

37. Przybylski, M., Weinmann, W. and Fligge, T.A., Mass spectrometry, in *Handbook of Spectroscopy*, G. Gauglitz and T. Vo-Dinh, Ed., Wiley-VCH, Vol. 1, Sec. 10, 317–362, 2003.

38. Hassell, C., Process mass spectrometry, in *Handbook of Spectroscopy*, G. Gauglitz and T. Vo-Dinh, Eds., Wiley-VCH, Vol. 2, Sec. 19, 316–335, 2003.

39. de Hoffman, S. and Stroubant, V., *Mass Spectrometry: Principles and Applications*, 2nd Edition, J.W. Wiley, New York, NY, 2004.

40. Dooley, K.C., Tandem mass spectrometry in the clinical chemistry laboratory, *Clin. Biochem.*, 36, 471, 2003.

41. Honour, K.C., Benchtop mass spectrometry in clinical biochemistry, *Ann. Clin. Biochem.*, 40, 628, 2003.

42. Baggerly, K.A., Morris, J.A., Wang, J. et al., A comprehensive approach to the analysis of matrix-assisted laser desorption/ionization-time-of-flight proteomics spectra from serum samples, *Electrophoresis*, 23, 1667, 2003.

43. Steen, H. and Mann, M., The ABC's (and XYZ's) of peptide sequencing, *Nat. Rev. Mol. Cell Biol.*, 5, 699, 2004.

44. Gevaert, K. and Vanderkerckhove, J., Protein identification methods in proteomics, *Electrophoresis*, 21, 1145, 2000.

45. Baldwin, M.A., Protein identification by mass spectrometry, *Mol. Cell. Proteomics*, 3, 1, 2004.

46. Wu, C.H., Huang, H., Yeh, L.-S.L. and Barker, W.C., Protein family classification and functional annotation, *Comp. Biol. Chem.*, 27, 37, 2003.

47. Boguski, M.S. and McIntosh, M.W., Biomedical informatics for proteomics, *Nature*, 422, 233, 2003.

48. Apweiler, R., Bairoch, A. and Wu, C.H., Protein sequence databases, *Curr. Opin. Chem. Biol.*, 8, 76, 2004.

49. López-Ferrer, D., Martinez-Bartolomé, S., Villar, M. et al., Statistical model for large-scale peptide identification in databases from tandem mass spectra usnig SEQUEST, *Anal. Chem.*, 76, 6853, 2004.

50. Smith, R.D., Paša-Toli, L., Lipton, M.S. et al., Rapid quantitative measurements of proteomes by Fourier transform ion cyclotron resonance mass spectrometry, *Electrophoresis*, 22, 1652, 2001.

51. Bergquist, J., FTICR mass spectrometry in proteomics, *Curr. Opin. Mol. Ther.*, 5, 310, 2003.

52. Page, J.S., Masselon, C.D. and Smith, R.D., FITCR mass spectrometry for qualitative and quantitative bioanalyses, *Curr. Opin. Biotech.*, 15, 3, 2004.

53. Zubarev, R.A., Electron-capture dissociation tandem mass spectrometry, *Curr. Opin. Biotech.*, 15, 12, 2004.

54. Ramstrom, M. and Bergquist, J., Miniaturized proteomics and peptidomics using capillary liquid separation and high resolution mass spectrometry, *FEBS Letters*, 567, 92, 2004.

55. Barrow, M.P., Burkitt, W.I. and Derrick, P.J., Principles of Fourier transform ion cyclotron resonance mass spectrometry and its application in structural biology, *Analyst*, 130, 18, 2005.

56. Bieman, K., Recent advances in protein sequencing by mass spectrometry. Introduction and overview, in *Methods in Protein Sequence Analysis*, K. Imahari and F. Sakiyama, Eds., Plenum Press, New York, NY, 119, 1993.

57. Burlingame, A.L., Baillie, T.A. and Derrick, P.J., Mass spectrometry, *Anal. Chem.*, 58, 165R, 1986.
58. Burlingame, A.L., Millington, D.S., Norwood, D.L. and Russell, D.H., Mass spectrometry, *Anal. Chem.*, 62, 268R, 1990.
59. Burlingame, A.L., Characterization of protein glycosylation by mass spectrometry, *Curr. Opin. Biotech.*, 7, 4, 1996.
60. Jimenez, C.R. and Burlingame, A.L., Ultramicroanalysis of peptide profiles in biological samples using MALDI mass spectrometry, *Exp. Neph.*, 6, 421, 1998.
61. Hirsch, J., Hansen, K.C., Burlingame, A.L. and Matthay, M.A., Proteomics: Current techniques and potential applications to lung disease, *Am. J. Phys. Lung Cell. Mol. Phys.*, 287, L1, 2004.
62. Glish, G.L. and Vachet, R.W., The basics of mass spectrometry in the twenty-first century, *Nat. Drug Disc.*, 2, 140, 2003.
63. Caprioli, R.M. and Suter, M.J.-F., Mass spectrometry, in *Introduction to Biophysical Methods for Protein and Nucleic Acid Research*, J.A. Glasel and M.P. Deutscher, Eds., Academic Press, San Diego, CA, Ch. 4, 157, 1995.
64. Conrads, T.P., Anderson, G.A., Veenstra, T.D., Pasa-Tolic, L. and Smith, R.D., Utility of accurate mass tags for proteome-wide protein identification, *Anal. Chem.*, 72, 3349, 2000.
65. Shen, Y., Tolic, N., Masselon, C. et al., Nanoscale proteomics, *Anal. Bioanal. Chem.*, 378, 1037, 2004.
66. Li, C., Hong, Y. and Tan, Y.X., Accurate qualitative and quantitative proteomic analysis of clinical hepatocellular carcinoma using laser capture microdissection coupled with isotope-coded affinty tag and two-dimensional liquid chromatography mass spectrometry, *Mol. Cell. Proteomics*, 3, 399, 2004.
67. Zaluzec, E.J., Gage, D.A. and Watson, J.T., Matrix-assisted laser desorption ionization mass spectrometry: Applications in peptide and protein characterization, *Protein Exp. Purif.*, 6, 109, 1995.
68. Kuster, B. and Mann, M., Identifying proteins and post-translational modifications by mass spectrometry, *Curr. Opin. Struct. Biol.*, 8, 393, 1998.
69. Wilm, M., Mass spectrometric analysis of proteins, *Adv. Protein Chem.*, 54, 1, 2000.
70. Veenstra, T.D., Proteome analysis of posttranslational modifications, *Adv. Protein Chem.*, 65, 195, 2003.
71. Khidekel, N. and Hsieh-Wilson, L.C., A 'molecular switchboard' — Covalent modifications to proteins and their impact on transcription, *Org. Biomol. Chem.*, 2, 1, 2004.
72. Metodiev, M.V., Timanova, A. and Stone, D.E., Differential phosphoproteome profiling by affinity capture and tandem matrix-assisted laser desorption/ionization mass spectrometry, *Proteomics*, 4, 1433, 2004.
73. Bonaldi, T., Imhof, A. and Regula, J.T., A combination of different mass spectroscopic techniques for the analysis of dynamic changes of histone modifications, *Proteomics*, 4, 1382, 2004.
74. Person, M.D., Monks, T.J. and Lau, S.S., An integrated approach to identifying chemically induced posttranslational modifications using comparative MALDI-MS and targeted HPLC-ESI-MS/MS, *Chem. Res. Tox.*, 16, 598, 2003.
75. Cantin, G.T. and Yates, J.R., III, Strategies for shotgun identification of post-translational modifications by mass spectrometry, *J. Chrom. A*, 1053, 7, 2004.
76. Mann, M. and Jensen, O.N., Proteomic analysis of post-translational modifications, *Nat. Biotech.*, 21, 255, 2003.

77. Biroccio, A., Urbani, A., Massoud, R., di Ilio, C., Sacchetta, P., Bernardini, S., Cortese, C. and Federici, G., A quantitative method for the analysis of glycated and glutathionylated hemoglobin by matrix-assisted laser desorption ionization-time of flight mass spectrometry, *Anal. Biochem.*, 336, 279, 2005.

78. Krishna, R.G. and Wold, F., Post-translational modification of proteins, *Adv. Enzymology*, 67, 265, 1993.

79. Krishna, R.G. and Wold, F., Post-translational modification of proteins, in *Methods in Protein Sequence Analysis*, K. Imahori and F. Sakami, Eds., Plenum Press, New York, NY, 167, 1993.

80. Bennett, K.L., Smith, S.V., Lambrecht, R.M., Truscott, R.J.W. and Sheil, M.M., Rapid characterization of chemically-modified proteins by electrospray mass spectrometry, *Bioconjugate Chem.*, 7, 16, 1996.

81. Fligge, T.A., Kast, J., Bruns, K. and Przybylski, M., Direct monitoring of protein-chemical reactions utilizing nanoelectrospray mass spectrometry, *J. Am. Soc. Mass Spec.*, 10, 112, 1999.

82. Apuy, J.L., Chen, X., Russell, D.H., Baldwin, T.O. and Giedroc, D.P., Ratiometric pulsed alkylation/mass spectrometry of the cysteine pairs in individual zinc fingers of MRE-binding transcription factor-1 (MTF-1) as a probe of zinc chelate stability, *Biochemistry*, 40, 15164, 2001.

83. Jahn, O., Hofmann, B., Brauns, O., Spiess, J. and Eckart, K., The use of multiple ion chromatograms in on-line HPLC-MS for the characterization of post-translational and chemical modification of proteins, *Intl. J. Mass Spec.*, 214, 37, 2002.

84. Kuyama, H., Watanabe, M., Toda, C., Ando, E., Tanaka, K. and Nishimura, O., An approach to quantitative proteome analysis by labeling tryptophan residues, *Rapid Comm. Mass Spec.*, 17, 1642, 2003.

85. Apuy, J.L., Busenlehner, L.S., Russell, D.H. and Giedroc, D.P., Ratiometric pulsed alkylation mass spectrometry as a probe of thiolate reactivity in different metallo-derivatives of *Staphylococcus aureus* pI258 CadC, *Biochemistry*, 43, 3824, 2004.

86. LeRiche, T., Skorey, K., Roy, P., McKay, D. and Bateman, K.P., Using mass spectrometry to study the photo-affinity labeling of protein tyrosine phosphatases 1B, *Intl. J. Mass Spec.*, 238, 99, 2004.

87. Beardsley, R.L., Karty, J.A. and Reilly, J.P., Enhancing the intensities of lysine-terminated tryptic peptide ions in matrix-assisted laser desorption/ionization mass spectrometry, *Rapid Comm. Mass Spec.*, 14, 2147, 2000.

88. Keough, T., Lacey, M.P. and Youngquist, R.S., Derivatization procedures to facility *de novo* sequencing of lysine-terminated tryptic peptides using postsource decay matrix-assisted laser desorption/ionization mass spectrometry, *Rapid Comm. Mass Spec.*, 14, 2348, 2000.

89. Keough, T., Youngquist, R.S. and Lacey, M.P., A method for high-sensitivity peptide sequencing using postsource decay matrix-assisted laser desorption ionization mass spectrometry, *Proc. Natl. Acad. Sci. USA*, 96, 7131, 1999.

90. Samyn, B., Debyser, G., Sergeant, K., Devreese, B. and Van Beeumen, J., A case study of *de novo* sequence analysis of *N*-sulfonated peptides by MALDI TOF/TOF mass spectrometry, *J. Am. Soc. Mass. Spec.*, 15, 1838, 2004.

91. Zappacosta, F. and Annan, R.S., *N*-terminal isotope tagging strategy for quantitative proteomics: Results-driven analysis of protein abundance changes, *Anal. Chem.*, 76, 6618, 2004.

92. Beardsley, R.L. and Reilly, J.P., Quantitation using enhanced signal tags: A technique for comparative proteomics, *J. Proteome Res.*, 2, 15, 2003.

93. Hagman, C., Ramström, M., Håkansson, P. and Bergquist, J., Quantitative analysis of tryptic protein mixtures using electrospray ionization Fourier transform cyclotron resonance mass spectrometry, *J. Proteome Res.*, 3, 587, 2004.
94. Matthiesen, R., Bauw, G. and Welinder, K.G., Use of performic acid oxidation to expand the mass distribution of tryptic peptides, *Anal. Chem.*, 76, 6848, 2004.
95. Hsu, J.-L., Huang, S.-Y., Chow, N.-H. and Chen, S.-H., Stable-isotope dimethyl labeling for quantitative proteomics, *Anal. Chem.*, 75, 6843, 2003.
96. Hsu, J.-L., Huang, S.-Y., Shiea, J.-T. et al., Beyond quantitative proteomics: Signal enhancement of the a_1 ion as a mass trap for peptide sequencing using dimethyl labeling, *J. Proteome Res.*, 4, 101, 2005.
97. Majors, R.E., Sample preparation in analytical chemistry (Organic Analysis), in *Handbook of Instrumental Techniques for Analytical Chemistry*, F. Settle, Ed., Prentice Hall PTR, Upper Saddle River, NJ, Ch. 2, 17, 1997.
98. *Handbook of Instrumental Techniques for Analytical Chemistry*, F. Settle, Ed., Prentice Hall PTR, Upper Saddle River, NJ, 1997.
99. *Analytical Chemistry*, R. Kellner, J.M. Mermet, M. Oth and H.M. Widmer, Eds., Wiley-VCH, Weinheim, Germany, 1998.
100. Two-Hybrid Systems. Methods and Protocols, P.N. MacDonald, Ed., *Methods in Molecular Biology* (J. Walker, Series Editor), Vol. 177, Humana Press, Towata, NJ, 2001.
101. Fields, S. and Bartel, P.L., The two-hybrid system. A personal view, in *Two-Hybrid Systems: Methods and Protocols*, P.N. MacDonald, Ed., Humana Press, Towata, NJ, Ch. 1, 3–8, 2001.
102. Xia, Y., Yu, H., Jansen, R., Seringhaus, M., Bxter, S., Greenbaum, D., Zhao, H. and Gerskin, M., Analyzing cellular biochemistry in terms of molecular networks, *Ann. Rev. Biochem.*, 73, 1051, 2004.
103. Walhout, A.J.M. and Vidal, M., Protein interaction maps for model organisms, *Nat. Rev. Mol. Cell. Biol.*, 2, 55, 2001.
104. Applications of chimeric genes and hybrid proteins, Part C, protein–protein interactions and genomes, J. Thorner, S.D. Emr and J.N. Abelson, Eds., *Meth. Enzymology*, 238, Academic Press, San Diego, CA, 2000.
105. Rigaut, G., Shevchenko, A., Rutz, B., Wilm, M., Mann, M. and Séraphin, B., A generic protein purification method for protein complex characterization and proteome exploration, *Nat. Biotech.*, 17, 1030, 1999.
106. Shevchenko, A., Schaft, D., Rogoer, A. et al., Diciphering portein complexes and protein interaction networks by tandem affinity purification and mass spectrometry: Analytical perspective, *Mol. Cell. Proteomics*, 1, 204, 2002.
107. Forler, D., Kocher, T., Rode, M. et al., An efficient protein complex purification method for functional proteomics in higher eukaryotes, *Nat. Biotech.*, 21, 89, 2003.
108. Li, Q., Dai, Z.Q., Shen, P.Y. et al., A modified mammalian tandem affinity purification procedures to prepare functiona polycytin-2 channels, *FEBS Letters*, 576, 231, 2004.
109. Drakss, R., Prisco, M. and Baserge, R., A modified tandem affinity purification tag technique for the purification of protein complexes in mammalian cells, *Proteomics*, 5, 132, 2005.
110. Vasilescu, J., Guo, X. and Kast, J., Identification of protein–protein interactions using *in vivo* crosslinking and mass spectrometry, *Proteomics*, 4, 3845, 2004.
111. Dihazi, G.H. and Sinz, A., Mapping low-resolution three-dimensional protein structures using chemical cross-linking and Fourier transform ion-cyclotron resonance mass spectrometry, *Rapid Comm. Mass Spec.*, 17, 2005, 2000.

112. Heinemeyer, J., Eubel, H., Wehmhaner, D., Jansch, L. and Braun, H.P., Proteomic approach to characterize the supermolecular organization of photosystems in higher plants, *Phytochemistry*, 65, 1683, 2004.

113. Shimura, K., Recent advances in capillary isoelectric focusing: 1997–2001, *Electrophoresis*, 23, 3847, 2002.

114. Blagoev, B., Kratchmarova, I., Ong, S.-E. et al., A proteomic study to elucidate functional protein–protein interactions applied to EGF signalling, *Nat. Biotech.*, 21, 315, 2003.

115. Leonaudakis, D., Conti, L.R., Anderson, L.R. et al., Protein trafficking and anchoring complexes revealed by proteomic analysis of inward rectifier potassium channel (Kir2-x)-associated proteins, *J. Biol. Chem.*, 279, 22331, 2004.

116. Ransome, L., Detection of protein–protein interactions by coimmunoprecipitation and dimerization, *Meth. Enzymology*, 254, 491, 1995.

117. Wong, C. and Naumouski, L., Methods to screen for relevant yeast two-hybrid derived clones by coimmunoprecipitation and localization of epitope-tagged fragments — Application to Bel-xl, *Anal. Biochem.*, 252, 33, 1997.

118. Ren, L., Emery, D., Kaboord, B., Chang, E. and Qoroonflen, M.W., Improved immunomatrix methods to detect protein:protein interactions, *J. Biochem. Biophys. Meth.*, 57, 143, 2003.

119. Barnovin, K., Two-dimensional gel electrophoresis for analysis of protein complexes, *Meth. Mol. Biol.*, 261, 479, 2004.

120. Yamada, K., Osawa, H. and Granner, D.K., Identification of proteins that interact with NF-YA, *FEBS Letters*, 460, 41, 1999.

121. Hryciw, D.H., Wang, Y., Devuyst, O. et al., Cofilin interacts with C1C-5 and regulates albumin uptake in proximal tubule cell lines, *J. Biol. Chem.*, 278, 40169, 2003.

122. Schmitt, E.K., Hoff, B. and Kuck, U., AcFKH1, a novel member of the forkhead family, associates with the RFX transcription factor CPCR1 in the cephalosporin C-producing fungus *Acremonium chrysogenum*, *Gene*, 342, 269, 2004.

123. Graumann, J., Dunipace, L.A., Seo, J.H. et al., Applicability of tandem affinity purification MudPIT to pathway proteomics in yeast, *Mol. Cell. Proteomics*, 3, 226, 2004.

124. Sasakura, Y., Kanda, K., Yoshimura-Suzuki, T. et al., Protein microarray system for detecting protein–protein interactions using an anti-his-tag antibody and fluorescence scanning: Effects of the heme redox state of protein–protein interactions of heme-regulated phosphodiesterase from *Escherichia coli*, *Anal. Chem.*, 76, 6521, 2004.

125. Figeys, D., Novel approaches to map protein interactions, *Curr. Opin. Biotech.*, 14, 119, 2003.

126. Cox, D.M., Du, M., Guo, X. et al., Tandem affinity purification of protein complexes from mammalian cells, *Biotechniques*, 33, 267, 2002.

127. Rohila, J.S., Chen, M., Cerny, R. and Fromm, M.E., Improved tandem affinity purification tag and methods for isolation of protein heterocomplexes from plants, *Plant J.*, 38, 172, 2004.

128. Ghosh, D., Krokhin, O., Antonovich, M. et al., Lectin affinity as an approach to the proteomic analysis of membrane glycoproteins, *J. Proteome Res.*, 3, 841, 2004.

129. Li, Q., Dai, X.Q., Shen, P.Y. et al., A modified mammalian tandem affinity purification procedure to prepare functional polycystin-2 channel, *FEBS Letters*, 576, 231, 2004.

130. Hunt, A.N., Skippen, A.J., Koster, G. et al., Acyl chain-based molecular selectivity for HL60 cellular phosphatidylinositol and of phosphatidyl choline by phosphatidylinositol transfer protein alpha, *Biochim. Biophys. Acta*, 1686, 50, 2004.

131. Theide, B., Hohenwarter, W., Krah, A., Mattow, J., Schmid, M., Schmidt, F. and Jungblut, P.R., Peptide mass fingerprinting, *Methods*, 35, 237–247, 2005.

132. Blueggel, M., Chamrad, D. and Meyer, H.E., Bioinformatics in proteomics, *Curr. Pharm. Biotech.*, 5(1), 79, Feb. 2004.

133. Lanne, B. and Panfilov, O., Protein staining influences the quality of mass spectra obtained by peptide mass fingerprinting after separation on 2-D gels. A comparison of staining with Coomassie Brilliant Blue and Sypro Ruby, *J. Proteome Res.*, 4, 175, 2005.

134. Ingram, V.M., Abnormal human haemoglobins I. The comparison of normal human and sickle cell haemoglobin by "fingerprinting," *Biochim. Biophys. Acta*, 28, 539, 1958.

135. Katz, A.M., Dreyer, W.J. and Anfinsen, C.B., Peptide separation by two-dimensional chromatography and electrophoresis, *J. Biol. Chem.*, 234, 2897, 1959.

136. Görg, A., Weiss, W. and Dunn, M.J., Current two-dimensional electrophoresis technology for proteomics, *Proteomics*, 4, 3665, 2004.

137. O'Farrell, P.H., High resolution two-dimensional electrophoresis of proteins, *J. Biol. Chem.*, 250, 4007, 1975.

138. Görg, A., Obermaier, C., Boguth, G. et al., The current status of two-dimensional electrophoresis with immobilized pH gradients, *Electrophoresis*, 21, 1037, 2000.

139. Ong, S.-E. and Pandey, A., An evaluation of the use of two-dimensional gel electrophoresis in proteomics, *Biomol. Eng.*, 18, 195, 2001.

140. Patton, W.F., Schulenberg, B. and Steinberg, T.H., Two-dimensional gel electrophoresis; better than a poke in the ICAT, *Curr. Opin. Biotech.*, 13, 321, 2002.

141. Molloy, M.P., Brzezinski, E.E., Hang, J., McDowell, M.T. and VanBogelen, R.A., Overcoming technical variation and biological variation in quantitative proteomics, *Proteomics*, 3, 1912, 2003.

142. Choe, L.H. and Lee, K.H., Quantitative and qualitative measure of intralaboratory two-dimensional protein gel reproducibility and the effects of sample preparation, sample load, and image analysis, *Electrophoresis*, 24, 3500, 2003.

143. Stasyk, T. and Huber, L.A., Zooming in: Fractionation strategies in proteomics, *Proteomics*, 4, 3704, 2004.

144. Zuo, X. and Speicher, D.W., Comprehensive analysis of complex proteomes using microscale solution isoelectricfocusing prior to narrow pH range two-dimensional electrophoresis, *Proteomics*, 2, 58, 2002.

145. Oh-Ishi, M. and Maeda, T., Separation techniques for high-molecular-mass proteins, *J. Chrom. B*, 771, 49, 2002.

146. Doherty, N.S., Littman, B.H., Reilly, K. et al., Analysis of changes in acute-phase plasma proteins in an acute-phase plasma proteins in an acute inflammatory response and in rheumatoid arthritis using two-dimensional gel electrophoresis, *Electrophoresis*, 20, 355, 1998.

147. Krah, A., Schmidt, F., Becher, D. et al., Analysis of automatically generated peptide mass fingerprints of cellular proteins and antigens form *Helicobacter pylori* 26695 separated by two-dimensional electrophoresis, *Mol. Cell. Proteomics*, 2, 1271, 2003.

148. Nishihara, J.C. and Champion, K.M., Quantitative evaluation of proteins in one- and two-dimensional polyacrylamide gels using a fluorescent stain, *Electrophoresis*, 23, 2203, 2002.

149. Ong, S.-E. and Pandey, A., An evaluation of the use of two-dimensional gel electrophoresis in proteomics, *Biomol. Eng.*, 18, 195, 2001.

150. Choe, L.H. and Lee, K.H., Quantitative and qualitative measure of intralabortory two-dimensional protein gel reproducibility and the effects of sample preparation, sample load, and image analysis, *Electrophoresis*, 24, 3500, 2003.

151. White, I.R., Pickford, R., Wood, J., Skehel, J.M., Gandadharan, B. and Cutler, P., A statistical comparison of silver and SYPRO Ruby staining for proteomic analysis, *Electrophoresis*, 25, 3048, 2004.

152. Tonge, R., Shaw, J., Middleton, B. et al., Validation and development of fluorescence two-dimensional differential gel electrophoresis proteomics, *Proteomics*, 1, 377, 2001.
153. Shaw, J., Rowlinson, R., Nickson, J. et al., Evaluation of saturation labelling two-dimensional difference gel electrophoresis fluorescent dyes, *Proteomics*, 3, 1181, 2003.
154. Karp, N.A., Kreil, D.P. and Lilley, K.S., Determining a significant change in protein expression with DeCyder™ during a pair-wise comparison using two-dimensional difference gel electrophoresis, *Proteomics*, 4, 1421, 2004.
155. Tomonaga, T., Matsushita, K., Yamaguchi, S. et al., Identification of altered protein expression and post-translational modifications in primary colorectal cancer by using agarose two-dimensional gel electrophoresis, *Clin. Cancer Res.*, 10, 2007, 2004.
156. McCormick, A.L., Schieltz, D.M., Goode, B. et al., Direct analysis and identification of proteins in mixtures by LC/MS/MS and database searching at the low-femtomole level, *Anal. Chem.*, 69, 767, 1997.
157. Wolters, D.A., Washburn, M.P. and Yates, J.R., III, An automated multidimensional protein identification technology for shotgun proteomics, *Anal. Chem.*, 73, 5683, 2001.
158. Yates, J.R., III, Eng, J.K., McCormack, A.L. and Schletz, D., Method to correlate tandem mass spectra of modified peptides to amino acid sequences in the protein database, *Anal. Chem.*, 67, 1426, 1995.
159. Yates, J.R., III, Eng, J.K. and McCormack, A.L., Mining genomics: Correlating tandem mass spectra of modified and unmodified peptides to sequences in nucleic acid databases, *Anal. Chem.*, 67, 3202, 1995.
160. Tabb, D.L., Scraf, A. and Yates, J.R., III, Gutentag: high-throughput sequence tagging via an empirically derived fragmentation model, *Anal. Chem.*, 75, 6415, 2003.
161. Cagney, G., Ami, S., Premawaradura, T., Lindo, M. and Emili, A., *In silico* protoeome analysis to facilitate proteomics experiments using mass spectrometry, *Proteome Sci.*, 1:5, 2003, http://www.Proteomesci.com/content/1/1/5.
162. Liu, H., Sadygov, R.G. and Yates, J.R., III, A model for random sampling and estimation of relative protein abundance in shotgun proteomics, *Anal. Chem.*, 76, 4193, 2004.
163. Issaq, H.J., The role of separation science in proteomics research, *Electrophoresis*, 22, 3629, 2001.
164. McDonald, W.H. and Yates, J.R., III, Shotgun proteomics and biomarker discovery, *Disease Markers*, 18, 99, 2002.
165. Liu, H., Lin, D. and Yates, J.R., III, Multidimensional separations for protein/peptide analysis in the post-genomic era, *BioTechniques*, 32, 989, 2002.
166. Jahn, O., Hofmann, B., Brauns, O., Speiss, J. and Eckart, K., The use of multiple ion chromatograms in on-line HPLC-MS for the characterization of post-translational and chemical modification of proteins, *Intl. J. Mass Spec.*, 214, 37, 2002.
167. Guttman, A., Varoglu, M. and Khandurina, J., Multidimensional separations in the pharmaceutical arena, *Drug Disc. Today*, 9, 136, 2004.
168. Wu, C.C. and MacCoss, M.J., Shotgun proteomics: Tools for the analysis of complex biological systems, *Curr. Opin. Mol. Ther.*, 4, 242, 2002.
169. Washburn, M.P., Wolters, D. and Yates, J.R., III, Large-scale analysis of the yeast proteome by multidimensional protein identification technology, *Nat. Biotech.*, 19, 242, 2001.
170. Link, A.J., Eng, S., Schieltz, D.M. et al., Direct analysis of protein complexes using mass spectrometry, *Nat. Biotech.*, 17, 676, 1999.
171. Jenõ, P., Mini, T., Moes, S., Hintermann, E. and Horst, M., Internal sequences from proteins digested in polyacrylamide gels, *Anal. Biochem.*, 224, 75, 1995.

172. Wada, Y. and Kadoya, M., In-gel digestion with endoproteinase LysC, *J. Mass Spec.*, 38, 117, 2003.
173. Zhang, W., Marzilli, L.A., Rouse, J.C. and Czupryn, M.J., Complete disulfide bond assignment of a recombinant immunoglobulin G4 monoclonal antibody, *Anal. Biochem.*, 311, 1, 2002.
174. Drubin, D.G., Miller, K.G. and Botstein, D., Yeast actin-binding proteins: Evidence for a role in morphogenesis, *J. Cell Biol.*, 107, 2551, 1988.
175. Hutchins, T.W. and Yip, T.-T., New desorption strategies for the mass spectrometric analysis of macromolecules, *Rapid Comm. Mass Spec.*, 7, 576, 1993.
176. Yip, T.-T., Van de Water, J., Gershwin, M.E., Coppel, R.L. and Hutchens, T.W., Cryptic antigenic determinants on the extracellular pyruvate dehydrogenase complex/mimetope found in primary biliary cirrhosis, *J. Biol. Chem.*, 271, 32825, 1996.
177. Rudiger, A.H., Rudiger, M., Carl, U.D., Chakraborty, T., Roepstorff, P. and Wehland, J., Affinity mass spectrometry-based approaches for the analysis of protein–protein interaction and complex mixture of peptide-ligands, *Anal. Biochem.*, 275, 162, 1999.
178. Dick, L.W., Jr. and McGown, L.B., Aptamer-enhanced laser desorption/ionization for affinity mass spectrometry, *Anal. Chem.*, 76, 3037, 2004.
179. Meng, J.C., Siuzdak, G. and Finn, M.G., Affinity mass spectrometry from a tailored porous silicon surface, *Chem. Comm.*, 21, 2108, 2004.
180. Merchant, M. and Weinberg, S.R., Recent advancements in surface-enhanced laser desorption/ionization time-of-flight mass spectrometry, *Electrophoresis*, 21, 1164, 2000.
181. Davies, H.A., The ProteinChip system from Ciphergen: A new technique for rapid, micro-scale protein biology, *J. Mol. Med.*, 78, B29, 2000.
182. Rubin, R.B. and Merchant, M., A rapid protein profiling system that speeds study of cancer and other diseases, *Am. Clin. Lab.*, 19, 28, 2000.
183. Fung, E.T., Thulasiranam, V., Weinberger, S.R. and Dalmasso, E.A., Protein biochips for differential profiling, *Curr. Opin. Biotech.*, 12, 65, 2001.
184. Weinberger, S.R., Dalmasso, E.W. and Fung, E.T., Current achievements using ProteinChip array technology, *Curr. Opin. Chem. Biol.*, 6, 86, 2002.
185. Chapman, K., The ProteinChip biomarker system from Ciphergen Biosystems: A novel proteomics platform for rapid biomarker discovery and validation, *Biochem. Soc. Trans.*, 30, 82, 2002.
186. Tang, N., Tornatore, P. and Weinberger, S.R., Current developments in SELDI affinity technology, *Mass Spec. Rev.*, 23, 34, 2004.
187. Zhu, H. and Snyder, M., Protein chip technology, *Curr. Opin. Chem. Biol.*, 7, 55, 2003.
188. Koopmann, J., Zhang, Z. and White, N., Serum diagnosis of pancreatic adenocarcinoma using surface-enhanced laser desorption and ionization mass spectrometry, *Clin. Chem. Res.*, 10, 860, 2004.
189. Warren, E.N., Elms, P.J., Parker, C.E. and Borchers, C.H., Development of a protein chip: A MS-based method for quantitation of protein expression and modification levels using an immunoaffinity approach. *Anal. Chem.*, 76, 4082, 2004.
190. Warren, E.N., Elms, P.J., Parker, C.E. and Borchers, C.H., Development of a protein protein chip: A MS-based method for quantitation of protein expression and modification of levels using an immunoaffinity approach, *Anal. Chem.*, 76, 4082, 2004.
191. Linke, T., Ross, A.C. and Harrison, E.H., Profiling of rat plasma by surface-enhanced laser desorption/ionization time-of-flight mass spectrometry, a novel tool for biomarker discovery in nutrition research, *J. Chrom. A*, 1043, 65, 2004.
192. Morris, D.L., Jr., Sutton, J.N., Harper, R.G. and Timperman, A.T., Reversed-phase HPLC separation of human serum employing a novel saw-tooth gradient: Toward multidimensional proteome analysis, *J. Proteome Res.*, 3, 1149, 2004.

193. Jock, C.A., Paulaukis, J.D., Baker, D., Olle, E., Bleavins, M.R., Johnson, K.T. and Heard, P.L., Influence of matrix application timing on spectral reproducibility and quality in SELDI-TOF-MS, *BioTechniques*, 37, 30, 2004.
194. Aldred, S., Grant, M.M. and Griffiths, H.R., The use of proteomics for the assessment of clinical samples in research, *Clin. Biochem.*, 37, 943, 2004.
195. Petricoin, E.F. and Liotta, L.A., SELDI-TOF-based serum proteomic pattern diagnostics for early detection of cancer, *Curr. Opin. Biotech.*, 15, 24, 2004.
196. Caputo, E., Moharram, R. and Martin, B.M., Methods for on-chip protein analysis, *Anal. Biochem.*, 321, 116, 2003.
197. Spencer, D.I.R., Robson, L., Purdy, D. et al., A strategy for mapping and neutralizing conformational immunogenic sites on protein therapeutics, *Proteomics*, 2, 271, 2002.

7 Clinical Proteomics

CONTENTS

INTRODUCTION

Clinical proteomics can be described as the use of proteomic technologies to address issues in the diagnosis and treatment of disease. While a role for proteomics is suggested in drug development,[1-4] the application of proteomics in the drug discovery process is somewhat less clear. Expression proteomics, defined in this specific context as the use of proteomic technology to study the effect of external perturbations on protein expression as measured by stable isotope labeling,[5] will possibly be useful in drug discovery; it may be more useful in the study of the effect of new drugs on cellular function as part of a preclinical toxicology program. In March 2005, 233 citations were obtained in a PUBMED search using "proteomic" and "therapeutic" as the descriptor while 605 citations were obtained for "proteomic" and "diagnostic." This result suggests that the major driver for clinical proteomics is diagnostic and prognostic use[6,7] with such activity focused on the identification of biomarkers. For the purpose of the present discussion, biomarkers are proteins that can be used for the study of a specific disease process.[8,9] The diagnostics developed as a result of biomarker identification are expected to be useful in patient treatment.[10]

While it is somewhat premature to expect substantial results from the application of proteomic technologies to clinical areas, some successes exist as demonstrated by the work of Glocker, Kekow and coworkers[11-13] on developing a useful diagnostic for differentiating rheumatoid arthritis and osteoarthritis. Briefly, these investigators found increased levels of calprotectin in plasma from rheumatoid arthritis patients but not osteoarthritis patients. Their work resulted in an immunoassay that can identify patients who would not benefit from anti-TNFα thus avoiding the complications of an expensive therapeutic that would not be effective. Rheumatoid arthritis, osteoarthritis and related joint diseases have been of considerable interest for the development of laboratory diagnostic tests. A small number of recent studies on the development and application of laboratory diagnostic tests are presented in Table 7.1. Development of new assays can come from classical approaches as well as proteomic technologies.

Clinical laboratory test results are one of several aids that a physician uses in diagnosing and treating a disease.[14,15] Few reference texts exist on the development of a diagnostic test; the key from a business perspective is that the diagnostic/prognostic test should have a positive effect on treatment[10] either by providing a timely diagnosis permitting early definitive treatment or a prognostic effect in therapy management.[16] A model for the use of proteomic technologies in clinical assay development can be provided by consideration of the use of nucleic acid technology in infectious disease.[17,18]

The vast majority of proteomic research on biomarker identification has focused on oncology with particular emphasis on ovarian cancer, bladder cancer and lung cancer. These areas are where an early specific diagnostic screening test is not available and such a test would be useful in decreasing the associated morbidity and mortality.[19–21] The reader is directed to an article by Schwartz discussing the current status of tumor markers published a decade ago.[22] Despite the passage of time, the considerations raised in this article are as timely today as they were in 1995. Schwartz defines tumor markers as "substances, measured by laboratory techniques, which are useful in the management of patients with cancer." Critical issues include the development of new markers and new assays for old markers, reference materials and standards, nomenclature standardization and the introduction of quality assurance programs that would include proficiency studies. These comments are possibly more critical today than they were in 1995. One of the more critical observations reported was the increase in assay specificity and assay sensitivity with the use of three markers as opposed to a single marker in breast cancer. A more recent study

TABLE 7.1
Studies on the Development of Diagnostics for Rheumatoid Arthritis

Study	Reference
ELISA for serum type IIa procollagen N-terminal propeptide/joint diseases	1
ELISA for autoantibodies to chromatin/juvenile rheumatoid arthritis	2
ELISA for anticyclic citrullinated peptide antibodies/juvenile idiopathic arthritis	3
General review of prognostic laboratory markers for rheumatoid arthritis	4
ELISA/serum fas/rheumatoid arthritis	5
ELISA/CD-26(dipeptidyl peptidase IV); decreases in rheumatoid arthritis compared to osteoarthritis; inversely correlated with CRP	6
ELISA/collagen 2-1 peptide and nitrated collagen 2-1 peptide/rheumatoid arthritis	7
Protein microarrays/autoimmune disease	8
2-Dimensional gel electrophoresis/MS/idenfication of potential biomarkers for rheumatoid arthritis	9
2-Dimensional LC/MS/identification of potential biomarkers for rheumatoid arthritis	10
2-Dimensional gel electrophoresis/Western blotting; identified autoantibodies to triosephosphate isomerase in rheumatoid arthritis	11

REFERENCES FOR TABLE 7.1

1. Rosseau, J.C., Sangdell, L.C., Demas, P.D. and Garnero, P., Development and clinical application in arthritis of a new immunoassay for serm type IIa procollagen NH_2 propeptide, *Meth. Mol. Med.*, 101, 25–37, 2004.
2. Ingelgnoli, F., Del Papa, N., Comina, D.P. et al., Autoantibodies to chromatin: Prevelance and clinical significance in juvenile rheumatoid arthritis, *Clin. Exptl. Rheum.*, 22, 499–501, 2004.
3. Low, J.M., Chaven, A.K., Kietz, D.A. et al., Determination of anti-cyclic citrullinated peptide antibodies in the sera of patients with juvenile idiopathic arthritis, *J. Rheum.*, 31, 1829–1833, 2004.
4. Lindqvist, E., Eberhardt, K., Bendtzen, K. et al., Prognostic laboratory markers of joint damage in rheumatoid arthritis, *Ann. Rheum. Dis.*, 64, 196–201, 2005.
5. Ates, A., Kihikli, G., Turgay, M. and Durnan, N., The levels of serum-soluble Fas in patients with rheumatoid arthritis and systemic sclerosis, *Clin. Rheum.*, 23, 421–425, 2004.
6. Busso, N., Wagtmann, N., Herling, C. et al., Circulating CD26 is negatively associated with inflammation in human and experimental arthritis, *Am. J. Path.*, 166, 433–442, 2005.
7. Deberg, M., Labasse, A., Christgua, S. et al., New serum biochemical markers (Coll 2-1 and Coll 2-1 NO_2) for studying oxidative-related type II collagen network degradation in patients with osteoarthritis and rheumatoid arthritis, *Osteo. Cart.*, 13, 258–265, 2005.
8. Hueber, W., Utz, P.J., Steinman, L. and Robinson, W.H., Autoantibody profiling for the study and treatment of autoimmune disease, *Arth. Res.*, 4, 290–295, 2005.
9. Detzlau, H., Schultz, M., Eggert, M. and Neeck, G., A pattern of protein expression in peripheral blood mononuclear cells distinguishes rheumatoid arthritis patients from healthy individuals, *Biochim. Biophys. Acta*, 1696, 121–129, 2004.
10. Liao, H., Wu. J., Kohn, E. et al., Use of mass spectrometry to identify protein biomarkers of disease severity in the synovial fluid and serum of patients with rheumatoid arthritis, *Arth. Rheum.*, 50, 2792–3803, 2004.
11. Xiang, Y., Sekine, T., Nakamura, H. et al., Idenfication of triose phosphate isomerase as an autoantigen in patients with osteoarthritis, *Arth. Rheum.*, 50, 1511–1521, 2004

by Zhang and coworkers[22a] demonstrates an increase in specificity while maintaining sensitivity when additional biomarkers, identified by proteomic analysis, were combined with CA125 for the diagnosis of ovarian cancer.

The current emphasis on early diagnosis should not in any way diminish the need for the development of useful prognostic tests.[16] Prognosis is a critical part of patient management in oncology, including cost control.[23] Some current prognostic markers include serum thyroglobulin in differentiated thyroid carcinoma, serum lactic dehydrogenase in nasopharyngeal carcinoma and alpha-fetoprotein in germ-cell cancer. This rapidly moving area deserves more attention in proteomic research.

Most clinical laboratory tests are based on the assay of an analyte in a biological fluid,[24] most often blood. The use of serological markers for diagnosis and prognosis in oncology is not new[25] and new investigators should consult earlier reference texts in this area[26,27] before "reinventing the wheel." More attention placed on the identification of biomarkers in precancerous conditions would also be useful.[28]

Next is the complex relationship between tumors and inflammation.[29] Chronic inflammatory conditions exist such as pancreatitis (pancreatic cancer), Crohn's disease (colon cancer) and lichen planus (oral cancer) that predispose to tumor development. The inflammatory response might also support or suppress tumor growth.[30-32] Thus, given the relationship between inflammation and tumor development, finding calprotectin, a marker of inflammation,[33,34] in a variety of cancerous and precancerous conditions is not surprising (Table 7.2). Thus, while calprotectin is a useful marker for the differentiation between rheumatoid arthritis, an inflammatory disease, and osteoarthritis, a disease without inflammation, its utility as a sole biomarker is not evident; it could be useful as a member of a panel of biomarkers.

While considerable promise exists in the application of proteomic technologies to study oncology, considerable controversy also persists about the value of proteomics in the development of diagnostics as demonstrated by the recent discussions about the use of protein profiling for ovarian cancer.[35-38] Here it would appear that the initial report[39] was somewhat overly enthusiastic about the promise of multi-analyte screening with considerable criticism of this study since its publication.[40-43] On balance, proteomics will be of value in the development of some novel and useful diagnostic and prognostic tests[36] but such work will clearly require time, hard effort and close collaboration between basic scientists and clinicians.[13] It will also clearly benefit from the use of multiple technical approaches as illustrated by work from von Eggeling and colleagues.[44] A certain amount of current concern exists that there be "products" resulting from proteomics in the short term.[45-47] This is not a trivial consideration and, at the risk of being redundant, the reader should consider carefully the comment of Loyda[10] regarding consideration in the development of diagnostic assays.

TABLE 7.2
The Promiscuity of Calprotectin as a Biomarker; Various Indications for Calprotectin as a Biomarker

Pathology	Reference
Differentiation of rheumatoid arthritis and osteoarthritis	1, 2
Reactive arthritis	3
Juvenile idiopathic arthritis	4
Ovarian carcinoma	5
Colorectal carcinoma	6
Intra-amniotic infection	7
Head and neck tumors	8
Nasal lavage fluid from chemical (dimethylbenzylamine, epoxy) exposure	9
Inflammatory bowel disease	10
Type 1 diabetes	11
Periodontal disease	12

REFERENCES FOR TABLE 7.2

1. Berntzen, H.B., Ölmez, Ü., Fagerhol, M.K. and Munthe, E., The leukocyte protein L1 in plasma and synovial fluid from patients rheumatoid arthritis and osteoarthritis, *Scand. J. Rheum.*, 20, 74–82, 1991.

2. Sinz, A., Bantscheff, M., Mikkat, S. et al., Mass spectrometric proteome analyses of synovial fluids and plasmas from patients suffering from rheumatoid arthritis and comparison to reactive arthritis or osteoarthritis, *Electrophoresis*, 23, 3445–3456, 2002.

2b. Uchida, T., Fukawa, A., Uchida, M., Fujita, K. and Saito, K., Application of a novel protein biochip technology for detection and identification of rheumatoid arthritis biomarkers in synovial fluids, *J. Proteome Res.*, 1, 495–499, 2002.

2b. Drynda, S., Ringel, B., Kekow, M. et al., Proteome analysis reveals disease-associated marker proteins to differentiate RA patients from other inflammatory joint diseases with the potential to monitor anti-TNF-α therapy, *Path. Res. Prac.*, 200, 165–171, 2004.

3. Hammer, H.B., Kvien, T.K., Glennås, A. and Melby, K., A longitudinal study of calprotectin as an inflammatory marker in patients with reactive arthritis, *Clin. Exptl. Rheum.*, 13, 59–64, 1995.

4. zur Wiesch, A.S., Foell, D., Frosch, M., Vogl, T., Sorg, C. and Roth, J., Myeloid related proteins MRP8/MRP14 may predict disease flares in juvenile idiopathic arthritis, *Clin. Exptl. Rheum.*, 22, 368–373, 2004.

5. Ott, H.W., Lindner, H., Sarg, B. et al., Calgranulins in cystic fluid and serum from patients with ovarian carcinomas, *Cancer Res.*, 63, 7507–7514, 2003.

6. Stulik, J., Osterreicher, J., Koupilova, K. et al., The analysis of S100A9 and S100A8 expression in matched sets of macroscopically normal colon mucosa and colorectal carcinoma: The S100A9 and S100A8 positive cells underlie and invade tumor mass, *Electrophoresis*, 30, 1047–1054, 1999.

7. Gravett, M.G., Novy, M.J., Rosenfeld, R.G. et al., Diagnosis of intra-amniotic infection by proteomic profiling and identification of novel biomarkers, *JAMA*, 292, 462–469, 2004.

8. Melle, C., Ernst, G., Schimmel, B. et al., A technical triade for proteomic identification and characterization of cancer biomarkers, *Cancer Res.*, 64, 4099–4104, 2004.

9. Lindahl, M., Irander, K., Tagesson, C. and Stahlbom, B., Nasal lavage fluid and proteomics as means to identify the effects of the irritating epoxy chemical dimethylbenzylamine, *Biomarkers*, 9, 56–70, 2004.

10. Vermeire, S., Van Assche, G. and Rutgeerts, P., C-Reactive protein as a marker for inflammatory bowel disease, *Inflamm. Bowel Dis.*, 10, 661–665, 2004.

11. Bouma, G., Lam-Tse, W.K., Wierenga-Wolf, A.F., Drexhage, H.A. and Versnel, M.A., Increased serum levels of MRP-8/14 in type 1 diabetes induce an increased expression of CD11b and an enhanced adhesion of circulating monocytes to fibronectin, *Diabetes*, 53, 1979–1986, 2004.

12. Lundy, F.T., Chalk, R., Lamey, P.J., Shaw, C. and Linden, G.J., Quantitative analysis of MRP-8 in gingival crevicular fluid in periodontal health and disease using microbore HPLC, *J. Clin. Periodon.*, 28, 1172–1177, 2001.

CLINICAL PROTEOMICS — TRENDS —
DIAGNOSTICS — BIOMARKERS

The above was intended as a brief overview of the area of clinical proteomics. Frankly, the term is not terribly useful and consideration should be given to its use. First, if you perform a search on PUBMED using "clinical proteomics" as the search term, you will find almost 497 citations (March 2005) including references where both clinical and proteomics are index terms ("clinical proteomics" or "clinical" and "proteomics"). If you confine the search to "clinical proteomics" where the two words are linked in a single expression (this is accomplished by enclosing clinical proteomics within quotation marks), this number is reduced to 70 citations of which 38 are from a single laboratory. A search with "proteomics" as the lone search term yields 3813 citations. This morass, combined with the variety of technical activities included within proteomics, make it critical that investigators and editors carefully select index terms. It is not at all clear that, despite the title of this chapter, clinical proteomics should be a search term.

Since, for reasons stated above, it is not clear how proteomics will contribute to the development of therapeutics in the short-term, the remainder of this discussion will focus on the use of proteomics to develop diagnostics and specifically on the identification and use of biomarkers. As biomarkers are identified that can be reliably used to diagnose and monitor the disease process, new therapeutic regimes will likely be identified but this does not necessarily mean that new drugs are required. Having disposed of the possibility of therapeutic development, the remainder of our discussion will focus on the use of proteomics to identify biomarkers that can be used as the basis for new clinical laboratory assays.

As we discuss the use of proteomic technology for the development of diagnostics, it is absolutely essential to very clearly understand that the use of proteomic technology does not change the basic process of the development and validation of a diagnostic agent. A cursory listing of the various steps is presented below:

- Identification of the disease/pathology.
- Systemic or organ/tissue localized.
- Identification of useful protein analyte (biomarker).
- If organ-specific disorder, is biomarker secreted/excreted into a useful biological fluid?
- Development of assay for biomarker that can be used in a clinical laboratory setting.
- Demonstration that the test will influence the treatment regime for the patient.
- Validation of clinical assay, including beta-testing.
- Approval by responsible regulatory agencies such as the FDA (USA) or MCA (UK).

The purpose of the above process is designed to provide an approved assay for use within a licensed clinical laboratory, most likely within a hospital setting. This process will allow the cost of the test to be reimbursed by a third-party payer. If the cost of the test is not reimbursed, the test will not be used. It is imperative that the test be useful in that the results of the test will influence the treatment of the patient.[10] Next, the ability of the assay to "fit" into the existing laboratory framework is a major consideration. The ability both to use existing instrumentation, such as microplate readers, and to present results in a form that will assimilate into the existing laboratory information system is extremely important in the marketing of a new assay. Finally, peer-reviewed publications that support the assay are extremely useful.[48,49]

Even if the target (biomarker) is not known at the start, a significant amount of literature research prior to starting the experimental work is absolutely critical. Zolg and Langen have presented an overview[50] of biomarker research from an industrial perspective. These investigators emphasize the need for the biomarker to be useful in the sense of fulfilling an obvious medical need as mentioned above[10] as well as clearly defining what biomarkers may exist for this specific situation. The work of Glocker, Kekow and colleagues was mentioned above and appears that a useful diagnostic has been developed. However, the proteomic research that identified calprotectin unfortunately duplicated earlier research performed with more conventional technologies.[51]

Nothing intrinsically unique exists in the use of proteomics for the development of diagnostics. In the final analysis, proteomics is mostly a set of advanced analytical techniques providing enhanced resolution and sensitivity to the study of proteins in mixtures. Proteomics does provide a sensitive approach for the identification of biomarkers and provides the basis for a useful collaboration between anatomic pathologists, biochemists and clinical pathologists. This collaboration is illustrated by the studies of von Eggeling and associates, as cited above.[44] Our model therefore is the identification of a biomarker by analytical technology, establishment of an assay for the biomarker by existing technology, such as an immunoassay, and validation of the assay in clinical practice. Validation is a term that is used extensively in the proteomics literature in the sense of validating a biomarker as opposed to validating an assay. A difference exists between the two concepts: The validation of a biomarker is an activity defined as establishing a firm relationship between a biomarker and the pathology. Controversy about this activity is a hallmark of the discussion of the assay for ovarian cancer discussed above.[37,38,42] The use of the term validation in this sense is unrelated to the concept of the validation of the assay for the biomarker, which is concerned with the accuracy, reproducibility and robustness of the assay as well as its applicability for the determination of the particular analyte.[52,53] Thus, first is the identification of a suitable analyte, the validation of the analyte/biomarker in a sufficiently large population[54–57] and finally the validation of the assay in a clinical setting. The differentiation between pathology and normal is not a trivial issue and can require a large population.[58–61] From an operational point of view, an early transition to an immunoassay is suggested for evaluation of the utility of the biomarker for diagnosis, prognosis and monitoring.

RHEUMATOID ARTHRITIS AND OSTEOARTHRITIS

In studies cited above, Glocker, Kekow and colleagues used proteomic technology to identify a biomarker that could differentiate between rheumatoid arthritis and osteoarthritis. In the first cited study,[11] blood plasma and synovial fluid samples were obtained from individuals with rheumatoid arthritis and osteoarthritis, and blood plasma samples were also obtained from normal subjects. The samples were fractionation by two-dimensional electrophoresis (the first dimension, isoelectric focusing, was performed in the pH range of 4.0 to 7.0). The electrophoretograms were stained with Coomassie Blue and "spots" of interest excised and digested with proteases. The digested samples were analyzed by MALDI-TOF mass spectrometry and the data analyzed with the SWISS-PROT database for the identification of proteins. The evaluation of differences in the electrophoretograms was assisted by the use of retinol-binding protein as an "internal standard." Clear differences were found in certain regions between the various samples (rheumatoid arthritis, osteo-arthritis, normal); differences were seen in fibrinogen degradation products and calprotectin (calgranulin A/calgranulin B, S100A9/S100A8, MRP8/MRP14). In subsequent work,[13] these investigators developed an immunoassay that could be used to differentiate between rheumatoid arthritis and osteoarthritis, permitting the identification of patients of rheumatoid arthritis who would benefit from anti-TNFα therapy. This type of assay can save a large amount of money in the health care system since the yearly cost of anti-TNFα therapy (Entercept, Embrel®) is approxi-mately $30,000 US per year. Another group of investigators have used ProteinChip® technology developed by Ciphergen (Fremont, CA)[62,63] to identify a protein found in synovial fluid from patients with rheumatoid arthritis and not in osteoarthritis.[64] These studies were performed with a hydrophobic (C_{16}) matrix and a strong anion exchange (SAX) matrix. These studies identified a protein consistent with the prop-erties of MRP8(S100A8) as described above. The apparent novelty of these obser-vations is possibly obscured by the failure to recognize the previous studies of Berntzen and coworkers[51] who reported elevated levels of leukocyte protein L1 in plasma and synovial fluids from patients with rheumatoid arthritis while normal levels of protein L1 are found in patients with osteoarthritis. More recently, MRP8/MRP14 has been suggested as a marker for relapse in inactive juvenile idiopathic arthritis[65] based on its previous identification in rheumatoid arthritis.[51,66,67] The terms S100A8 and S100A9 are recognized as new terminology for MRP8 and MRP14 and calgranulin A and calgranulin B;[67–69] thus, calprotectin is the complex of S100A8 (MRP8) and S100A9 (MRP14). Although the linage is a little difficult, the parent term appears to be L1 (L1 heavy chain and L1 light chain) that is released from leukocytes following immunological injury and is thought to be involved with leukocyte trafficking.[70] Elevated levels of these proteins is seen in the plasma/serum from individuals with a variety of inflammatory diseases.[71–73] This later study[73] used proteomic technology to identify calgranulin B (S100A9) as a potential marker for intraamniotic infection. Amniotic fluid was reduced, carboxamidomethylated and fractionation on SDS-PAGE and proteins were detected by staining with Coomassie Blue. The stained protein bands were excised and digested in situ with trypsin. The resulting digests were subjected to RP-HPLC and the fractions analyzed by mass spectrometry. Proteins were identified by use of the SEQUEST database.[74] In addi-tion to calgranulin B, insulin-growth-factor-binding-protein-1 was also differentially expressed with intraamniotic infection. The identity of these proteins was subse-quently confirmed with Western blots.

HEAD AND NECK CANCER

The other study cited above[44] in the introductory statement also used ProteinChip® technology. In this study, von Eggeling and coworkers[44,75] used proteomic technology to identify a biomarker for head and neck cancers. Using laser microdissection as a prefractionation technique, tissue samples from tumor tissue and normal tissue were obtained for analysis. Head and neck tumor samples ($n = 57$) and matched normal mucosa ($n = 44$) were obtained after surgical resection. Laser microdissection was used to obtain a sample of 3000 to 5000 cells. This sample was extracted with a lysis buffer (100 mM sodium phosphate, pH 7.5) containing CHAPS, EDTA, 2-mercaptoethanol, leupeptin and PMSF. The supernatant fractions were analyzed by surface-enhanced laser desorption/ionization (SELDI-MS) using a SAX chip. Immunodepletion with specific antibodies before ProteinChip® analysis suggested that there were decreased or absent levels of calgranulin A and calgranulin B. Two-dimensional gel electrophoresis was performed using undissected tumor and normal tissue samples. Protein "spots" were identified by Coomassie Blue staining. "Spots" that were consistently differentially expressed were excised, digested *in situ* with trypsin and analyzed using a C_{16} hydrophobic protein chip. Immunodepletion of the samples prior to the ProteinChip® analysis suggested that there were decreased or absent levels of calgranulin A and calgranulin B. Analysis of the samples by two-dimensional electrophoresis also demonstrated differences in these proteins. Monoclonal antibodies to calgranulin A and calgranulin B were used to demonstrate the presence of these proteins in the tumor tissue (squamous cell carcinoma). Thus, a combination of immunohistochemistry, histology and protein analysis was used to identify a putative biomarker for head and neck cancers. The relationship of calgranulin A, calgranulin B and calprotectin has been discussed above. The protein marker calprotectin, identified as a biomarker in the above two studies, is a migratory inhibitory factor-related protein that is a member of the S100 family and involved in leukocyte trafficking.[70] Elevated levels of these proteins in the plasma/serum from individuals with a variety of inflammatory diseases have been previously noted.

The above two studies were discussed in some detail for several reasons. One reason is that the experimental approach was nicely designed and the laboratory work of high quality. The second reason is that both studies benefited from previous studies that suggested calprotectin would be a useful target.[51,67,76,77] Stulik and coworkers[77] observed an increase in the S100 proteins in intestinal mucosal tumor cells reflecting invasion of the tumors as well as the relationship between the inflammatory response and tumors. This interesting biology provides for immense complication in identifying a specific and sensitive biomarker for oncology screening.

SPECIFICITY AND SENSITIVITY — CA125 AND OTHER MARKERS

The promiscuity of calprotectin has been discussed above. The other side of the coin is the finding of multiple markers for similar pathology in multiple studies. Head and neck (squamous cell) tumors provide a somewhat discrete sample for this issue.

The finding of calprotectin by von Eggeling and colleagues[44,75] is discussed above. Several other studies on head and neck tumors exist that have identified different markers by using proteomic technology. One group[78] suggested that serum amyloid A protein is useful for following relapse in nasopharyngeal cancer. These investigators used SELDI (ProteinChip®) technology to analysis serum samples from 31 nasopharyngeal cancer patients who were enrolled in a Phase II clinical study with a gemcitabine and cisplatin combination. Serum amyloid A protein increased in patients who relapsed during the trial; initial identification of serum amyloid A protein was made via SELDI technology and confirmed by immunoassay. Another group[79] showed an increase in annexin I and annexin II in cell cultures of two head and cell cancer cell lines (UMSCC10A and UMSCC10B). Koong and coworkers[80] suggest that 1κB kinase ß is an endogenous marker of tumor hypoxia in patients with head and neck squamous cell carcinomas. These investigators used a combination of two-dimensional gel electrophoresis and PowerBlot Western blot technology. This latter approach permits a more extensive, high-throughput immunoblotting process.[81,82]

Specificity for a disease would appear to a problem with the use of calprotectin. Sensitivity would also appear to be an issue with the use of CA125 as a biomarker for ovarian cancer (Table 7.3). However, CA125 has been used for the study of ovarian cancer for some time[84,85] with considerable success. Current work would suggest that sensitivity can be increased by using CA125 in combination with other potential biomarkers identified with proteomic technology. At the current time, CA125 appears to be useful for the monitoring of chemotherapy in ovarian cancer[85] and is therefore useful as a prognostic marker. Investigators would be well advised to consider the comments of Droegemueller[86] regarding the issues concerning the early diagnosis of ovarian cancer. With a specificity of 99.6%, nine false-positives would appear for every one positive diagnosis. More recently, Jacobs and Menon,[87] while emphasizing the value of proteomic technology in identifying new biomar-

---

TABLE 7.3
Observations on the Use of CA125 as a Biomarker for Ovarian Cancer

Study Description	Reference
Failure of CA125 to detect early stage ovarian cancer.	1, 2
Identification of a new biomarker complementary to CA125.	3
Value of pattern recognition in diagnostics.	4
Haptoglobin α-subunit complementary to CA125 in the diagnosis of ovarian cancer.	5
Value of other biomarkers such as CA19.9, CA15.3, and TAG.72 in the diagnosis of ovarian cancer. CA19.9 has a high sensitivity for the mucinous histotype. CA125 is also useful for endometrial cancer. CA125 is useful for the monitoring of therapy. Elevated CA125 may also be elevated in cervical cancer.	6
Development of a multivariate model with three additional biomarkers a polipoprotein A1, down-regulated; a truncated form of transyretin, down-regulated; inter-alpha-trypsin inhibitor heavy chain H4, up-regulated.	7
Combination of soluble mesothelial related marker with CA125 for the diagnosis of ovarian cancer.	8

REFERENCES FOR TABLE 7.3

1. Jones, M.B., Krutzsch, H., Shu, H. et al., Proteomic analysis and identification of new biomarkers and therapeutic targets for invasive ovarian cancer, *Proteomics*, 2, 76–84, 2002.
2. Ardekami, A.M., Liotta, L.A. and Petricoin, E.F., III, Clinical protential of proteomics in the diagnosis of ovarian cancer, *Expt. Rev. Mol. Diag.*, 2, 313–320, 2002.
3. Rai, A.J., Zhang, Z., Rosenzweig, J. et al., Proteomic approaches to tumor marker discovery, *Arch. Path. Lab. Med.*, 126, 1518–1526, 2002.
4. Steinert, R., van Hogen, P., Fels, L.M., Gunther, K., Lippert, H. and Reymond, M.A., Proteomic prediction of disease outcome in cancer: Clinical framework and current status, *Am. J. Pharm.*, 3, 107–115, 2003.
5. Ye, B., Cramer, D.W., Skates, S.J. et al., Haptoglobin-alpha subunit as potential serum biomarker in ovarian cancer: Identification and characterization using proteomic profiling and mass spectrometry, *Clin. Cancer Res.*, 9, 2904–2911, 2003.
6. Gadducci, A., Cosier, S., Carpi, A., Nicolini, A. and Genazzani, A.R., Serum tumor markers in the management of ovarian endometrial and cervical cancer, *Biomed. Pharm.*, 58, 24–38, 2004.
7. Zhang, Z., Bast, R.C., Jr., Yu, Y. et al., Three biomarkers identified from serum proteomic analysis for the detection of early stage ovarian cancer, *Cancer Res.*, 64, 5882–5890, 2004.
8. McIntosh, M.W., Drescher, C., Karlan, B., Schella, N., Urban, N., Hellstrom, K.E. and Hellstrom, I., Combining CA 125 and SMR serum markers for diagnosis and early detection of ovarian cancer, *Gynecol. Oncol.*, 95, 9–15, 2004.

kers, note that the same specificity would be required. This latter reference[87] provides an excellent overview of current studies on the development of new diagnostics. However, at present there is no test that merits the screening of the general population for ovarian cancer.[88] Without being pessimistic, the investigators note that the price of a false positive is a laparotomy or laparoscopy to rule out ovarian cancer.

Notwithstanding the potential issues of specificity and sensitivity, the identification of new biomarkers is necessary for the early diagnosis of diseases such as ovarian cancer and bladder cancer, for toxicology and environmental health studies and for personalized medicine. Identification of new biomarkers will also be useful for preclinical studies in drug development. Proteomic technologies will likely form the basis for studies designed for the discovery of new biomarkers. The most obvious studies will involve the identification of new protein biomarkers in serum; less obvious studies will utilize chemical proteomics to identify new drugs. Table 7.4 provides a list of some studies on the use of proteomics to discover protein biomarkers, and that the majority of work has been done with cancer is easily observed. Within oncology, most of the work has been done with ovarian cancer and to a lesser extent with lung cancer, prostate cancer and bladder cancer. These diseases are where the development of a diagnostic that would provide early identification of a tumor could be an advance in therapy.[89–92]

The most public work on biomarker development in oncology seems to focus on ovarian cancer. Table 7.5 describes work on the development of biomarkers for

TABLE 7.4
Studies on the Validation of Biomarkers Identified Using Proteomic Technology

Study Description	Reference
Biomarker identification for use in environmental health.	1, 5
Comparison of protein profiles (laser capture microdissection) for invasive versus low malignancy potential ovarian cancer.	2
Biomarkers in cancer screening/detection.	3
Protein profiling with surface-enhanced laser desorption/ionization in breast cancer.	4
Tumor cells and tumor biomarkers in small cell lung cancer; multiple biomarkers including circulating tumor cells.	6
Biomarker selection and validation; clinical endpoint evaluation and application.	7, 9
Haptoglobin α-subunit and ovarian cancer.	8
Colorectal cancer.	10
Drug development and toxicology.	11, 12
Identification and validation of biomarkers for lung cancer.	13
Protein profiling in intraamniotic infection.	14
Up- and down-regulation of biomarkers in ovarian cancer; development and validation of a multi-variate model.	15

REFERENCES FOR TABLE 7.4

1. Albertini, R.J., Developing sustainable studies on environmental health, *Mut. Res.*, 480–481, 317–331, 2001.
2. Jones, M.B., Krutzsch, H., Shu, H. et al., Proteomic analysis and identification of new biomarkers and therapeutic targets for invasive ovarian cancer, *Proteomics*, 2, 76–84, 2002.
3. Negm, R.S., Verma, M. and Srivastava, S., The promise of biomarkers in cancer screening and detection, *Trends Mol. Med.*, 8, 288–293, 2002.
4. Li, J., Zhang, Z., Rosenzweig, J., Wang, Y.Y. and Chan, D.W., Proteomics and bioinformatics to detect breast cancer, *Clin. Chem.*, 48, 1296–1304, 2002.
5. Goodsaid, F.M., Genomic markers of toxicity, *Curr. Opin. Drug Disc. Dev.*, 6, 41–49, 2003.
6. Ma, P.C., Blaszkowsky, L., Bhati, A. et al., Circulating tumor cells and serum tumor biomarkers in small cell lung cancer, *Anticancer Res.*, 23(1A), 49–62, 2003.
7. Colburn, W.A., Biomarkers in drug discovery and development: from target identification through drug marketing, *J. Clin. Pharm.*, 43, 329–341, 2003.
8. Ye, B., Cramer, D.W., Skates, S.J. et al., Haptoglobin-alpha subunit as potential serum biomarker in ovarian cancer: identification and characterization using proteomic profiling and mass spectrometry, *Clin. Cancer Res.*, 9, 2904–2911, 2003.
9. Colburn, W.A. and Lee, J.W., Biomarker validation and pharmacokinetic — Pharmacodynamic modeling, *Clin. Pharm.*, 42, 997–1022, 2003.
10. Sanderson, P., Johnson, I.T., Mathers, J.C. et al., Emergy diet-related surrogate end points for colorectal cancer: UK Food Standards Agency diet and colonic health workshop reports, *Brit. J. Nutrition*, 91, 315–323, 2004.

11. Walgren, J.L. and Thompson, D.C., Application of proteomic technologies to the drug development process, *Tox. Letters*, 149, 377–385, 2004.
12. Floyd, E. and McShane, T.M., Development and use of biomarkers in oncology drug development, *Tox. Path.*, Suppl. 1, 32, 106–115, 2004.
13. Chanin, T.D., Merrick, D.T., Franklin, W.A. and Hirsch, F.R., Recent developments on biomarkers for the early detection of lung cancer: Perspectives based on publications 2003 to present, *Curr. Opin. Pulm. Med.*, 10, 242–247, 2004.
14. Gravett, M.G., Novy, M.J., Rosenfeld, R.G. et al., Diagnosis of intra-amniotic infection by proteomic profiling and identification of novel biomarkers, *JAMA*, 292, 462–469, 2004.
15. Zhang, Z., Bast, R.C., Yu, Y. et al., Three biomarkers identified from serum proteomic analysis for detection of early stage ovarian cancer, *Cancer Res.*, 64, 5882–5890, 2004.

TABLE 7.5
Biomarkers for Ovarian Cancer

Study[a]	Biomarker(s)	Reference
Serum proteomics analysis, SELDI[b]	CA125; CA125 + apolipoprotein A1, truncated transthyretin, inter-α-trypsin inhibitor heavy chain H4	1
Serum proteomics analysis, SELDI	CA125 + Haptoglobin α-subunit	2
Serum proteomic analysis, SELDI	CA125 + 4 other biomarkers	3
Serum proteomic analysis, two-dimensional gel electrophoresis	52 kDa FK506 binding protein, Rho G-protein dissociation inhibitor (RhoGD1), glyoxalase 1	4
Serum proteomic analysis	CA125 + soluble mesothelin-related marker	5
Serum protein analysis	Hemoglobin	6
Immunohistocytochemistry	Type IV collagen, CD44v6, P53, Ki-67	7
Review	CA125, tumor-associated trypsin inhibitor	8
Two-dimensional gel electrophoresis	OP-18, PCNA, triosephosphate isomerase, elongation factor-2, GST all upregulated; TM2 and laminin c downregulated	9, 10

[a] Biomarkers are separate when separated by a comma; combined with + sign to give profiling.
[b] SELDI, Surface-Enhanced Laser Desorption/Ionization.

REFERENCES FOR TABLE 7.5

1. Zhang, Z., Bast, R.C., Jr., Yu, Y. et al., Three biomarkers from serum proteomic analysis for detection of early stage ovarian cancer, *Cancer Res.*, 64, 5882–5890, 2004.
2. Ye, B., Craner, D.W., Skates, S.J. et al., Haptoglobin-alpha subunit as potential serum biomarkers in ovarian cancer: Identification and characterization using proteomic profiling and mass spectrometry, *Clin. Cancer Res.*, 9, 2904–2811, 2003.

3. Rai, A.J., Zhang, Z., Rosenzweig, J. et al., Proteomic approaches to tumor marker discovery, *Arch. Path. Lab. Med.*, 126, 1518–1526, 2002.
4. Janes, M.B., Krutzsch, H., Shu, H. et al., Proteomic analysis and identification of new biomarkers and therapeutic targets for invasive ovarian cancer, *Proteomics*, 2, 76–84, 2002.
5. McIntosh, M.W., Drescer, C., Karlan, B., Scheller, N., Urban, N., Hellstrom, K.F. and Hellstrom, I., Combining CA 125 and SMR serum marker for diagnosis and early detection of ovarian cancer, *Gynecol. Oncol.*, 95, 9–15, 2004.
6. Ferrero, A., Zola, P., Mazzola, S. et al., Pretreatment serum hemoglobin level and a preliminary investigation of intratumoral microvessel density in advanced ovarian cancer, *Gynecol. Oncol.*, 95, 323–329, 2004.
7. Bar, J.K., Grelewski, P., Popiela, A., Naga, L. and Rabxzyski, J., Type IV collagen and CD44v6 expression in benign, malignant primary and metastatic ovarian tumors: Correlation with Ki-67 and p53 immunreactivity, *Gynecol. Oncol.*, 95, 23–31, 2004.
8. Saksela, E., Prognostic markers in epithelial ovarian cancer, *Intl. J. Gynecol. Path.*, 12, 156–161, 1993.
9. Alaiya, A.A., Franzén, B., Fujioka, K., Moberger, B., Schedoins, K., Silfversuärd, C., Linder, S. and Aver, G., Phenotypic analysis of ovarian carcinoma: Polypeptide expression in benign, borderline, and malignant tumors, *Intl. J. Cancer*, 73, 678–683, 1997.
10. Franzén, B., Okuzawa, K., Linder, S., Katz, H. and Aver, G., Non-enzymatic extraction of cells from clinical tumor material for analysis of gene expression by two-dimensional gel electrophoresis, *Electrophoresis*, 14, 382–390, 1993.

ovarian cancer. The most extensively used biomarker for ovarian cancer in CA125, first described in 1981 by Bast and colleagues[93] who describe a monoclonal antibody, OC125, that reacted with each of six epithelial ovarian cancer cell lines with tissue from ovarian cancer patients but not with nonmalignant tissues including fetal and adult ovary. Evidence was found to support specificity for epithelial ovarian cancer as OC125 did not react with a pancreatic cancer cell line, a breast cancer cell line or carcinoembryonic antigen (CEA). The reactive epitope for OC125 was found to be located on a high molecular weight glycoprotein expressed in coelomic epithelium during embryonic development and was designated CA125[a],[94] A radio-immunoassay was subsequently developed for diagnostic use[83],[95] that has tseen considerable use in monitoring ovarian cancer. The fact was recognized early that while the measurement of CA125 was useful in monitoring the treatment of ovarian cancer, using it for early diagnosis was not possible because of a lack of specificity.[96] In this latter paper,[95] the specificity of the assay for CA125 antigen assay was 83% (RIA) for ovarian tumors; high reactivity (66%) was also found with cervical tumors and uterine tumors (56%) with much less reactivity (20%) for colon cancer.

Other studies exist which demonstrate the lack of sensitivity of CA125 for the diagnosis of ovarian cancer. A report exists of elevation of CA125 in a patient with right heart failure[97] due to an atrial septal defect where it was suggested that this was due to increased secretion of CA125 from activated peritoneal mesothelium. CA125 is reported to be elevated in benign lesions such as large endometrioma.[98] Krediet[99] has shown that CA125 is a mesothelial cell marker that can be measured in peritoneal effluent and used as a measure of *in vivo* biocompatibility of dialysis solutions. DiBaise and Donovan[100] reported elevated CA125 in hepatic cirrhosis. These latter observations have been confirmed and extended by Xiao and Liu[101] and Kim and coworkers.[102]

CLINICAL PROTEOMICS BY TECHNICAL APPROACH

In the rest of this discussion, we will address the topic of clinical proteomics by a technical approach rather than by pathology. First, while it may seem somewhat redundant, we need to review several definitions that are useful in considering clinical proteomics by technical approach:

- *proteome*: "the complete set of proteins produced from the information encoded in a genome."[103]
- *proteomics*: "the qualitative and quantitative study of the proteome under various conditions, including protein expression, modification, localization, function, and protein–protein interactions as a means of understanding biologicals processes."[103]
- *clinical proteomics*: the use of proteomic technologies to study clinical problems; most frequently used for studies of the discovery, description and characterization of biomarkers.
- *biomarker*: proteins that can be used to diagnose and monitor a specific disease process including the study of the effect of therapeutic intervention. Changes in biomarker expression in fluids, cells or tissues are used as surrogate markers.[54,56] Biomarkers are not unique entities that are only discovered by proteomic technology;[104] biomarkers can be described as analytes used in clinical laboratories.[105,106]
- *surrogate marker*: a characteristic that can be used as a substitute for a clinical endpoint.[107]

The oldest of the techniques and likely also the most frequently used is two-dimensional electrophoresis. The first dimension — isoelectric focusing and separation — is based on the charge of the protein (isoelectric point); the second dimension is performed in the presence of sodium dodecyl sulfate and separation is based on molecular size.[108,109] This system is complex but has been demonstrated to be reproducible to the extent of being able to identify potential biomarkers.[110,111] Depending on the detection system, the number of protein "spots" can number more than 5000. Identification of biomarkers by two-dimensional electrophoresis is based on the comparison of diseased tissue versus normal tissue. In most studies, the protein "spots" are identified on the electrophoretograms by staining with Coomassie Blue, straining with ammoniacal silver or with a fluorescent dye such as SYPRO, the gels compared, the difference "spots" excised either manually or with the aid of a robotic process and the proteins either digested *in situ* with proteolytic enzymes or analyzed directly by mass spectrometry.[112,113] Another option is the derivatization of the protein sample with fluorescent dyes prior to electrophoresis; selection of different fluorophores allows the more facile comparison of several electrophoretograms.[114,115] As noted above, this approach can be very useful if the work is carefully performed as demonstrated by the calprotectin studies mentioned above and the use of a combination of techniques. A partial list of biomarker studies based on two-dimensional gel electrophoresis technology is given in Table 7.6.

While different separation modalities are used, the general concept of multi-dimensional protein identification technologies (MuDPIT)[116,117] is quite similar in concept to the two-dimensional electrophoretic approach described above. This approach is described as the use of orthogonal techniques; for example, an ion-exchange column could the first "phase" and a reversed-phase column would be the second "phase." Mass spectrometry of the intact proteins or digests thereof is used for analysis and identification using the same databases as described above. This technology has seen limited use in the identification of clinical biomarkers, as shown in Table 7.7.

The third most extensively used procedure has properties of both of the above procedures. Surface-enhanced laser desorption/ionization-time of flight mass spectrometry (SELDI-TOF-MS) was developed as affinity mass spectrometry.[118] The approach has been commercialized by Cipergen (Fremont, CA) as ProteinChip® technology.[119,120] A partial list of studies using SELDI-based technology for biomarker identification is shown in Table 7.8. The SELDI-TOF (ProteinChip®) appears to be the most popular approach. However, as indicated in the discussion above, the problems with the rigorous validation of the technology are legion. From a technical perspective, this process would appear to be the least demanding, which is perhaps both a blessing and a curse.

CONCLUSIONS

At present, we are pessimistic about the promise of proteomics delivering clinically significant results in the short term any more rapidly than the more traditional disciplines such as analytical biochemistry and immunohistochemistry. Good and perhaps even great science will come from the thoughtful application of proteomics. Without a new paradigm, it is unlikely to address the complete description of a proteome. As such, a move from the strict discovery-oriented approach to a more hypothesis-directed approach might be useful. In this regard, the application of proteomic technologies to the definition of prognostic markers will be more rewarding that the current search for the "Holy Grail" of a novel screening diagnostic. Also imperative is that investigators do a diligent literature search. A prime example of the necessity of such diligence is provided by the above cited example of the "discovery" of calprotectin as a biomarker for rheumatoid arthritis.

TABLE 7.6
Selected Applications of Two- Dimensional Electrophoresis in Clinical Proteomics

Study	Pathology
Gharib, T.G., Chen, G., Huang, C.C. et al., Genomic and proteomic analysis of vascular endothelial growth factor and insulin-like growth factor-binding protein 3 in lung adenocarcinomas, *Clin. Lung Cancer*, 5, 307–319, 2004.	lung cancer
Kim, C.H., Kim, do K., Choi, S.J. et al., Proteomic and transcriptomic analysis of interleukin-1-beta treated lung carcinoma cell line, *Proteomics*, 3, 2454–2471, 2003.[a]	lung cancer
MacKeigan, J.P., Clements, C.M., Lich, J.D., Pepe, R.M., Hod, Y. and Ting, J.P., Proteomic profiling drug-induced apoptosis in non-small cell lung carcinoma: Identification of RS/DJ-1 and RhoGD1alpha, *Cancer Res.*, 63, 6928–6934, 2003.	lung cancer
Li, C., Chen, Z., Ziao, Z. et al., Comparative proteomic analysis of human lung squamous carcinoma, *Biochem. Biophys. Res. Comm.*, 309, 253–260, 2003.	lung cancer
Chen, G., Gharib, T.G., Thomas, D.G. et al., Proteomic analysis of eIF-5A in lung adenocarcinomas, *Proteomics*, 3, 496–504, 2003.	lung cancer
Brichary, F., Beer, D., Le Naour, F., Giondono, T. and Hanash, S., Proteomics-based identification of protein gene product 9.5 as a tumor antigen in lung cancer, *Cancer Res.*, 61, 7908–7912, 2001.	lung cancer
Friedman, D.B., Hill, S., Kella, J.W., Merchant, N.B., Levy, S.E., Coffey, R.J. and Caprioli, R.M., Proteome analysis of human colon cancer by two-dimensional difference gel electrophoresis and mass spectrometry, *Proteomics*, 4, 793–811, 2004.	colon cancer
Celis, J.E. and Gromov, P., Proteomics in translational cancer research: Toward an integrated approach, *Cancer Cell.*, 3, 9–15, 2003.	bladder cancer
Celis, J.E., Gromova, I., Rank, F. and Gromov, P., Proteomics in bladder cancer, in *Oncogenomics: Molecular Approaches to Cancer*, C. Brenner and D. Duggan, Eds., John Wiley & Sons, Hoboken, NJ, Ch. 9, 157–184, 2004.	
Ahran, M., Flaig, M.J., Gillespie, J.W. et al., Evaluation of ethanol-fixed paraffin embedded tissues for proteomic applications, *Proteomics*, 3, 412–421, 2003.	prostate cancer
Meehan, K.L., Holland, J.W. and Dawkins, H.J., Proteomic analysis of normal and malignant prostate tissue to identify novel proteins lost in cancer, *Prostate*, 50, 54–63, 2002.	prostate cancer
Cheng, X., Activation of antioxidant pathways in ras-mediated oncogenic transformation of human surface ovarian epithelial cells revealed by functional proteomics and mass spectrometry, *Cancer Res.*, 64, 4577–4584, 2004.	ovarian cancer
Ahmed, N., Barker, G., Olivo, K.T. et al., Proteomic-based identification of haptoglobin-1 precursor as a novel circulating biomarker of ovarian cancer, *Brit. J. Cancer*, 9, 129–140, 2004.	ovarian cancer
Kageyama, S., Isono, T., Iweki, H. et al., Identification by proteomic analysis of calreticulum as a marker for bladder cancer and evaluation of diagnostic accuracy of its detection in urine, *Clin. Chem.*, 50, 857–866, 2004.	bladder cancer
Kuruma, H., Egawa, S., On-Ishi, M., Kodera, Y. and Maeda, T., Proteome analysis of prostate cancer, *Prostate Cancer Prostatic Disease*, 2004.	prostate cancer

[a] Study shows that transcript levels were of little value in predicting the level of protein expression.

TABLE 7.7
Selected Applications of Multi- Dimensional Protein Identification Technology to Clinical Proteomics

Study	Pathology
Beer, I., Barnea, E., Ziv. T. and Adman, A., Improving large-scale proteomics by clustering of mass spectrometry data, *Proteomics*, 4, 950–960, 2004.	lung cancer
Kaiser, T., Kamel, H., Park, A. et al., Proteomics applied to the clinical follow-up of patients after allogeneic hematopoietic stem cell transplantation, *Blood*, 104, 340–349, 2001.	GVHD
Wang, H., Kachman, M.T., Schwartz, D.R., Cho, K.R. and Lubman, D.M., A protein molecular weight map of ES2 clear cell ovarian cancer cells using a two-dimensional liquid separation/mass mapping technique, *Electrophoresis*, 23, 3168–3181, 2002.	ovarian cancer

TABLE 7.8
Selected Applications of ProteinChip® Technology[a] in Clinical Proteomics

Study	Pathology
Rai, A.J. and Chan, D.W., Cancer proteomics. Serum diagnostics for tumor marker discovery, *Ann. N.Y. Acad. Sci.*, 1022, 286–294, 2003.	cancer
Johann, D.J., Jr., McGuigan, M.D., Patel, A.R. et al., Clinical proteomics and biomarker discovery, *Ann. N.Y. Acad. Sci.*, 1022, 295–305, 2003.	general
Gerton, C.L., Fan, X.J., Chittams, J. et al., A serum proteomics approach to the diagnosis of ectopic pregnancy, *Ann. N.Y. Acad. Sci.*, 1022, 306–316, 2003.	ectopic pregnancy
Wilson, L.L., Tran, L., Morton, D.L. and Hoon, D.S.B., Detection of differentially expressed proteins in early stage melanoma patients using SELDI-TOF mass spectrometry, *Ann. N.Y. Acad. Sci.*, 1022, 371–322, 2003.	melanoma
Xiao, Y., Liu, D., Tang, Y. et al., Development of proteomic patterns for detecting lung cancer, *Disease Markers*, 19, 33–39, 2003–2004.	lung cancer
Cho, W.C., Yip, T.T., Yip, C. et al., Identification of serum amyloid a protein as a potentially useful biomarker to monitor relapse of nasopharyngeal cancer by serum proteomic profiling, *Clin. Cancer Res.*, 10, 43–52, 2004.	naso-pharyngeal cancer
Ornstein, D.K., Rayford, W., Fusaro, V.A. et al., Serum proteomic profiling can discriminate prostate cancer from benign prostates in men with total prostate specific antigen levels between 2.5 and 15.0 ng/ml, *J. Urology*, 172, 1302–1305, 2004.	prostate cancer
Gretzer, M.B., Chan, D.W., von Rootselaar, C.L. et al., Proteome analysis of dunning prostate cancer cell lines with variable metastatic potential using SELDI-TOF, *Prostate*, 60, 325–331, 2004.	prostate cancer
Li, J., White, N., Zhang, Z. et al., Detection of prostate cancer using serum proteomic pattern in a histologically confirmed population, *J. Urology*, 171, 1782–1787, 2004.	prostate cancer
Zheng, Y., Xu, Y., Ye, B. et al., Prostate carcinoma tissue proteomics for biomarker discovery, *Cancer*, 98, 2576–2582, 2003.	prostate cancer

(Continued)

TABLE 7.8
Selected Applications of ProteinChip® Technology[a] in Clinical Proteomics (*Continued*)

Yasui, Y., Pepe, M., Thompson, M.L. et al., A data-analytic strategy for protein biomarker discovery: Profiling of high-dimensional proteomic data for cancer detection, *Biostatistics*, 4, 449–463, 2003.	prostate cancer
Adam, B.L., Qu, Y., Davis, J.W. et al., Serum protein fingerprinting coupled with a pattern-matching algorithm distinguishes prostate cancer from benign prostate hyperplasia and healthy men, *Cancer Res.*, 62, 2614–3609, 2002.	prostate cancer
Zhang, Z., Bast, R.C., Jr., Yu, Y. et al., Three biomarkers identified from serum proteomic analysis for the detection of early stage ovarian cancer, *Cancer Res.*, 64, 5882–5890, 2004.	ovarian cancer
Ye, B., Cramer, D.W., Skates, S.J. et al., Haptoglobin alpha subunit as potential serum biomarker in ovarian cancer: Identification and characterization using proteomic profiling and mass spectrometry, *Clin. Cancer Res.*, 9, 2904–2911, 2003.	ovarian cancer
Rai, A.J., Zhang, Z., Rosenzweig, J. et al., Proteomic approaches to tumor marker discovery. Identification of biomarkers for ovarian cancer, *Arch. Path. Lab. Med.*, 126, 1518–1526, 2002.	ovarian cancer
Oehr, P., Proteomics as a tool for detection of nuclear matrix proteins and new biomarkers for screening of early tumor stages, *Anticancer Res.*, 23, 805–812, 2003.	ovarian cancer
Yip, T.T., Chan, J.W., Cho, W.C. et al., Protein chip array profiling analysis in patients with severe acute respiratory syndrome identified serum amyloid A protein as a biomarker potentially useful in marking the extent of pneumonia, *Clin. Chem*, 51, 47–55, 2005.	pneumonia
Zhu, X.D., Zhang, W.H. and Li, C.L., New serum biomarkers for detection of HBV-induced liver cirrhosis using SELDI protein technology, *World J. Gastro.*, 10, 2327–2329, 2004.	hbv-induced liver cirrhosis
Wilson, L.L., Tran, L., Morton, D.L. and Hoon, D.S., Detection of differentially expressed proteins in early-stage melanoma patients using SELDI-TOF mass spectrometry, *Ann. N.Y. Acad. Sci.*, 1002, 317–322, 2003.	melanoma
Sun, Z., Fu, X., Zhang, L., Xang, X., Liu, F. and Hu, G., A protein chip system for parallel analysis of multi-tumor markers and its application in cancer detection, *Anticancer Res.*, 24, 1159–1165, 2004.	multiple tumors
Balachandra, K., Laisupasin, P., Dhepakson, P. et al., Preliminary clinical evaluation of a protein chip for tumor marker serodiagnosis of various cancers, *Asian Pac. J. Allergy Immunol.*, 21, 171–178, 2003.	multiple tumors
Koopman, J., Zhang, Z., White, N. et al., Serum diagnosis of pancreatic adenocarcinoma using surface-enhanced laser desorption and ionization mass spectrometry, *Clin. Cancer Res.*, 10, 860–868, 2004.	pancreatic cancer
Won, Y., Song, H.J., Kay, T.W. et al., Pattern analysis of serum proteome distinguishes renal cell carcinoma form other urologic diseases and healthy persons, *Proteomics*, 3, 2310–2316, 2003.	renal cell carcinoma
Carrette, O., Demalte, I., Scherl, A. et al., A panel of cerebrospinal fluid potential biomarkers for the diagnosis of Alzheimer's disease, *Proteomics*, 3, 1486–1494, 2003.	Alzheimer's disease

(*Continued*)

TABLE 7.8
Selected Applications of ProteinChip® Technology[a] in Clinical Proteomics (Continued)

Lehrer, S., Roboz, J. and Ding, H., Putative protein markers in the sera of men with   prostate cancer prostatic neoplasms, *BJU Intl.*, 92, 223–225, 2003.

[a] ProteinChip® technology is a commercial term for surface-enhanced laser desorption/ionization as technique for mass spectrometry developed by Ciphergen (Fremont, CA). This technique was originally developed as affinity mass spectrometry at the University of California at Davis. SELDI — Surface-enhanced laser desorption/ionization; Gilbert, K., Figueredo, S., Meng, X.-Y., Yip, C. and Fung, E.T., Serum protein-expression profiling using the ProteinChip® biomarker system, in *Protein Array: Methods and Protocols* (*Methods in Molecular Biology*, John Walker, Ed., Vol. 364), E.T. Fung, Ed., Humana Press, Towata, NJ, Ch. 22, 259–269, 2004.

REFERENCES

1. Chen, S. T., Pan, T.L., Tsai, Y.C. and Huang, C.M., Proteomics reveals protein profile changes in doxorubicin-treated MCF-7 human breast cancer cells, *Cancer Letters*, 181, 95, 2002.
2. Godl, K., Wissing, J., Kurtenbach, A. et al., An efficient proteomics method to identify the cellular targets of protein kinase inhibitors, *Proc. Natl. Acad. Sci. USA*, 100, 15434, 2003.
3. Verhoeckx, K.C.M., Bijlsma, S., Jespersen, S. et al., Characterization of anti-inflammatory compounds using transcriptomics, proteomics, and metabolomics in combination with multivariate data analysis, *Intl. Immunopharm.*, 4, 1499, 2004.
4. Rosario, L.A., Biomarkers in the development of oncology drug products: Can they make a difference? *Preclinica*, 2, 382, 2004.
5. Goshe, M.B. and Smith, R.D., Stable isotope-coded proteomic mass spectrometry, *Curr. Opin. Biotech.*, 14, 101, 2003, 1995.
6. Rodland, K.D., Proteomics and cancer diagnosis: the potential of mass spectrometry, *Clin. Biochem.*, 37, 579, 2004.
7. Gao, J. Garulacan, L.-A. and Storm, S.M., Biomarker discovery in biological fluids, *Methods*, 35, 291, 2005.
8. Biomarker Definitions Working Group, Biomarkers and surrogate endpoints: Preferred definitions and conceptual framework, *Clin. Pharm. Ther.*, 69, 89, 2001.
9. Ilyin, S.E., Belkowski, S.M. and Plala-Salamón, C.R., Biomarker discovery and validation: Technologies and integrative approaches, *Trends Biotech.*, 22, 411, 2004.
10. Loyda, H.-J., Putting new assays to the test, *IVD Technology*, 38, March 2004.
11. Sinz, A., Bantscheff, M., Nikkat, S. et al., Mass spectrometric proteome analyses of synovial fluids and plasmas from patients suffering from rheumatoid arthritis and comparison to reactive arthritis or osteoarthritis, *Electrophoresis*, 23, 3445, 2002.
12. Lorenz, P., Bantscheff, M., Ibarhim, S.M. et al., Proteome analysis of diseased joints from mice suffering from collagen-induced arthritis, *Clin. Chem. Lab. Med.*, 41, 1622, 2003.
13. Drynda, S., Ringel, B., Kekow, M. et al., Proteome analysis reveals disease-associated marker proteins to differentiate RA patients from other inflammatory joint diseases with the potential to monitor anti-TNFα therapy, *Path. Res. Prac.*, 200, 165, 2004.

14. *Illustrated Guide to Diagnostic Tests*, Springhouse Corporation, Springhouse, PA, 1994.
15. *2004 Lange Current Medical Diagnosis and Treatment*, L.M. Tierney, Jr., S.J. McPhee and M.A. Papadaleis, Eds., Lange/McGraw-Hill, New York, NY, 2004.
16. *Prognostic Factors in Cancer*, 2nd Edition, M.K. Gospodarowicz, D.E. Hensen, R.V.P. Hutter, B. O'Sullivan, L.H. Sobin and C. Wittekind, Eds., Wiley-Liss, New York, NY, 2001.
17. Hacker, J. and Heesemann, J., Molecular diagnosis and epidemiology, in *Molecular Infectious Biology: Interactions between Microorganisms and Cells*, J. Hacker and J. Heesemann, Eds., Wiley-Liss/Spectrum, Heidelberg, Germany, Ch. 18, 195, 2002.
18. *Molecular Diagnosis of Infectious Disease*, Humana Press, Towata, NJ, 1998.
19. Schalken, J., Molecular diagnostics and therapy of prostate cancer: New avenues, *Eur. Urology*, Suppl. 3, 34, 1998.
20. Field, J.K., Brambilla, C., Caporaso, N. et al., Consensus statements from the Second International Lung Cancer Molecular Biomarkers Workshop: A European strategy for developing lung cancer molecular diagnostics in high risk populations, *Intl. J. Oncol.*, 21, 369, 2002.
21. Tsukishiro, S., Suzumori, N., Nishikawa, H. et al., Use of serum secretory leukocyte protease inhibitor levels in patients to improve specificity of ovarian cancer diagnosis, *Gynecol. Oncol.*, 96, 516, 2005.
22. Schwartz, M.K., Current status of tumour markers, *Scand. J. Clin. Lab. Invest.*, Suppl. 221, 5, 1995.
22a. Zhang, Z., Bast, R.C., Jr., Yu, Y. et al., Three biomarkers identified from serum proteomic analysis for the detection of early stage ovarian cancer, *Cancer Res.*, 64, 5882, 2004.
23. MacKillop, W.J., The importance of prognosis in cancer medicine, in *Prognostic Factors in Cancer*, 2nd Edition, M.K. Gospodarowicz, D.E. Hensen, R.V.P. Hutter, B. O'Sullivan, L.H. Sobin and C. Wittekind, Eds., Wiley-Liss, New York, NY, Ch. 1, 3–16, 2001.
24. Kratz, A., Ferraro, M., Sluss, P.M. and Lewandrowski, K.B., Laboratory reference values, *N. Eng. J. Med.*, 351, 1548, 2004.
25. Jones, H.B., On a new substance occurring in the urine of a patient with mollities ossium, *Phil. Trans. Royal Soc.*, London, United Kingdom, 138, 55, 1848.
26. Roulston, J.E. and Leonard, R.C.F., *Serological Tumour Markers, An Introduction*, Churchill Livingstone, Edinburgh, United Kingdom, 1993.
27. *Serological Tumor Markers*, S. Sell, Ed., Humana Press, Towota, NJ, 1992.
28. *Precancer: Biology, Importance and Possible Prevention*, J. Pontén, Ed., Cold Spring Harbor Laboratory Press, Cold Spring Harbor, NY, 1998.
29. *Cancer and Inflammation*, Novartis Foundation Symposium, John Wiley & Sons, Chichester, United Kingdom, 2004.
30. Balkwill, F. and Mantovani, A., Inflammation and cancer: Back to Virchow, *Lancet*, 357, 539, 2001.
31. Coussens, L.M. and Werb, Z., Inflammatory cells and cancer: Think different! *J. Exptl. Med.*, 193, F23, 2001.
32. Blankenstein, T., The role of inflammation in tumour growth and tumour suppression, in *Cancer and Inflammation*, Novartis Foundation Symposium, John Wiley & Sons, Chichester, United Kingdom, 205–214, 2004.
33. Roth, J., Vogl, T., Sorg, C. and Sunderkötten, C., Phagocyte-specific S100 proteins: A novel group of proinflammatory molecules, *Trends Immunol.*, 24, 155, 2003.
34. Ott, H.W., Lindner, H., Sorg, B. et al., Calgranulins in cystic fluid and serum from patients with ovarian carcinomas, *Cancer Res.*, 63, 7507, 2003.

35. Check, E., Running before we can walk? *Nature*, 429, 496–497, 2004.
36. Kolch, W., Mischak, H., Chalmers, M.J. et al., Clinical Proteomics: A question of technology, *Rapid Comm. Mass Spec.*, 18, 2365, 2004.
37. Baggerly, K.A., Morris, J.S., Edmonson, S.R. and Coombes, K.R., Signal in noise: Evaluating reported reproducibility of serum proteomic tests for ovarian cancer, *J. Natl. Cancer Inst.*, 97, 307, 2005.
38. Ransohoff, D.F., Lessons from controversy: Ovarian cancer screening and serum proteomics, *J. Natl. Cancer Inst.*, 97, 315, 2005.
39. Petricoin, E.F., III, Ardekani, A.M., Hitt, B.A. et al., Use of proteomic patterns in serum to identify ovarian cancer, *Lancet*, 359, 572, 2002.
40. Rockhill, B., et al., Comments, Proteomic patterns in serum and identification of ovarian cancer, *Lancet*, 360, 169, 2002.
41. Diamandis, E.P., Analysis of serum proteomic patterns for early cancer diagnosis: Drawing attention to potential problems, *J. Natl. Cancer Inst.*, 96, 353, 2004.
42. Baggerly, K.A., Morris, J.A. and Coombes, K.R., Reproducibility of SELDI-TOF protein patterns in serum: Comparing data sets from different experiments, *Bioinformatics*, 20, 777, 2004.
43. Rockhill, B., Importance of disclosure of patent applications, *Lancet*, 364, 577, 2004.
44. Melle, C., Ernst, G., Schimmel, B. et al., A technical triade for proteomic identification and characterization of cancer biomarkers, *Cancer Res.*, 64, 4099, 2004.
45. Harry, J.L., Wilkins, M.R., Herbert, B.R., Packer, N.H., Gooley, A.A. and Williams, K.L., Proteomics: Capacity versus utility, *Electrophoresis*, 21, 1071, 2000.
46. Humprey-Smith, I., A human proteome project with a beginning and an end, *Proteomics*, 4, 2519, 2004.
47. Bodovitz, S. and Joos, T., The proteomics bottleneck: strategies for preliminary validation of potential biomarkers and drug targets, *Trends Biotech.*, 22, 4, 2004.
48. Luce, B.R. and Brown, R.E., The use of technology assessment by hospitals, health maintenance organizations, and third-party payers in the United States, *Intl. J. Tech. Assess.*, 11, 79, 1995.
49. Zarkowsky, H., Managed care oranization's assessment of reimbursement for new technology, procedures, and drugs, *Arch. Path. Lab. Med.*, 123, 677, 1999.
50. Zolg, J.W. and Langen, H., How industry is approaching the search for new diagnostic markers and biomarkers, *Mol. Cell. Proteomics*, 3, 345, 2004.
51. Berntzen, H.B., Olmez, U., Fagerhol, M.K. and Monthe, E., The leukocyte protein L1 in plasma and synovial fluid from patients with rheumatoid arthritis and osteoarthritis, *Scand. J. Rheum.*, 20, 74, 1991.
52. Lundblad, R.L., Approach to assay validation for the development of biopharmaceuticals, *Biotech. Appl. Biochem.*, 34, 195, 2001.
53. Lundblad, R.L. and Wagner, P., Ruminations on the issue of assay validation for biopharmaceuticals, *Preclinica*, 2, 1, 2004.
54. Colburn, W.A., Surrogate markers and clinical pharmacology, *J. Clin. Pharm.*, 35, 441, 1995.
55. Bielokova, B. and Martin, R., Development of biomarkers in multiple sclerosis, *Brain*, 127, 1463, 2004.
56. Colburn, W.A., Selecting and validating biological markers for drug development, *J. Clin. Pharm.*, 27, 355, 1997.
57. Patterson, S., Selecting targets for therapeutic validation through differential protein expression using chromatography using chromatography-mass spectrometry, *Bioinformatics*, Suppl. 2, 18, 181, 2002.

58. Liu, S.C., Sauter, E.R., Clapper, M.L. et al., Markers of cell proliferation in normal epithelia and dsyplastic leukoplakes of the oral cavity, *Cancer Epid. Biomarkers Prev.*, 7, 597, 1998.

59. Ahmed, M.I., Abd-Emalelib, F., Ziada, N.A. and Khalifa, A., Evaluation of some tissue and serum biomarkers in prostatic carcinoma among Egyptian males, *Clin. Biochem.*, 32, 439, 1999.

60. Yasui, Y., Pepe, M., Thompson, M.L., et al,, A data-analytic strategy for protein biomarker discovery: Profiling of high-dimensional proteomic data for cancer detection, *Biostatistics*, 4, 449, 2003.

61. Eissa, S., Swellam, M. and el-Mosallamy, H., Diagnostic value of urinary molecular markers in bladder cancer, *Anticancer Res.*, 23, 4347, 2003.

62. Merchant, M. and Weinberger, S.R., Recent advancements in surface-enriched laser desorption ionization time-of-flight-mass spectrometry, *Electrophoresis*, 21, 1164, 2000.

63. Tang, N., Tornatore, P. and Weinberger, S.R., Current developments in SELDI affinity technology, *Mass Spec. Rev.*, 23, 34, 2004.

64. Uchida, T., Fukawa, A., Uchida, M., Fujito, K. and Saito, K., Application of a novel protein Biochip technology for detection and identification of rheumatoid arthritis biomarkers in synovial fluid, *J. Proteome Res.*, 1, 495, 2002.

65. zur Weish, A.S., Foell, D., Frosch, M. et al., Myeloid related proteins MRP8/MRP14 may predict disease flares in juvenile idiopathic arthritis, *Clin. Exptl. Rheum.*, 22, 368, 2004.

66. Brun, J.G., Haga, H.J., Bøe, E. et al., Calprotectiin in patients with rheumatoid arthritis: Relation to clinical and laboratory variable of disease activity, *J. Rheum.*, 19, 859, 1992.

67. Hammer, H.B., Kvien, T.K., Glennås, A. and Melby, K., A longitudinal study of calprotectin as an inflammatory marker in patients with reactive arthritis, *Clin. Exptl. Rheum.*, 13, 59, 1995.

68. Yui, S., Nakatoni, Y. and Mikami, M., Calprotectin (S100A8/S100A9), an inflammatory protein complex from neutrophils with a broad apoptosis-inducing activity, *Biol. Pharm. Bull.*, 26, 754, 2003.

69. Passey, R.J., Xu, K., Hume, D.A. and Gecsy, C.L., S100A8; emerging functions and regulation, *J. Leuk. Biol.*, 66, 549, 1999.

70. Fagerhol, M.K., Dale, I. and Andersson, T., Release and quantitation of a leukocyte derived protein L1, *Scand. J. Haematol.*, 24, 393, 1980.

71. Vermeine, S., Van Assche, G. and Rutgeerts, P., C-Reactive protein as a marker for inflammatory bowel disease, *Inflamm. Bowel Dis.*, 10, 661, 2004.

72. Bruzzese, E., Rais, V., Gaudiello, G. et al., Intestinal inflammation in a frequent feature of cystic fibrosis and is reduced by probiotic administration, *Aliment. Pharm. Ther.*, 20, 813, 2004.

73. Gravatt, M.G., Novy, M.J., Rosenfeld, R.G. et al., Diagnosis of intra-amniotic infection by proteomic profilng and identification of novel biomarkers, *JAMA*, 292, 462, 2004.

74. Yates, J.R., III, Carmock, E., Hays, L. et al., Automated protein identification using microcolumn liquid chromatography-tandem mass spectrometry, *Meth. Mol. Biol.*, 112, 553, 1999.

75. Melle, C., Ernst, G., Schimmel, B. et al., Biomarker discovery and identification in laser microdissected head and neck squamous cell carcinoma with ProteinChip® technology, two-dimensional electrophoresis, tandem mass spectrometry, and immunohistochemistry, *Mol. Cell. Proteomics*, 2, 443, 2003.

76. Schafer, T., Sachse, G.E. and Gassen, H.G., The calcium-binding protein MRP-8 is produced by human pulmonary epithelial cells, *Biol. Chem. Hoppe Seyler*, 372, 1, 1991.

77. Stulik, J., Osterreicher, J., Koupilova, K. et al., The analysis of S100A9 and S100A8 expression in matched sets of macroscopically normal colon mucosa and colorrectal carcinoma: The S100A9 and S100A8 positive cells underlie and invade tumor mass, *Electrophoresis*, 20, 1047, 1999.

78. Cho, W.C., Yip, T.T., Yip, C. et al., Identification of serum amyloid a protein as a potentially useful biomarker to monitor relapse of nasopharyngeal cancer by serum proteomic profiling, *Clin. Cancer Res.*, 10, 43, 2004.

79. Wu, W., Tang, X., Hu, W. et al., Identification and validation of metastasis-associated proteins in head and neck cancer lines by two-dimensional electrophoresis and mass spectrometry, *Clin. Exptl. Metastasis*, 19, 319, 2002.

80. Chen, Y., Shi, G., Xia, W. et al., Identification of hypoxia-regulated proteins in head and neck cancer by proteomics and tissue array profiling, *Cancer Res.*, 64, 7302, 2004.

81. Docetaxel induced gene expression patterns in head and neck squamous cell carcinoma using cDNA microarray and PowerBlot, *Clin. Cancer Res.*, 8, 3910, 2002.

82. Kim, H.J. and Lotan, R., Identification of retinoid-modulated proteins in squamous cells using high-throughput immunoblotting, *Cancer Res.*, 64, 2439, 2004.

83. Bast, R.C., Jr., Klug, T.L., St. John, E. et al., A radioimmunoassay using a monoclonal antibody to monitor the course of epithelial ovary cancer, *N. Eng. J. Med.*, 309, 883, 1983.

84. Cannistra, S.A., Cancer of the ovary, *N. Eng. J. Med.*, 351, 2519, 2004.

85. Gronlund, B., Hogdall, C., Hilden, J., Engelholm, S.A., Høgdall, E.V.S. and Hansen, H.H., Should CA-125 response criteria be preferred to response criteria in solid tumors (RECIST) for prognostication during second-line chemotherapy of ovarian carcinoma, *J. Clin. Oncol.*, 22, 4051, 2004.

86. Droegemueller, W., Screening for ovarian carcinoma: Hopeful and wistful thinking, *Am. J. Obstet. Gynecol.*, 170, 1095, 1994.

87. Jacobs, I.J. and Menon, U., Progress and challenges in screening for early detection of ovarian cancer, *Mol. Cell. Proteomics*, 3, 355, 2004.

88. Hensley, M.L., Alektior, K.M. and Chi, D.S., Ovarian and fallopian-tube cancer in gynecologic oncology, 2nd Edition, R.P. Borakat, M.W. Bevers, D.M. Gershenson and W.J. Hoskin, Eds., Martin Dunitz Ltd., London, United Kingdom, Ch. 14, 249–269, 2000.

89. Sanchez-Carbayo, M., Recent advances in bladder cancer diagnosis, *Clin. Biochem.*, 37, 562, 2004.

90. Mobley, J.A., Lam, Y.W., Lau, K.M. et al., Monitoring the serological proteome: The latest modality in prostate cancer detection, *J. Urology*, 172, 331, 2004.

91. Skates, S.J., Menon, U., McDonald, N. et al., Calculation of the risk of ovarian cancer from serial CA-125 values for preclinical detection in postmenopausal women, *J. Clin. Oncol.*, Suppl. 21, 206S, 2003.

92. Jett, J.R. and Midthun, D.E., Screening for lung cancer: current status and future direction, Thomas A. Neff Lecture, *Chest*, Suppl. 125, 158S, 2004.

93. Bast, R.C., Jr., Feeney, M., Lazarus, H. et al., Reactivity of a monoclonal antibody with human ovarian carcinoma, *J. Clin. Invest.*, 68, 1331, 1981.

94. Kabaway, S.E., Bast, R.C., Bhan, A.K. et al., Tissue distribution of a coelomic epithelium-related antigen recognized by the monoclonal antibody OC125, *Lab. Invest.*, 48, 42A, 1983.

95. Nilhoff, J.M., Klug, T.L., Schaetzl, E. et al., Elevation of serum CA125 in carcinomas of the fallopian tube, endometrium, and endocervix, *Am. J. Obstet. Gynecol.*, 148, 1057, 1984.

96. Canney, P.A., Moore, M., Wilkinson, P.M. and Janes, R.D., Ovarian cancer antigen CA125: A prospective clinical assessment of its role as a tumor marker, *Brit. J. Cancer*, 50, 765, 1984.

97. Matthew, B., Bhatia, V., Mahy, I.R., Ahmed, I. and Francis, L., Elevation of the tumor marker CA125 in right heart failure, *South. Med. J.*, 97, 1013, 2004.

98. Phopong, V., Chen, O. and Ultchaswadi, P., High level of CA125 due to large endometrioma, *J. Med. Assoc. Thai.*, 87, 1108, 2004.

99. Krediet, R.T., Dialyzate cancer antigen 125 concentration as marker of periotoneal membrane status in patients treated with chronic peritoneal dialysis, *Periotoneal Dialysis Intl.*, 21, 560, 2001.

100. DiBaise, J.K. and Donovan, J.P., Markedly elevated CA125 in hepatic cirrhosis: Two case illustrations and review of the literature, *J. Clin. Gastro.*, 28, 159, 1999.

101. Xiao, W.B. and Liu, Y.L., Elevation of serum and ascites cancer antigen 125 levels in patients with liver cirrhosis, *J. Gastro. Hepatol.*, 18, 1315, 2003.

102. Kim, Y.S., Kim, D.Y. and Ryu, K.H., Clinical significance of serum CA 125 in patients with chronic liver diseases, *Korean J. Gastro.*, 42, 409, 2003.

103. *Dorland's Illustrated Medical Dictionary*, 30th Edition, Saunders (Elsevier), Philadelphia, PA, 2003.

104. Anderson, N.L., Polenski, M., Pieper, R. et al., The human proteome. A nonredundant list developed by combination of four separate sources, *Mol. Cell. Proteomics*, 3, 311, 2004.

105. Crescenzi, G., Cedrati, V., Landoni, G. et al., Cardiac biomarker release after CABG with different surgical techniques, *J. Cardio. Vasc. Anesth.*, 18, 34, 2004.

106. Diamandis, E.P., Mass spectrometry as a diagnostic and cancer biomarker discovery tool. Opportunities and potential limitations, *Mol. Cell. Proteomics*, 3, 367, 2004.

107. Downing, G.J., et al., Biomarkers Definition Group, Biomarkers and surrogate endpoints: Preferred definitions and conceptual framework, *Clin. Pharm. Ther.*, 69, 89, 2000.

108. O'Farrell, P.H., High resolution two-dimensional electrophoresis of proteins, *J. Biol. Chem.*, 250, 4007, 1975.

109. Görg, A., Obermaier, C., Boguth, G., Harder, A., Scheibe, B., Wildgruber, R. and Weiss, W., The current status of two-dimensional electrophoresis with immobilized pH gradients, *Electrophoresis*, 21, 1037, 2000.

110. Molloy, M.P., Brzezinski, E.E., Hang, J. et al., Overcoming technical variation and biological variation in quantitative proteomics, *Proteomics*, 3, 1912, 2003.

111. Choe, L.H. and Lee, K.H., Quantitative and qualitative measure of intralaboratory two-dimensional protein gel reproducibility and the effects of sample preparation, sample load, and image analysis, *Electrophoresis*, 24, 3500, 2003.

112. Doherty, N.S., Littman, B.H., Reilly, K. et al., Analysis of changes in acute-phase plasma proteins in an acute-phase plasma proteins in an acute inflammatory response and in rheumatoid arthritis using two-dimensional gel electrophoresis, *Electrophoresis*, 20, 355, 1998.

113. Krah, A., Schmidt, F., Becher, D. et al., Analysis of automatically generated peptide mass fingerprints of cellular proteins and antigens form *Helicobacter pylori* 26695 separated by two-dimensional electrophoresis, *Mol. Cell. Proteomics*, 2, 1271, 2003.

114. White, I.R., Pickford, R., Wood, J. et al., Statistical comparison of silver and SYPRO Ruby staining for proteomic analysis, *Electrophoresis*, 25, 3048, 2004.

115. Tonge, R., Shaw, J., Middleton, B. et al., Validation and development of fluorescence two-dimensional differential gel electrophoresis proteomics, *Proteomics*, 1, 377, 2001.

116. Wolters, D.A., Washburn, M.P. and Yates, J.R., III, Automated multidimensional protein identification technology for shotgun proteomics, *Anal. Chem.*, 73, 5683, 2001.

117. Fujii, K., Nakana, T., Kawamura, T. et al., Multidimensional protein profiling technology and its application to human plasma proteome, *J. Proteome Res.*, 3, 712, 2004.

118. Hutchins, T.W. and Yip, T.-T., New desorption strategies for the mass spectrometric analysis of macromolecules, *Rapid Comm. Mass Spec.*, 7, 576, 1993.

119. Merchant, M. and Weinberg, S.R., Recent advancements in surface-enhanced laser desorption/ionization time-of-flight mass spectrometry, *Electrophoresis*, 21, 1164, 2000.

120. Tang, N., Tornatore, P. and Weinberger, S.R., Current developments in SELDI affinity technology, *Mass Spec. Rev.*, 23, 34, 2004.

8 Validation Issues in Proteomics and the Transition from Research to Diagnostic Development

CONTENTS

INTRODUCTION

The bias of this book is toward the development of diagnostics using proteomic technology. The development of new diagnostics will be driven by the identification of unique biomarkers that can be used as an analyte (biomarker) and can be demonstrated to be useful for the management of a disease.[1-3] Recently, Loyda[1] emphasized the necessity of a new diagnostic impacting the treatment. In an earlier paper, Sandberg and Thue[2] discussed issues impinging on the development strategy for a diagnostic product emphasizing that this process must be driven by physician need; clinical studies need to demonstrate what changes in concentration of a specific analyte result in a change in patient treatment. Powers and Greenberg[3] emphasized the need for analytical quality specifications in the development of diagnostic products; here again, the message is that the development of a diagnostic product must be driven by medical need. Kenny and others[4] have published a consensus statement from a conference on *Strategies to Set Global Quality Specifications in Laboratory Medicine*. Several recommendations emerged that must be considered in the development of a diagnostic including "evaluation of the effect of analytical performance on clinical decisions in general" and the "evaluation of the effect of analytical performance on clinical outcomes in specific clinical settings."

Major practical considerations exist that might be considered in the transition from basic proteomic research to a commercial product. When the goal is a useful commercial product as opposed to a "boutique" assay, the investigators should have a clear idea as to what the final product is going to measure and how the measurement will influence treatment. An article by Schwartz in 1995[5] reviews the status of cancer tumor markers and is an excellent article that makes several points just as valid today as they were a decade ago. First, multiple markers have increased statistical specificity and sensitivity. Second, all components of quality assurance must be considered including a preanalytical phase, which includes the process from the ordering of the test to the preparation of the sample, the analytical phase and the post-analytical phase, which includes the use of the data in patient management.[6] The next step from a business perspective is to establish the site of assay performance; a clinical laboratory associated with a hospital, a free-standing clinic or physician's office or a point-of-care (POC), or all of these. This latter consideration is, in part, driven by the technology required. Basic proteomic technologies utilize mass spectrometry as the basic analytical instrumentation. While mass spectrometry is gaining value as an analytical technique in the clinical laboratory,[7,8] it is not clear as to whether such technology (1) is suitable for proteomics, and (2) can be realistically pursued outside of a large institution. As will be mentioned several times in this discussion, proteomic technology will be of value in the identification of biomarkers but the laboratory assays will be based on immunoassays. Issues also exist of identification of biomarkers used in assay development for susceptibility, disease diagnosis and treatment prognosis. The issues in the commercial development of proteomic technologies for diagnosis are not dissimilar for molecular diagnostics, defined as the use of nucleic acid technologies for diagnostics.[9,10]

Proteomic technology may have the most value in the identification of biomarkers that can be used for the early diagnosis of diseases such as ovarian cancer and bladder cancer.[11–14] The importance of publishing the results in the peer-reviewed literature with rigorous review cannot be overemphasized; documentation is found suggesting that information appearing in peer-reviewed literature is of great value in obtaining third-party reimbursement[15,16] and resulting use in general patient care. Zarkowsky[16] discusses the importance of having well-executed clinical studies that will clearly demonstrate a substantial improvement in health care. A new diagnostic product will only be successful if there is reimbursement for the use of the product and it is not unreasonable to suggest that such reimbursement will only be forthcoming if the product is useful.

ISSUES AFFECTING ASSAY VARIABILITY

In what may appear to be a diversion, we will spend some time on issues influencing assay variability, as such issues are not only critical to the final diagnostic product but deserve attention during the development phase to truly assess the significance of the observations. The transition from a basic science observation to a clinical assay with multicenter use is challenging.[17,18] Markó-Varga and Fehniger[18] make several interesting points in their review. The first, discussed below in somewhat greater detail and in other chapters when necessary, is differentiation between health (normal) and disease. The second point, likely far important for the development of therapeutics than for diagnostics, is degeneracy[19] and its effect on the uniqueness of biomarkers. Edelman and Gally[19] review degeneracy in biological systems and its importance in the maintenance of homeostasis. The existence of multiple systems, such as multiple codons for protein synthesis or multiplicity of structurally distinct antibodies that bind to the same antigen, is useful in a more complex organism. One example cited is the existence of a population deficient in albumin;[20] albumin is generally considered absolutely critical for the maintenance of intravascular fluid dynamics.[21,22] Edelman and Gally also reference the problem of "knockout" mice that appear phenotypically normal, again suggesting multiplicity of pathways for some homeostatic functions. It is unlikely that degeneracy would influence the identification of unique biomarker; however, situations exist were a single biomarker is elevated in a variety of disease states. An example is provided by calprotectin as discussed in Chapter 7 (Table 2). Degeneracy would likely influence protein profiling studies.[23,24] However, the effect on the use of profiling studies could be positive in that degeneracy might result in multiple markers for a single event, thus requiring the use of a panel rather than a single analyte.

INDEX OF VARIABILITY

The success of a clinical assay depends on the ability to distinguish normal from abnormal and to reliably use an analyte as a surrogate marker for the abnormal situation.[25,26] A surrogate marker can be defined as an analyte (or biomarker) that substitutes for a clinical endpoint. One of the better examples is the use of the partial thromboplastin time to measure blood coagulation in hemophilia, where the correction of a prolonged clotting time is associated with a decrease in bleeding. The accuracy of a clinical assay depends on the variability in assay that can be separated into analytical variability and individual variability; individual variability can be further separated into intraindividual variability and interindividual variability.[27–29] Other work distinguishes physiologic variation from analytic variation,[30] resulting in the development of the "index of individuality."[31] The index of variability has proved to be a useful tool in evaluating the utility of reference values for a normal population.[32] The index of individuality is derived from analytical variability, intraindividual variability and interindividual variability. Analytical variability is the variability introduced by sample collection/preparation (preanalytical variability) and variability in the analytical procedure (analytical variability). Intraindividual variability is the random fluctuation of a laboratory value around a homeostatic set point which can be influenced by a variety of environmental factors.[33–36] While not specifically considered by Baggerly and coworkers[24] in their analysis of the data used for the profiling of serum proteins in ovarian cancer,[37] the intraindividual variability should be considered together with analytical variability. This latter point is raised by Diamandis and others[38,39] in their criticism of the ovarian cancer SELDI data.

In 1956, R. Williams[40] provided evidence for substantial differences in analytes between healthy individuals. In 1998, Williams refined this concept,[41] noting that "practically every human being is a deviate in some way." G. Williams extended this concept in 1970[42] by observing the individuality of a clinical biochemical pattern and its importance in preventive medicine; "For every clinical evident degenerative state there must be a long chain of related chemical, physiological, and biological events...which constitute individual changes that can be measured and documented by modern methods." Williams further observed that time is important and that for a chronic disease, "repeated studies on different adult population groups over long periods of time will be necessary." A limited number of studies are found that review this type of information. Ricós and colleagues[33] have reviewed a number of analytes in an extremely thorough study. Some intraindividual variations (CV_w) are similar to interindividual variation (CV_b) (for albumin, $CV_w = 3.1$; $CV_b = 4.1$), while other intraindividual variations are much smaller than interindividual variations (for CA125, $CV_w = 13.6$; $CV_b = 46.5$). In studies on cholesterol, the lowest values for CV_w were obtained for multiple samples taken during a single day; longer study periods had higher values for CV_w. Harris and coworkers[28] observed that a long-term study of serum analytes was assisted by the concomitant assay of a constant "pooled" serum. The results of these various studies suggests that the assay of a given analyte is most useful when the combination of the personal and analytical variability is substantially less that the interindividual variability (index of individuality; $(CV_A + CV_I)^{1/2} / CV_G$, where CV_A is the analytical coefficient of variation, CV_I is the within-subject coefficient of variation and CV_G is the between-subject coefficient of variation).[31,32,43,44] The complete understanding of this concept requires knowledge of the concept of the reference interval.[45,46] Horn and coworkers[46] proposed a robust approach to reference interval estimation and evaluation. These investigators state that "the concept of a reference interval in medicine is based on determining a set of values within which some percentage, 95% for example, of the values of a particular analyte in a health population would fall." The following statements are derived from the analysis presented by Petersen and coworkers[47] following the earlier concepts of Harris:[31]

- If the index of individuality is less than 0.6, there is some benefit to using a subject-specific reference interval as when the index of individuality is low, the concentration of a signal biomarker would have to increase several times to exceed the upper reference interval.
- If the index of individuality is greater that 1.4, the reference interval can be used to identify individual deviation.

The reader is directed to several recent studies that have examined the use of the index of individuality for the validation of clinical laboratory assays.[48–51]

Somewhat remarkable is that the above studies are not cited or considered to a greater extent in the burgeoning literature on clinical proteomics. An exception is the review by Anderson and Anderson on plasma proteomics[52] where these workers do suggest caution in the interpretation of biomarker identification and use. Just because a biomarker is discovered with proteomic technology, it is not of higher value than a biomarker discovered with conventional technology and certainly does not imply that the above considerations would not hold for the development of a diagnostic based on a biomarker discovered with proteomic technology. Seriously considering a concept such as the index of individuality in clinical proteomics may not be necessary but it would not be unreasonable to consider some longitudinal studies to explore the possibility of analytical variability and intraindividual variability.

PERSONALIZED MEDICINE

A somewhat longer term goal is the use of proteomics in the development of personalized medicine.[53–58] This is not a new concept as G. Williams[59] stressed the importance of the individuality of biochemical patterns in preventive medicine. Thus, while the sophistication of the technical approach has markedly increased, the basic concept has not changed. Thus, the concept of personalized medicine is at least 40 years old but it would seem that fruition is still in the distance,[58] as Nebert and coworkers suggest that clinical test based on proteomics will not be available for the next five to ten years. This is probably overly pessimistic and success is possible in an earlier time frame; it will, however, require more focus than has been exhibited by most studies. Implicit is the use of diagnostics/prognostics for personalized medicine or theranostics.[60] The effectiveness of personalized medicine will definitely require a thorough understanding of intrasubject variance as described above and less of an understanding of intersubject variance and the reference interval. The utility of personalized medicine will also require a rigorous cost-analysis to justify third-party reimbursement.[15,16]

ASSAY VALIDATION

The key to the use of biomarkers in the development of effective diagnostics/prognostics will depend on the "validation" of their use. Validation is defined as the process of establishing something as valid, with valid being defined as being both relevant and meaningful.[61] Within the context of the validation of biomarkers, the first stage of validation is establishing the biomarker and being at once relevant and meaningful with respect to a specific pathology. The second stage would be the validation of a diagnostic to accurately measure the analyte. The third stage would be the validation of the assay in the measurement of the analyte in clinical samples that address the issue of the index of individuality as described above. This process is shown Table 8.1.

Sensitivity and specificity are issues in the validation of proteomic assays. A difference exists in the definition of specificity and sensitivity between the statistician and the analytical chemist. Validation of biomarkers in the first stage as described above is based on two considerations: sensitivity and specificity. Sensitivity for a diagnostic test can be defined as the proportion of positive tests as part of the total number of positive cases. Specificity can be defined as the proportion of negative tests to the total number of negative individuals.[62] Predictive probabilities

---

TABLE 8.1
Steps in the Identification and Validation of a Biomarker

Identification of Biomarker: This would involve a limited study that would identify a biomarker in a
small number of patients. There should be very careful consideration of the concept of a surrogate
biomarker/marker at this point. The impact of the assay on patient treatment should be clear. Unless
this issue is answered, the assay will provide information about the pathology but would not be of
commercial value.

Validation of Biomarker: This would be a somewhat larger study that would determine the specificity
of the biomarker with a specific pathology. An example is provided by troponin and myocardial
infarction (Koerbin, G., Tate, J.R., Potter, J.M. and Hickman, PE, The comparative analytical
performance of four troponin I assays at low concentration, *Ann. Clin. Biochem.*, 42, 19–23, 2005).
At this point, there should be an understanding as to whether the assay is diagnostic or prognostic.

Development of the Assay: This would be the process of assay development that would involve the
selection of the method, instrument and endpoint.

Validation of the Assay: This would be the validation of the assay in the clinical surroundings. This
would, for example, involve matrix effects. The consideration in validation is presented in Table 2.

Commercialization of the Assay: This is a complex process involving technical and business decisions.
Paramount here will be the clear demonstration that the assay will have value in patient treatment.

---

can be derived from specificity and sensitivity information. In the aforementioned
review,[62] Sharples presents a short and remarkably intelligible discussion of the
application of statistics to the selection and validation of a biomarker. Regulatory
agencies and laboratory accrediting groups have a different definition or understanding of the terms specificity and sensitivity.[5] In laboratory analysis, specificity refers
to the ability of an assay to identify a specific analyte within an assay matrix such
as blood plasma or urine. Sensitivity is described as the lowest level of an analyte
that can be accurately measured (lower assay limit) as opposed to the limit of
detection, which can only determine if the analyte is present. These definitions of
sensitivity and specificity are critical considerations for the validation of the assay.

BIOMARKERS IN TOXICITY AND ENVIRONMENTAL HEALTH

The above comments are concerned with the development of biomarkers for use in
the diagnosis, prognosis and monitoring the therapeutic course of a specific disease.
However, biomarkers are not the province of a specific discipline such as oncology
or a technology such as proteomics. The study of biomarkers has been a long-time
interesting the fields of toxicology and environmental health.[63,64] The major driving
force in the use of biomarker development in environmental health has been risk
assessment. The definition of a biomarker is similar to that described above as any
material, structure or process that can be measured in an organism and that influences
or predicts the outcome of a disease.[65] In the field of environmental health, biomarkers may be classified as indices to exposure, susceptibility or effect. Validation
is concerned with the study of issues such as the pharmacokinetics of the biomarker,
temporal relevance and background variability in the study population in addition

to the issue of the validation of the laboratory assay.[66] Another definition of bio-marker within the field of environmental health suggests division into the observable properties of an organism that can be used to (1) identify the organism, (2) record the exposure of an organism to an external stimuli that would include toxic chemi-cals, (3) provide surrogate markers associated with a disease process, or (4) define the susceptibility of an organism.[67,68] Issues related to the validation of a biomarker in environmental health have been discussed by Henderson.[69] A biomarker of expo-sure is not necessarily a biomarker for the disease process.[70] However, biomarkers of exposure exist that are indeed biomarkers for the disease process, demonstrated by the mutation of p53 secondary to vinyl chloride exposure and p53 mutations in angiosarcomas.[71]

The study of toxicity biomarkers with the use of expression profiling using proteomic analysis has been discussed by Steiner and Anderson.[72] The ability to analyze changes in the expression patterns of relevant cells, tissues and organisms exposed to toxic chemicals shall enable the definition of exposure markers and perhaps also early biomarkers of disease. Also noted is that proteomic technology will be of great value in the preclinical study of new drugs where more information can be obtained with the study of expression profiling in tissue culture and cell culture. The use of surrogate tissue analysis[73-76] will fit nicely into the use of expression profiling to study toxicity where peripheral blood leukocytes (peripheral blood mononuclear cells) can be used as a surrogate tissue for endovascular chlamy-dial infection,[73] oxidative DNA damage[74] and hepatocellular carcinoma.[75] The work by Rockett and coworkers[76] reviews the use of surrogate tissues with emphasis on peripheral blood leukocytes/peripheral blood mononuclear cells. The above discus-sion leads to several conclusions:

- Surrogate markers and surrogate tissues can be useful for the diagnosis and prognosis of disease.
- Validation of a biomarker for a disease state will be a process complicated by choosing a biomarker to identify.
- Use of the proteomic technologies (SELDI, two-dimensional gel electro-phoresis, MuDPIT) used for the identification of the biomarker is unlikely for the validation of the biomarker as there exists, with current technology, a lack of reproducibility/repeatability (Table 8.2) of mass spectrometry. The second issue concerns the necessity to validate the sample preparation procedure. An immunoassay will likely be developed based on biomarker identification that can then be used for a multi-institutional validation.
- Investigators should carefully evaluate prior studies, lest they miss impor-tant observations; just because a biomarker was not discovered by proteomic technology does not invalidate the work.
- The question exists on whether an unvalidated assay can be used to validate a diagnostic target. Granted that this can become a circular argu-ment, it is worth thinking about as part of the overall regulatory process.
- Experimental details such as temperature, time and solvents need to be rigorously defined in the text of manuscripts. Methods for the determina-tion of protein concentration need to be described other than by vendor

source; the choice of standard should be defined and defended. Only with the rigorous description of experimental conditions will it be possible to reproduce data between various laboratories. While assay validation between various laboratories may not be possible at this time, target validation between several laboratories is critical.

DESIGN CONTROL

Design control is a quality concept accepted on a global basis in a manner similar to the ISO (International Organization for Standardization) 9000 series. The ISO 9000 series represents international quality standards for product design, manufacturing and distribution. Design control is an interrelated set of practices and procedures that are incorporated into the design and development process.[77] Design control is also described as an iterative process where product development is driven by user needs. The application of design control to the development of diagnostic products has been reviewed by Powers and Greenberg.[3] In their model, the product requirements are defined by the "customer," most likely the clinical laboratory, whose requirements in turn are driven by physician need. The physician need is driven by the requirement to improve patient treatment either by improved diagnosis or improved prognosis. While the same analyte might serve for both purposes, that is not always the situation. For example, CA125 as an isolated marker is not particularly good for the early diagnosis of ovarian cancer but it is a good marker for following treatment. Thus, it would be useful if the objective is established prior to initiation of the program. This, of course, is a bit of an issue since proteomics is discovery-driven rather than hypothesis-driven. The early work with proteomic technology would be expected to provide support for statistical specificity and sensitivity, while analytical specificity and sensitivity would be addressed during the development of the clinical laboratory assay. When a potentially useful analyte is identified, an immunoassay will be developed such as done by Glocker, Drinz and coworkers[78,79] after identification/validation of the biomarker, calprotectin (leukocyte L protein, MRP8/MRP/14, S100A8/S100A14).

The process of assay validation will not be discussed at this time as such discussion is not of particular value. Some detail is presented in Table 8.2, and investigators are encouraged to consider these issues as they become involved in the process of identifying biomarkers.

CONCLUSIONS

Proteomic technology will result in the identification of biomarkers that will be useful either in diagnosis or prognosis. Most current work is discovery-driven rather than hypothesis-driven and only selected portions of a proteome are subjected to analysis. Most studies would clearly benefit from a consideration of basic issues in clinical laboratory medicine such as the index of individuality. Differentiation must be made between target validation and assay validation.

TABLE 8.2
Issues in the Validation of an Assay[a]

Selectivity — The ability of an assay to differentiate and quantify the analyte in the presence of other components in the sample.

Accuracy — Describes the closeness of mean test results obtained by the method to the true value (concentration) of the analyte.
  • Detection Limit
  • Quantitation Limit

Precision — Describes the closeness of individual measures of an analyte when the procedure is applied repeatedly to multiple aliquots of a single homogeneous volume of biological matrix.
  • *Repeatibility* — intraanalyst, intrainstrument
  • *Reproducibility* — intra- and interlaboratory

Recovery — The detector response obtained from an amount of the analyte added to and extracted from the biological matrix, compared to the detector response obtained for the true concentration of the pure analytical standard. Also referred to as "spike recovery."

Calibration/Standard Curve — Relationship between instrument response and known concentration of the analyte.

Stability — Stability of the analyte in the biological matrix.
  • Robustness

[a] This material is taken from Guidance for Industry. Bioanalytical Method Validation, http://www.fda.gov/cder/guidance/4252fnl.htm. The reader is also recommended to an ICH Harmonised Tripartite Guideline, Q2B Validation of Analytical Procedures: Methodology, November 1996, http://www.ich.org.

REFERENCES

1. Loyda, H.-J., Putting new assays to the test, *IVD Technology*, 38–42, March 2004.
2. Sandberg, S. and Thue, G., Quality specifications derived from objective analyses based upon clinical needs, *Scand. J. Clin. Lab. Invest.*, 59, 531, 1999.
3. Powers, D.M. and Greenberg, N., Development and use of analytical quality specifications in the *in vitro* diagnostics medical device industry, *Scand. J. Clin. Lab. Invest.*, 59, 539, 1999.
4. Kenny, D., Fraser, C.G., Petersen, P.H. and Kallner, A., Consensus agreement, *Scand. J. Clin. Lab. Invest.*, 59, 585, 1999.
5. Schwartz, M.K., Current status of tumour markers, *Scand. J. Clin. Lab. Invest.*, Suppl. 221, 5, 1995.
6. Birkmayer, G.J.D., Schwartz, M.K. and Klavins, J.V., Quality assurance of tumour marker assays, *J. Tumor Marker Oncol.*, 5, 195, 1990.
7. Hounor, J.W., Benchtop mass spectrometry in clinical biochemistry, *Ann. Clin. Biochem.*, 40, 628, 2003.
8. Dooley, K.C., Tandem mass spectrometry in the clinical chemistry laboratory, *Clin. Biochem.*, 36, 471, 2003.
9. Kiechle, F.L. and Holland-Staley, C.A., Genomics, transcriptomics, proteomics, and numbers, *Arch. Path. Lab. Med.*, 127, 1089, 2003.

10. Fortina, P., Surrey, S. and Kricka, L.J., Molecular diagnostics: Hurdles for clinical implementation, *Trends Mol. Med.*, 8, 264, 2005.

11. Hu, W., Wu, W., Kobayashi, R. et al., Proteomics in cancer screening and management in gynecologic cancer, *Curr. Oncol. Rep.*, 6, 456, 2004.

12. Posadas, E.M., Davidson, B. and Kohn, E.C., Proteomics and ovarian cancer: Implications for diagnosis and treatment: A critical review of the recent literature, *Curr. Opin. Oncol.*, 16, 478, 2004.

13. Umezo-Goto, M., Tanyi, J., Lahad, J. et al., Lysophosphatidic acid production and action: Validated targets in cancer? *J. Cell. Biochem.*, 92, 1115, 2004.

14. Breedlove, G. and Busenhart, C., Screening and detection of ovarian cancer, *J. Midwifery Women's Health*, 50, 51, 2005.

15. Luce, B.R. and Brown, R.E., The use of technology assessment by hospitals, health maintenance organizations, and third-party payers in the United States, *Intl. J. Tech. Assess. Health Care*, 11, 79, 1995.

16. Zarkowsky, H., Managed care organizations' assessment of reimbursement for new technology, procedures, and drugs, *Arch. Path. Lab. Med.*, 123, 677, 1999.

17. Banks, R. and Selby, P., Clinical proteomics — Insights into pathologies and benefits for patients, *Lancet*, 362, 415, 2003.

18. Markó-Varga, G. and Fehniger, T.E., Proteomics and disease — The challenges for technology and discovery, *J. Proteome Res.*, 3, 179, 2004.

19. Edelman, G.M. and Gally, J.A., Degeneracy and complexity in biological systems, *Proc. Natl. Acad. Sci. USA*, 98, 13763, 2001.

20. Buehler, B.A., Hereditary disorders of albumin synthesis, *Ann. Clin. Lab. Sci.*, 8, 283, 1978.

21. Heyl, J.T. and Janeway, C.A., The use of human albumin in military medicine, *US Navy Med. Bull.*, 40, 785, 1942.

22. Brugnara, C. and Churchill, W.H., Plasma and component therapy, in *Thrombosis and Hemorrhage*, 2nd Edition, J. Loscalzo and A.I. Shafer, Eds., Williams & Wilkins, Baltimore, MD, Ch. 2, 1135–1162, 1998.

23. Yasui, Y., Pepe, M., Thompson, M.L. et al., A data-analytic strategy for protein biomarker discovery: Profiling of high-dimensional proteomic data for cancer detection, *Biostatistics*, 4, 449, 2003.

24. Baggerly, K.A., Morris, J.S. and Coombes, K.R., Reproducibility of SELDI-TOF protein patterns in serum: Comparing datasets from different experiments, *Bioinformatics*, 20, 777, 2004.

25. Colburn, W.A., Selecting and validating biologic markers for drug development, *J. Clin. Pharm.*, 37, 355, 1997.

26. Downing, G.J. et al., Biomarkers and surrogate endpoints: Preferred definitions and conceptual framework, *Clin. Pharm. Ther.*, 69, 89, 2001.

27. Williams, G.Z., Young, D.S., Stein, M.R. and Cotlove, E., Biological and analytic components of variation in long-term studies of serum constituents in normal subjects. I. Objectives, subject selection, laboratory procedures, and estimation of analytic deviation, *Clin. Chem.*, 16, 1016, 1970.

28. Harris, E.K., Kanofsky, P., Shakarji, G. and Cotlove, E., Biological and analytic components of variation in long-term studies of serum constituents in normal subjects. II. Estimating biological components of variation, *Clin. Chem.*, 16, 1022, 1970.

29. Cotlove, E., Harris, E.K. and Williams, G.Z., Biological and analytic components of variation in long-term studies of serum constituents in normal subjects, III. Physiological and medical implications, *Clin. Chem.*, 16, 1028, 1970.

30. Harris, E.K., Distinguishing physiologic variation from analytic variation, *J. Chron. Dis.*, 23, 469, 1970.
31. Harris, E.K., Effects of intra- and inter-individual variation on the appropriate use of normal ranges, *Clin. Chem.*, 20, 1535, 1974.
32. Petersen, P.H., Fraser, C.G., Sandberg, S. and Goldschmidt, H., The index of individuality is often misinterpreted quantity characteristic, *Clin. Chem. Lab. Med.*, 37, 655, 1999.
33. Ricós, C., Alvarez, V., Cava, F. et al., Current databases on biological variation: Pros, cons and progress, *Scand. J. Clin. Lab. Invest.*, 59, 591, 1999.
34. Fuentes-Arderiu, X., Variability of the biological variation, *Scand. J. Clin. Lab. Invest.*, 62, 561, 2002.
35. Petersen, P.H., Fraser, C.G., Jørgensen, L. et al., Combination of analytical quality specifications based on biological within- and between-subject variation, *Ann. Clin. Biochem.*, 39, 543, 2002.
36. Bolann, B.J. and Åsberg, A., Analytical quality goals derived from the total deviation from patients' homeostatic set points, with a margin for analytical errors, *Scand. J. Clin. Lab. Invest.*, 64, 443, 2004.
37. Petricoin, E.F, III, Ardekani, A.M., Hitt, B.A. et al., Use of proteomic patterns in serum to identify ovarian cancer, *Lancet*, 359, 572, 2002.
38. Diamandis, E.P., Analysis of serum proteomic patterns for early cancer diagnosis: Drawing attention to potential problems, *J. Natl. Cancer Inst.*, 96, 353, 2004.
39. Garber, K., Debate rages over proteomic patterns, *J. Natl. Cancer Inst.*, 96, 816, 2004.
40. Williams, R.J., *Biochemical Individuality, The Basis for the Genotrophic Concept*, University of Texas Press, Austin, TX, 1956.
41. Williams, R.J., Biochemical variation: Its significance in biology and medicine, in *Biochemical Individuality, The Basis for the Genotrophic Concept*, NTC Contemporary, New Canaan, CT, Ch. 1, 1998.
42. Williams, G.Z., Individuality of clinical biochemical patterns in preventive health maintenance, *J. Occup. Med.*, 9, 567, 1967.
43. Frasher, C.G. and Harris, E.K., Generation and application of data on biological variation in clinical chemistry, *Crit. Rev. Clin. Lab. Sci.*, 27, 409, 1989.
44. Petersen, P.H., Sandberg, S., Fraser, C.G. and Goldschmidt, H., Influence of index of individuality on false positives in repeated sampling from healthy individuals, *Clin. Chem. Lab. Med.*, 39, 160, 2001.
45. Stamm, D., A new concept of quality control of clinical laboratory investigations in the light of clinical requirements and based on reference method values, *J. Clin. Chem. Clin. Biochem.*, 20, 817, 1982.
46. Horn, P.S., Pesce, A.J. and Copeland, B.E., A robust approach to reference interval estimation and evaluation, *Clin. Chem.*, 44, 622, 1998.
47. Petersen, P.H., Fraser, C.G., Sandberg, S. and Goldschmidt, H., The index of individuality is often a misinterpreted quantity characteristic, *Clin. Chem. Lab. Med.*, 37, 655, 1999.
48. Tuxen, M.K., Sölétormos, G., Petersen, P.H., Schiøler, V. and Dombernowsky, P., Assessment of biological variation and analytical imprecision of CA 125, CEA, and TPA in relation to monitoring of ovarian cancer, *Gynecol. Oncol.*, 74, 12, 1999.
49. Plebani, M., Bernardi, D., Meneghetti, M.F., Ujka, F. and Zaninotto, M., Biological variability in assessing the clinical value of biochemical markers of bone turnover, *Clin. Chem. Acta*, 299, 77, 2000.
50. Møller, H.J., Petersen, P.H., Rejnmark, L. and Moestrup, S.K., Biological variation of soluble CD163, *Scand. J. Clin. Lab. Invest.*, 63, 15, 2003.

51. Chiu, S.K., Collier, C.P., Clark, A.F. and Wynn-Edwards, K.E., Salivary cortisol on ROCHE Elecsys immunoassay system: Pilot biological variation studies, *Clin. Biochem.*, 36, 211, 2003.

52. Anderson, N.L. and Anderson, N.G., The human plasma proteome. History, characteristics, and diagnostics products, *Mol. Cell. Proteomics*, 1, 845–867, 2002.

53. Califf, R.M., Defining the balance of risk and benefit in the era of genomics and proteomics, *Health Affairs*, 23, 77, 2004.

54. Weston, A.D. and Hood, L., Systems biology, proteomics, and the future of health care: Toward preventive and predictive medicine, *J. Proteome Res.*, 3, 179, 2004.

55. Jain, K.K., Role of pharmacoproteomics in the development of personalized medicine, *Pharmacogenomics*, 5, 331, 2004.

56. Utrecht, J.P., Is it possible to more accurately predict which drug candidates will cause an idiosyncratic drug reaction? *Curr. Drug. Metab.*, 1, 133, 2000.

57. Wilkins, M.R., What do we want from proteomics in the detection and avoidance of adverse drug reactions, *Tox. Letters*, 127, 245, 2002.

58. Nebert, D.W., Jorge-Nebert, L., and Vesell, E.S., Pharmacogenomics and "individualized drug therapy": high expectations and disappointing achievements, *Am. J. Pharm.*, 3, 361, 2003.

59. Williams, G.Z., Individuality of clinical biochemical patterns in preventive health maintenance, *J. Occup. Med.*, 9, 567, 1967.

60. Picard, F.J. and Bergeron, M.G., Rapid molecular theranostics in infectious diseases, *Drug Disc. Today*, 7, 1092, 2002.

61. *Webster's Ninth New Collegiate Dictionary*, Merriam-Webster, Springfield, MA, 1987.

62. Sharples, L.D., Statistical approaches to rational biomarker selection, in *Biomarkers in Disease: An Evidence-Based Approach*, A.K. Trull, L.M. Demers, D.W. Holt, A. Johnston, J.M. Tredger and C.P. Price, Eds., Cambridge University Press, Cambridge, United Kingdom, Ch. 3, 24–31, 2002.

63. *Biomarkers in Risk Assessment: Validity and Validation, Environmental Health Criteria 222.* International Programme on Chemical Safety, World Health Organization, Geneva, Switzerland, 2001.

64. *Biomarkers. Medical and Workplace Applications*, M.L. Mendelsohn, L.C. Mehr and J.P. Peekes, Eds., John Henry Press, Washington, DC, 1998.

65. *Biomarkers in Risk Assessment: Validity and Validation, Environmental Health Criteria 222.* International Programme on Chemical Safety, World Health Organization, Geneva, Switzerland, 1, 2001.

66. *Biomarkers in Risk Assessment: Validity and Validation, Environmental Health Criteria 222.* International Programme on Chemical Safety, World Health Organization, Geneva, Switzerland, 16–18, 2001.

67. Schulte, P.A. and Rothman, N., Epidemiological validation of biomarkers of early biological effect and susceptibility, in *Biomarkers: Medical and Workplace Applications*, M.L. Mendelsohn, L.C. Mehr and J.P. Peekes, Eds., John Henry Press, Washington, DC, 23–32, 1998.

68. Howe, G.R., Practical uses of bimarkers in population studies, in *Biomarkers: Medical and Workplace Applications*, M.L. Mendelsohn, L.C. Mehr and J.P. Peekes, Eds., John Henry Press, Washington, DC, 41–49, 1998.

69. Henderson, R.E., in *Biomarkers: Medical and Workplace Applications*, M.L. Mendelsohn, L.C. Mehr and J.P. Peekes, Eds., John Henry Press, Washington, DC, 33–39, 1998.

70. Albertini, R.J., Developing sustainable studies on environmental health, *Mut. Res.*, 480–481, 317, 2001.

71. Toraason, M., Albertini, R., Bayard, S. et al., Applying new biotechnologies to the study of occupational cancer — A workshop summary, *Env. Health Pers.*, 112, 412, 2004.

72. Steiner, S. and Anderson, N.L., Expression profiling in toxicology — Potentials and limitations, *Tox. Letters*, 112–113, 467, 2000.

73. Sessa, R., Di Pietro, M., Schiavoni, G. et al., *Chlamydia pneumoniae* DNA in patients with symptomatic carotid atherosclerotic disease, *J. Vasc. Surg.*, 37, 1027, 2003.

74. Liu, Z., Wang, L.E., Strom, S.S. et al., Overexpression of hMTH in peripheral lymphocytes and risk of prostate cancer: A case-control analysis, *Mol. Carcin.*, 36, 123, 2003.

75. Moustafa, M.A., Ogino, D., Nishimura, M. et al., Comparative analysis of ATP-binding cassette (ABC) transporter gene expression levels in peripheral blood leukocytes and in liver with hepatocellular carcinoma, *Cancer Sci.*, 95, 530, 2004.

76. Rockett, J.C., Burczynksi, M.E., Fornace, A.J., Jr., Hermann, P.C., Kravetz, S.A. and Dix, D.J., Surrogate tissue analysis: monitoring toxicant exposure and health status of inaccessible tissues through analysis of accessible tissues and cells, *Tox. Appl. Pharm.*, 194, 189, 2004.

77. Design Control Guidance for Medical Device Manufacturers, CRDH, FDA, http://www.fda.gov/crdh/comp/designd.html.

78. Sinz, A., Bantscheff, M., Mikkat, S. et al., Mass spectrometric proteome analyses of synovial fluids and plasmas from patients suffering from rheumatoid arthritis and comparison to reactive arthritis or osteoarthritis, *Electrophoresis*, 23, 3445, 2002.

79. Drynda, S., Ringel, B., Kekow, M. et al., Proteome analysis reveals disease-associated marker proteins to differentiate RA patients from other inflammatory joint diseases with the potential to monitor anti-TNF therapy, *Path. Res. Prac.*, 200, 165, 2004.

Index